International Automobile Scientific Forum (IASF-2018)

Intelligent Transport System Technologies and Components

IOP Conference Series: Materials Science and Engineering
Volume 534

Moscow, Russia
18 – 19 October 2018

ISBN: 978-1-5108-8919-4
ISSN: 1757-8981

Printed from e-media with permission by:

Curran Associates, Inc.
57 Morehouse Lane
Red Hook, NY 12571

Some format issues inherent in the e-media version may also appear in this print version.

Printed by Curran Associates, Inc. (2019)

For permission requests, please contact the Institute of Physics at the address below.

Institute of Physics
Dirac House, Temple Back
Bristol BS1 6BE UK

Phone: 44 1 17 929 7481
Fax: 44 1 17 920 0979

techtracking@iop.org

Additional copies of this publication are available from:

Curran Associates, Inc.
57 Morehouse Lane
Red Hook, NY 12571 USA
Phone: 845-758-0400
Fax: 845-758-2633
Email: curran@proceedings.com
Web: www.proceedings.com

Table of contents

Volume 534

International Automobile Scientific Forum (IASF-2018), Intelligent Transport System Technologies and Components

18–19 October 2018, Moscow, Russian Federation

Accepted papers received: 16 April 2019
Published online: 12 June 2019

Preface

Papers

IOP Conf. Series: Materials Science and Engineering **534** (2019) 011001 doi:10.1088/1757-899X/534/1/011001

Preface to IASF-2018

IASF-2018 "Intelligent Transport System Technologies and Components" was held on October 18-19, 2018. It was dedicated to the centenary of the State Research Center of the Russian Federation FSUE "NAMI".

Among the guests and participants of the Forum were representatives of the Ministry of Industry and Trade of the Russian Federation, the Ministry of Transport of the Russian Federation, representatives of automotive plants. The forum was also attended by representatives of leading universities, the Academy of Sciences of the Republic of Belarus, foreign universities and Russian and foreign companies. The total number of participants exceeded five hundred.

The forum was attended by regular partners: OICA, AVL List GMbH, FISITA, SAE International, Association of Russian Automakers, Association of Traffic Engineers (ATE), Publishing House AAE Press, National Technology Platform of the Russian Federation Environmentally Friendly Vehicles Green Car.

A welcoming speech addressed to the participants and guests of the forum was made by Alexander Morozov, Deputy Minister of Industry and Trade of the Russian Federation, Sergey Gaisin, Chief Executive Officer of FSUE NAMI, David Schutt, Chief Executive Officer of SAE International, FISITA Vice President, Murli Iyer, Vice President - Americas FISITA; Executive Advisor – Global Affairs SAE International and Valentin Gapanovich, President of Non-Commercial Partnership "Union of Industries of Railway Equipment".

Within four panel discussions "Regulatory and Legal Framework of Autonomous Driving and Social and Economic Problems of Autonomous Vehicle Implementation", "Prospects of Development of New Types of Intelligent Land Transport and Technological Vehicles", "Intelligent Transport System Components" and "Modern Trends in Advanced Driver Assistance Systems (ADAS)" by representatives and heads of companies, scientific and educational institutions, such as OICA, GLONASS Union, FSUE "NAMI", Moscow Automobile and Road Construction State Technical University (MADI), Institute for Transport Economics and Transport Policy Studies HSE, BMSTU (Bauman Moscow State Technical University), Federal Scientific Agroengineering Center VIM, FEV Polska sp. z o.o., United Engineering Center of GAZ Group, Safety Engineering Research sp.j., KAMAZ PJSC, National Instruments Rus LLC, Robert Bosh LLC, AVL List GMbH, Ford Sollers Holding LLC, The Joint Institute of Mechanical Engineering of the NAS of Belarus, LLC Ener Z, JSC "Scientific Design Bureau of Computing Systems", Rohde & Schwarz RUS LLC, Israel Institute of Technology, Russian State Agrarian University - Moscow Timiryazev Agricultural Academy reports were presented and discussed.

During the breaks, guests were shown videos devoted to the 100th anniversary of FSUE "NAMI . In addition, all participants could visit the exposition, which this year was

IOP Conf. Series: Materials Science and Engineering **534** (2019) 011001 doi:10.1088/1757-899X/534/1/011001

represented by automobiles for top government officials of different periods and historical cars NAMI-1 and KIM-10. The centerpiece of the exhibition was certainly automobiles based on a united modular platform under the AURUS brand, represented by Senat, Senat Limousine and Arsenal models. National Instruments, Rohde and Schwarz RUS, Keysight Ansys, Bosch and AVL stands were also presented.

The second day was held at the FSUE "NAMI" Testing Centre and began with a demonstration of vehicles equipped with unmanned vehicle controls in different driving conditions: emergency braking, parking, bridging a railway crossing, driving along a route, etc.

The second part of the day contained a joint session of 2 sections: "Technologies and components of modern intelligent transport systems" and "Scientific and technical developments of young scientists, post-graduate students and students." Within the sections, there were made presentations by representatives of foreign and Russian companies, universities: AVL List GMbH, Italdesign Giugiaro SpA, CSI SpA, FSUE "NAMI", ENERGO LLC, SINOPSYS LLC, Keysight Technologies Russia, Moscow Polytechnic University and BMSTU (Bauman Moscow State Technical University).

Poster presentations were made by BMSTU (Bauman Moscow State Technical University), Research Institute "MIVAR", FSUE "NAMI", Ryazan State Radio Engineering University (RSREU), Moscow Automobile and Road Construction State Technical University (MADI), Nizhny Novgorod State Technical University n.a. R.E. Alekseev and Moscow Polytechnic University.

Detailed information about the results of the International Automotive Scientific Forum (IASF-2018) "Intelligent Transport System Technologies and Components" are available on the forum website **http://iasf.nami.ru/**

Peer review statement

All papers published in this volume of *IOP Conference Series: Materials Science and Engineering* have been peer reviewed through processes administered by the proceedings Editors. Reviews were conducted by expert referees to the professional and scientific standards expected of a proceedings journal published by IOP Publishing.

IOP Conf. Series: Materials Science and Engineering **534** (2019) 012001 doi:10.1088/1757-899X/534/1/012001

Possible scenarios of autonomous vehicles' testing in Russia

A M Ivanov, S S Shadrin, S R Kristalniy and N V Popov

MADI, Moscow, Russian Federation

E-mail: ivanov-am@madi.ru

Abstract. This article represents out point of view on what is happening in Russia in the field of autonomous driving technologies development, testing and implementation. A method for addressing the issue of autonomous vehicles (AVs) testing and critical testing scenarios detection is proposed. Road tests of ADAS Autonomous Emergency Braking System (AEBS) were performed in winter conditions at a proving ground and in virtual environment with simulation techniques, practical results are presented.

1. Introduction

Autonomous driving is a global trend and no automakers, suppliers, or technological companies will be able to pass it by [1, 2, 3, 4]. The rapid worldwide attempts to implement autonomous driving technologies with ADAS (advanced driver assistance systems) functionality at the first stage and the lack of legal regulation in Russia have led to the situation when we have a lot of modern passenger cars with ADAS systems on public roads. These ADAS systems (autonomous emergency braking, lane keeping, automated parking, etc.) have not been tested and may not have been adapted for use in Russian conditions [5]. It's a potential threat to national road safety.

Russian government has established the association of "Autonet" in 2016 with the aim of new market consolidation (AVs, cooperative transport, electric vehicles, and transport services). This association has a road map for ADAS regulation deployment, and it consists of the next stages:

- A choice of appropriate ADAS systems to be tested;
- A development of national standards on safety requirements and testing methods;
- A creation of national AVs' testing system on the basis of previously developed standards;
- Amendments to the Technical Regulations of the Customs Union TR CU 018 [6] (type approval level).

The third item in the above list sounds good, but seems to be tremendously hard to be solved.

At the same time, deployment of automated and autonomous vehicles can lead to the next negative technical and social problems [5]:

- inadequate car's behavior due to imperfection of control algorithms;
- inadequate car's behavior in the events of failures, malfunctions, connection with infrastructure lost, after road accidents, vehicle's parameters change (tires change, software modification);
- malfunction due to the conflict of driver's actions and automated system operation;
- operation in conditions, which are not specified by car manufacturer;
- lack of cybersecurity;
- moral and ethical problems;
- contradiction with existing national legislation.

IOP Conf. Series: Materials Science and Engineering **534** (2019) 012001 doi:10.1088/1757-899X/534/1/012001

The above negative technical and social problems of AVs' implementation should be solved with appropriate national testing system.

2. Foreign Experience

The UNECE/WP.29 on automated/autonomous and connected vehicles discusses the following grading of AVs operating conditions with respect to different road types: highways, city roads, other roads, parking areas, closed areas. Testing methods in this case are divided into several categories: virtual and real testing of typical scenarios; tests on public roads; manufacturer's audit (ISO 26262); software validation; etc.

Another example is the list of road situations, developed by researchers in the USA [7, 8], to which car's automated control systems should respond adequately. AVs should:

- detect and respond to speed limit changes;
- perform high speed freeway merge;
- perform low speed merge;
- park on the shoulder or transition the vehicle to a minimal risk state;
- detect and respond to dangerous oncoming traffic;
- detect permitted and restricted zones for driving on a road, and perform maneuvering in permitted areas;
- perform car following including 'stop and go' and emergency braking;
- detect and respond to stopped vehicles;
- detect and respond to intended lane changes;
- detect and respond to static obstacles;
- detect and respond to traffic light signals, road signs of stopping and giving the way;
- navigate intersections and perform turns;
- perform driving at roundabouts;
- detect free parking slots and perform parking;
- detect and respond to prescriptive and prohibitive road signs;
- detect and respond to traffic-controller's signals;
- follow the priority rights of driving;
- respect other requirements of national traffic rules;
- follow other road requirements (temporary road signs, marking, etc.);
- detect and drive safely in road repair areas;
- react to the situation in the area of road accident;
- detect and respond to emergency vehicles;
- detect and respond to cars with activated light alarm;
- give way to pedestrians and cyclists on pedestrian crossings;
- ensure safe distance between vehicles, pedestrians, cyclists on roadside;
- detect and respond to temporary changes in traffic patterns.

This testing pattern can be extended with the next situations, actual for Russian operating conditions:

- perform overtaking on two-lane roads with oncoming traffic;
- choose a safe cornering speed;
- detect and respond to dangerous defects of road pavement (roughness, ruts, etc.);
- detect and respond to road areas with poor adhesion characteristics;
- navigate a vehicle on a road without lane markings or with lane markings of poor quality;
- perform operation in inclement weather conditions.

3. Proposals for Critical Testing Scenarios Detection

Fundamentally, the AVs' testing system should cover the whole life cycle of AVs, beginning from the stage of design and ending with utilization. At the same time, methods of testing consist of simulations, proving ground tests, public road tests and further data monitoring of automated systems in use.

IOP Conf. Series: Materials Science and Engineering **534** (2019) 012001 doi:10.1088/1757-899X/534/1/012001

The very important question is how to localize critical testing scenarios for proving ground and public road tests from the enormously great variety of all possible states, modes, internal and external conditions?

We suggest using the next sequence of actions for this case:

1. Definition of the test's purpose, which may include: the task of functional or operational safety monitoring; determination of operating conditions, reliability metrics, fault tolerance, performance variation between the same cars, as well as during the whole life cycle, etc.
2. Creation of test scenarios or selection from the existing ones (e.g. EuroNCAP, ADAC, ISO):
    a. Definition of ranges of changeable parameters of the system "driver—car—road—environment" [9].
    b. The choice of evaluation criteria and their permissible values.
3. Flow simulation of all possible variations of the developed test scenarios in a virtual environment performing:
    a. Critical testing scenarios detection with respect to the test's purpose.
4. Critical scenarios' field tests performing:
    a. Virtual model's parameters validation, the above cycle repetition.
5. Changing the design or control algorithms to achieve goals of testing.

4. Experimental Research – Road Tests

This section provides some experimental results of Autonomous Emergency Braking System (AEBS) testing in winter conditions [10]. Vehicle under test is Subaru XV of 2018'th model year with technical vision system based on stereo cameras (Subaru EyeSight). The following measurement equipment has been used: MSW – measurement steering wheel (Kistler); CPFTA – braking pedal sensor (Kistler); WPT – external wheel speed sensor (Kistler); TANS – Tri-Axial Navigational Sensor (Kistler); CS 1016 FAMOS Online – data acquisition system (IMC); GPS antennas (IMC, JAVAD); internal camera for dashboard AEBS warning signal detection; external camera for tailgate stop signals detection; on-board CAN bus data monitoring system [11]. Road tests were conducted at Dmitrovskiy proving ground (NAMI).

Figure 1 illustrates the results for AEBS winter testing on dry asphalt. Initial host car velocity was set to be approximately 35 km/h. The stationary target looked like a soft wall covered with thin reflective metallic coating. The first vertical line in Figure 1 indicates the in-vehicle AEBS warning signal, this moment corresponds to the distance of 28.8m between host car and stationary target, and TTC (time to collision) of 2.63s. The second vertical line indicates activation of the tailgate stop signals, this moment corresponds with the beginning of automated partial braking process with the approximate value of calculated TTC of 0.82s. We can mention, that then in the control strategy of tested vehicle the next stage of more intensive braking was activated with deceleration of up to 8 m/s^2 (see the graph of deceleration in Figure 1). The third vertical line indicates full vehicle stop in front of a stationary target with free space of 0.3m between them. Collision was avoided.

IOP Conf. Series: Materials Science and Engineering **534** (2019) 012001 doi:10.1088/1757-899X/534/1/012001

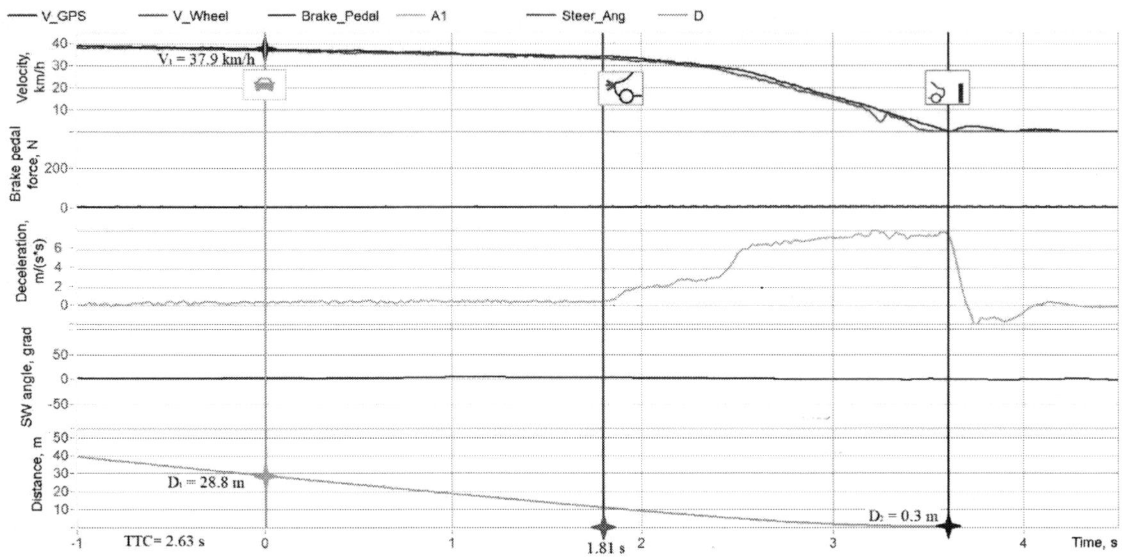

Figure 1. AEBS winter test – stationary target, dry asphalt, daylight.

Figure 2 illustrates the results for AEBS winter testing of the same car and the same scenario as above, but on snow road. Initial host car velocity was set to be approximately 35 km/h. The first vertical line indicates the in-vehicle AEBS warning signal, this moment corresponds to the distance of 37.7m between host car and stationary target, and TTC of 3.52s. We can mention, that these values of distance and TTC has been increased for snow conditions in comparison with dry asphalt. The second vertical line indicates activation of the tailgate stop signals, the braking process begins, it also has two stages and the maximum deceleration achieves values of 4 m/s^2, but it is also clear that it was the maximum possible deceleration on such a road (see the graph of wheel velocity in Figure 2 with ABS cycles). The TTC of braking process start approximately equals to 1.44s, and this value is also higher than the TTC value for dry asphalt (0.82s). The third vertical line indicates collision of a host car with stationary target. The velocity equals to the value of 6.8km/h, collision was not avoided.

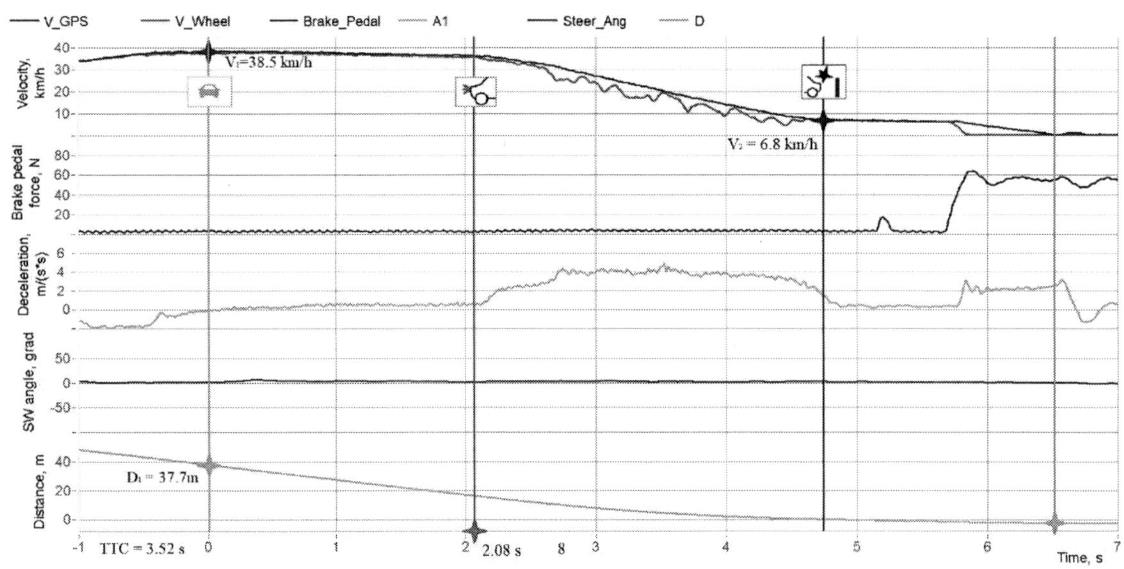

Figure 2. AEBS winter test – stationary target, snow road, daylight.

We can say, that it seems, that AEBS of Subaru XV adapts to the road friction conditions because TTC values of the above road tests are essentially different. The other side of this is that we don't know the actual control algorithms of vehicle under test, as well as the actual characteristics of vision recognition system. Perhaps in other lighting and weather conditions this AEBS will demonstrate other results. The task of AEBS control algorithms decompilation (reverse engineering) seems ambitious and non-effective for further virtual testing of all possible variations of test scenarios and weather conditions.

5. Experimental Research – Virtual Tests

We have used standard software for ADAS virtual testing. We simulated straight driving of a host car with initial velocity of 60 km/h with a stationary object (Euro NCAP car target) on a way. Ideal road conditions have been used (dry asphalt, daylight). The AEBS control algorithm gave a warning at 2.6s TTC, then it applied 40% of braking at 1.6s TTC, and 100% of braking at 0.6s TTC. Virtual vehicle under test was equipped with long range radar (LRR – 120m range, 9 degrees narrow beam) and short range radar (SRR – 30m range, 90 degrees wide beam). The test scenarios were as follows: 1) driving with idealized radar sensors; 2) driving with characteristics of real radars (Continental ARS510 as a LLR, and Continental SRR510 as a SRR with characteristics from the official web site).

Figure 3 illustrates virtual AEBS operation with idealized sensors and ideal road conditions. Collision was avoided.

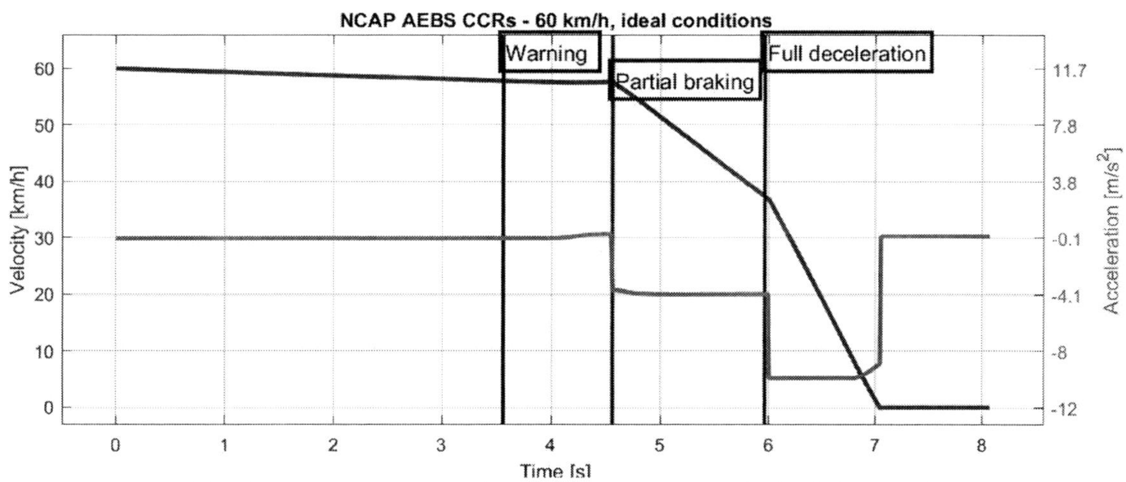

Figure 3. AEBS virtual testing – stationary target, dry asphalt, daylight, idealized sensors.

Figure 4 illustrates virtual AEBS operation with characteristics of real radars and ideal road conditions. Collision was not avoided, collision speed equals to 37.07km/h. Warning and partial braking TTC of the both experiments are the same, but in the second case we see malfunction of AEBS control algorithm (the graph of deceleration in Figure 4 – it stopped braking). Perhaps, this situation has occurred because of target tracking fail. In any case, we see that the results of virtual tests are essentially sensitive to the adequacy of ADAS control algorithms and sensors characteristics.

IOP Conf. Series: Materials Science and Engineering **534** (2019) 012001 doi:10.1088/1757-899X/534/1/012001

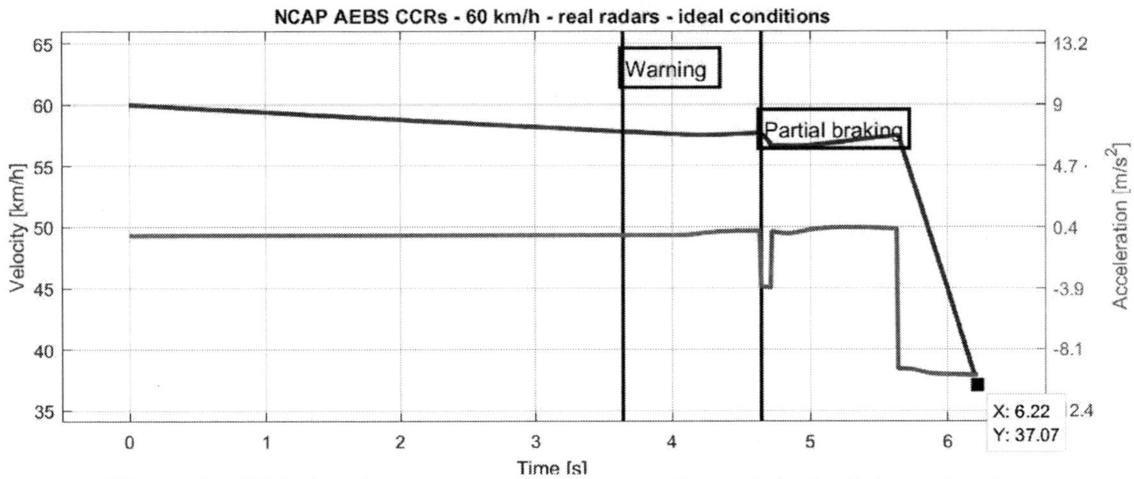

Figure 4. AEBS virtual testing – stationary target, dry asphalt, daylight, real radars.

The information of exact ADAS control algorithms and vision recognition system characteristics is very important for virtual tests and critical scenarios detection. This type of information can be provided by car manufactures or their OEMs only. Of course this information is essentially confident, but it can be provided as a hardware or software "black boxes" for testing purposes.

6. Conclusion

A comprehensive approach with the use of virtual and field tests is needed to assess the safety of autonomous road vehicles with respect to the peculiarities of road and weather-climatic conditions in Russia.

It is impossible to assess the complete ADAS systems safety without the use of exact control algorithms and vision recognition system characteristics, which can be provided by car manufactures or their OEMs as a hardware or software "black boxes" for testing purposes.

A clear statement of the acceptable operational conditions and the functionality of automated in-vehicle control systems should be declared in the instruction manual. AVs legal permission to access public roads should be done after comprehensive tests only.

We have proposed a method for addressing the issue of autonomous vehicles testing and critical testing scenarios detection.

References

[1] Eskandarian A 2012 *Handbook of Intelligent Vehicles* (Springer) p 1628
[2] Saikin A M, Bakhmutov S V, Terenchenko A S et al 2014 Tendency of creation of "driverless" vehicles abroad *Biosciences biotechnology research Asia* **11** (Spl. Edn.) pp 241-46
[3] Ruparel T, Yonten K and Eskandarian A 2014 Analyzing roadside safety implications of future vehicle designs *ASME International Mechanical Engineering Congress and Exposition Proceedings* (IMECE)
[4] V V Gaevskiy and A M Ivanov 2018 Problems of the application of intelligent driver assistance systems on a single-track vehicles *IOP Conf. Ser.: Mater. Sci. Eng.* **386** 012021
[5] A M Ivanov and S S Shadrin 2018 Development of autonomous vehicles' testing system *IOP Conf. Ser.: Mater. Sci. Eng.* **315** 012011 (doi:10.1088/1757-899X/315/1/012011)
[6] Technical regulations of the Customs Union "On the safety of wheeled vehicles" TR CU 018/2011
[7] Christopher Nowakowski et al 2014 Development of California Regulations to govern the testing and operation of automated driving systems *California PATH Program* (University of California, Berkeley, available at http://docs.trb.org/prp/15-2269.pdf)

[8] Federal Automated Vehicles Policy. Accelerating the next revolution in roadway safety *U.S. Department of Transportation* (NHTSA, Sept. 2016, 12507-091216-v9) p 116
[9] Shadrin Sergey Sergeevich and Ivanov Andrey Mikhailovich 2016 Technical aspects of external devices into vehicles' networks integration *International Journal of Applied Engineering Research* **11**(10) pp 7003-06
[10] Ivanov A M, Kristalniy S R, Popov N V, Toporkov M A and M I Isakova 2018 New testing methods of automatic emergency braking systems and the experience of their application *IOP Conf. Ser.: Mater. Sci. Eng.* **386** 012019
[11] Shadrin S S, Ivanov A M and Karpukhin K E 2016 Using data from multiplex networks on vehicles in road tests, in intelligent transportation systems, and in self-driving cars *Russian Engineering Research* **36**(10) pp 811–14

IOP Conf. Series: Materials Science and Engineering **534** (2019) 012002 doi:10.1088/1757-899X/534/1/012002

Perspectives of the unmanned cargo transport with electric traction development under the conditions of the Russian Federation

A M Saikin

Special programs department, FSUE "NAMI", Moscow, Russian Federation

E-mail: a.saykin@nami.ru

Abstract. The development of road transport in the world is highly connected with increased requirements for safety and cost-effectiveness of carrying out passenger and cargo transportation, reducing the negative impact of road transport on people and the environment, as well as the desire of countries that do not have direct access to energy sources, "get off the oil needle". To ensure the fulfillment of these requirements, the key development areas in the automotive industry of developed countries are unmanned vehicles and electric vehicles. Taking into account foreign experience, the article looks at the prospects and features of developing unmanned cargo vehicles with electric traction, at operation conditions of the Russian Federation, which differ markedly from the conditions of vehicle operation in Central Europe, North America and the countries of the Asian region.

1. Introduction

All leading automakers currently have projects aimed at the creation of autonomous unmanned vehicles (UV) and in 2020-2025 are planning the introduction of vehicles with autonomous control. However, in most cases, we are talking about 3rd-4th levels of traffic control systems automation in accordance with the standard SAE J3016 [1]. At the 3rd level of automation, the traffic control system (TCS) takes full control over the dynamics of the vehicle when the driver reacts appropriately to its signal to intervene. At the 4th level of automation, the TCS takes over the control of the vehicle's movement even if the driver ignores the requirement to intervene in the motion control process. At this level, the management solution is taken by the vehicle TCS.

In 2016, the European Union directive UE (347/2012 & 351/2012) on road safety came into force. The directive prescribes to equip all trucks registered or used within the borders of the EEC with special systems to improve road safety. Driver assistance systems (ADAS) are commonly used, they provide detection of pedestrians, drivers alert about an emergency, traffic assistants in traffic jams, braking, vehicle interaction systems with other traffic participants and road infrastructure.

In comparison with other areas of vehicle development, the creation of UV, ADAS systems and UV components is carried out at a faster pace. As an illustration of this fact, Figure 1 shows the patenting dynamics of developments abroad concerning land UV, ADAS systems and UV components.

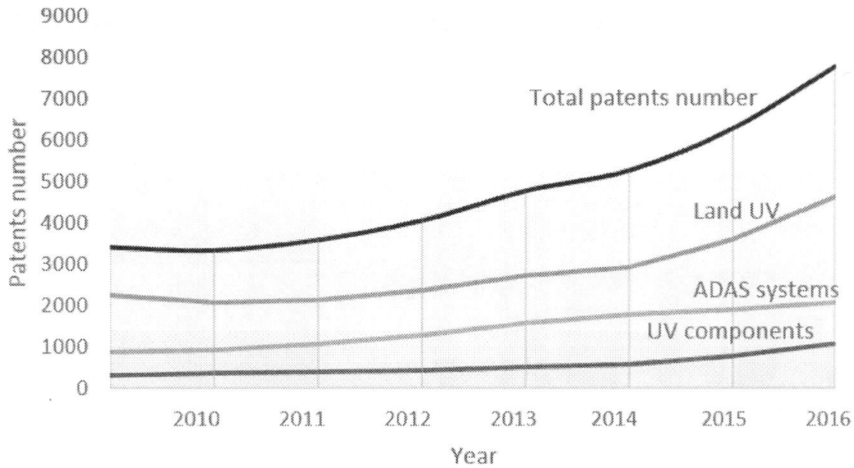

Figure 1. Patenting dynamics of land UV abroad, ADAS systems and UV components in general and separately for the period from 2010 to 2016.

It is considered that the implementation of the UVs will increase productivity and reduce the cost of cargo and passenger transportation, ensure road safety, use the road capacity more efficiently, expand the use of cars for people with disabilities, provide the ability to transport goods in hazardous areas, during natural and anthropogenic disasters or military actions, increase passenger comfort, reduce fuel consumption and travel time, lower harmful emissions and greenhouse gases into the atmosphere [2-7].

Another, the most important direction in promising vehicles creation, is the introduction of cargo electric vehicles [9-11], which do not emit harmful substances, including greenhouse gases, with the exhaust gases of internal combustion engines. The decision to decommission the vehicle with petrol and diesel engines is part of the reducing carbon dioxide emissions (CO_2) program adopted in the climate agreement in Paris in December 2015. According to the experts of this program, the vehicles are responsible for about one-fifth emissions of carbon dioxide into the atmosphere. Norway (2025), Germany (2040), Great Britain (2040), France (2040) and other countries announced plans for a complete ban on cars with internal combustion engines.

Constraints that prevent the wide use of electric vehicles are the need to create an infrastructure for recharging batteries, long battery charging time in comparison with refueling, relatively small mileage of electric vehicles on one battery charge and special requirements for energy accumulation and storage elements. The batteries should be explosion-proof and fire-safe, have small weight and size parameters, high efficiency of charge-discharge characteristics, have a wide range of operating temperatures, minimum self-discharge, long service life, be mechanically robust, easy to maintain and release minimum toxic gases.

2. Main part

The main trend in the development of promising cargo vehicles is the creation of UVs with electric drive. Most foreign developments on unmanned trucks, including ones with electric drive, are being carried out both through the serial cargo vehicles conversion and equipping them with motion control systems using a variety of vision systems as well as by developing fundamentally new designs, recently in a cabinless performance. The main companies engaged in the development of cargo UVs are Einride, Komatsu, Caterpillar, Volvo, Ford, Waymo, a subsidiary of Alphabet, a subsidiary of Google, Otto, a subsidiary of Uber, Daimler concern, etc. In recent years, at cargo UVs development, the emphasis is made on the UVs with an electric drive and without a cabin. Among domestic developers of UVs KAMAZ, UAZ, Cognitive Technologies, KB Aurora, Traft, etc. should be noted. Examples of foreign cargo vehicles developed by foreign automakers are shown in Figure 2.

a) T-pod Cargo UV b) Autonomous Haulage Vehicle (AHV) unmanned dump truck c) SURUS unmanned cargo platform d) Ford cargo platform.

Figure 2. Cargo UVs, developed by foreign automakers.

There is no driver's cabin in the Swedish promising UV T-pod with the Einride control system (Fig. 2a)) [2]. The UV is equipped with a remote control system and rectangular chassis with an electric drive. The remote control is performed by the operator and, according to the developer, is especially important when driving around the city. T-pod technical parameters: weight with full load is 20 tons, length is 7 m (23 feet), battery capacity is 200 kWh, mileage at one charge is 200 km (124 miles) [3]. According to the developer forecasts, the use of T-pod UVs will not only reduce environmental pollution, but also improve traffic safety, create new jobs and reduce transportation costs. According to the plans of 200 trucks, T-pod will be operated between the cities of Gothenburg and Helsingborg beginning in 2020 [4].

Unmanned mining dump truck Autonomous Haulage Vehicle (AHV) of the Japanese company Komatsu (Fig. 2b) is also designed without a driver's cabin [5]. Komatsu has been developing unmanned mining technologies since 2008 in cooperation with Rio Tinto. During this time, its autonomous unmanned dump trucks performed in classical design with the driver's cabin transported hundreds of millions tons of rock in the mines of Chile and Australia. At present, Komatsu company has a serious scientific and technical background and experience of operating UVs for the development of cabinless dump trucks. The AHV technical vision system provides scanning of the surrounding space from all sides of the machine, which allows the dump truck to move in any direction without the assistance of a driver or operator. Due to the lack of a driver's cabin, the bed volume is increased. All 4 wheels of the dump truck are steerable, it has increased maneuverability, which allows to abandon reversal sections on the movement routes in quarries and mines.

The absence of the cabin allowed achieving more uniform load distribution on the wheels: 250 tons of cargo are distributed along the axes almost equally. Komatsu AHV technical parameters: length is 15 m, width is 8 m, load capacity is 230 t, weight is 416 t, engine power is 2700 horsepower (2014 kW), maximum speed is 65 km/h regardless movement direction, tire size is 59/80R6, independent suspension of each wheel, individual transmission of each wheel, independent rotation of each wheel.

IOP Conf. Series: Materials Science and Engineering **534** (2019) 012002 doi:10.1088/1757-899X/534/1/012002

General Motors Company has developed a Silent Utility Rover Universal Superstructure (SURUS) unmanned cargo platform also without a driver's cabin [7]. SURUS uses second generation Hydrotec hydrogen fuel cells produced by General Motors, an independent electric drive of each wheel, all-wheel drive steering, a system of lithium-ion batteries (Fig. 1 c) [6]. The length of the platform is 5 m, width is 2.3 m. One hydrogen filling provides the car with a power for more than 400 miles (644 km). Several platforms can theoretically move in unmanned mode behind a leader car. The fuel cell can operate as a mobile generator with alternating current of 120 V, except for charging its own batteries. SURUS can be made in the form of a conventional truck or in the form of a cabinless platform. The dimensions of the chassis correspond to the dimensions of the sea container for the transport of goods: length is from 6 to 13 m, width is 2.4-2.5 m, height is 2.5-2.9 m.

According to GM forecast, SURUS UV can be used as a truck, mobile or emergency power source, in flexible cargo delivery systems, for commercial cargo transportation, as light and medium-tonnage military trucks, only for military use in the future.

The SURUS chassis is designed for military and civil use - in the army, police, rescue services and cargo transportation. The platform can be used as a mobile power plant charged by solar panels and as an ambulance for driving on rough terrain or as an unmanned cargo platform. In addition to absolute noiselessness, the chassis does not emit harmful substances into the atmosphere, fast battery charge is ensured. GM expects that SURUS UV will become the basis for the serial use of fuel cells and will be used in complex transport situations caused by natural disasters and military conflicts. SURUS can move both on the highway and off-road [7].

As part of the researching promising technologies program for a metropolis called "City of tomorrow", the Ford campaign introduced the concept of an autonomous delivery service for parcels auto delivery [8]. Appearance of the chassis is shown in figure 2d.

Autodelivery service includes cargo vans and drones. As part of the service, unmanned self-propelled chassis transports any cargo from groceries to medical supplies on the ground and drones deliver them by air at the final stage of transportation in order to ensure cargo delivery to destinations unavailable for UVs. The range of the drones does not exceed 15 m, which is sufficient to transport cargo from the UVs to, for example, private door. The concept "The City of Tomorrow" provides the solution to mobility problems in urban conditions, including traffic jams and air pollution. Ford plans to launch the Autodelivery service in 2021.

From the brief analysis it follows that foreign developments of promising UVs are being carried out almost for the entire range of cargo vehicles with an emphasis on UVs with electric drive and without driver's cabin.

Compared with UVs, the development of cargo electric vehicles began much earlier. Now they have found wide practical application. Constraints that prevent the large-scale use of electric vehicles claim the creation of an expensive infrastructure for recharging batteries, electric vehicles are characterized by a longer battery charging time compared to fueling, a relatively small mileage of electric vehicles on one battery charge and special requirements for energy storage and accumulation elements. The batteries should be explosion-proof and fire-safe, have small weight and size parameters, high efficiency of charge-discharge characteristics, have a wide range of operating temperatures, minimum self-discharge, long service life, be mechanically robust, easy to maintain and release minimum toxic gases. However, priority is given to electric vehicles due to the parameters of the electric drive, the battery and its components constant improvement and the transition to more mass production.

Among the latest developments of cargo electric vehicles [9-12] the heavy truck Urban eTruck from Mercedes, launched in production; automobile tractor class 8 Nikola One from American company Nikola Motor; port tractor YT202-EV from the leading European manufacturer of port (terminal) tractors - the company Terberg Benschop BV should be noted.

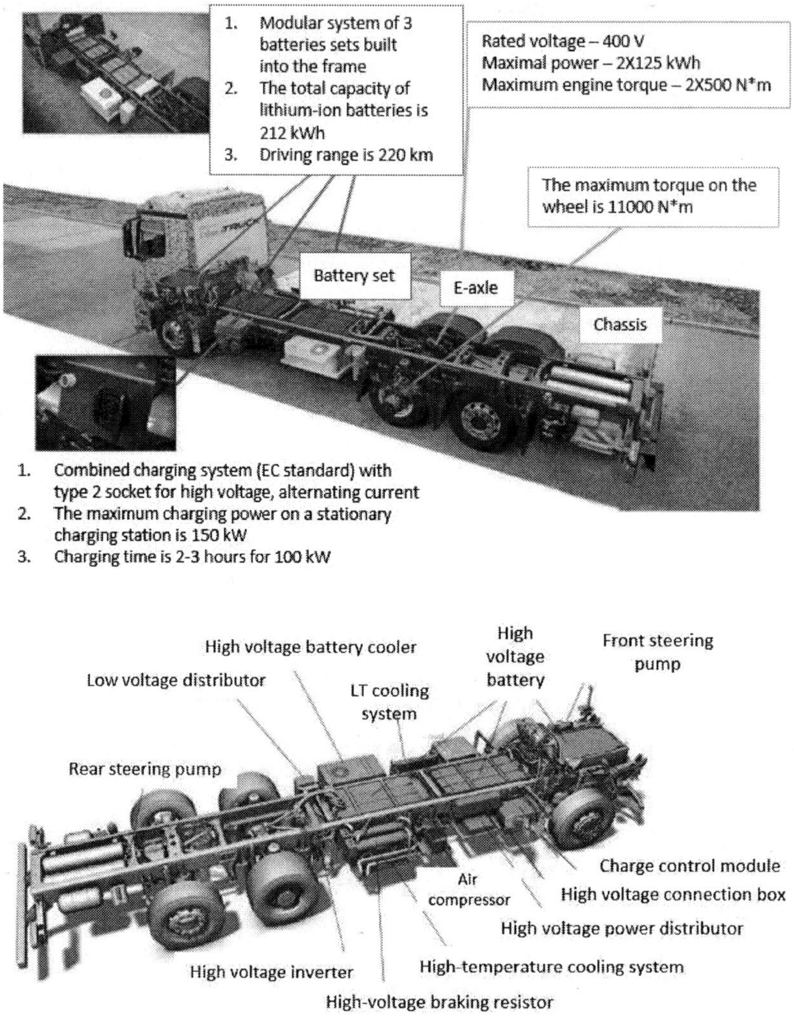

Figure 3. Composition, parameters and layout of the Urban eTruck electric drive main elements on the chassis.

The Urban eTruck truck maximum permissible mass is 26 tons. The Urban eTruck electric drive includes two electric motors with a total capacity of 250 kW located on each side of the car. Their power supply is provided by two modular rechargeable lithium-ion batteries with capacity of 212 kW/h and the regenerative braking system feeds the batteries during braking. The mileage without recharging is 200 km. The FleetBoard telematics system allows the driver to communicate with the traffic controller, allowing the planning of the delivery route. All information about the current state of the machine is displayed on a 12.3-inch display located on the front of the driver's cabin. The cargo vehicle Urban eTruck is a modular type. Serial production of the machine is planned to begin after 2020.

Truck Nikola One can work on electricity, natural gas and other types of fuel. It can carry cargo weighing about 36 287 kg and at the same time pass 1609 km without a single stop. The battery capacity is 32 kWh and it is charged through the turbine, as well as from regenerative braking. The car is equipped with six electric motors with a total output capacity of 2000 horsepower and a torque of 5017 N·m. According to the developer, the operating costs of Nikola One are 2 times less than in case of conventional diesel trucks.

The Terberg YT202-EV, designed for public roads, is equipped with a lithium-iron-phosphate battery with a voltage of 614 V. The mileage of road train, including a single tractor and a standard tilt semitrailer, is 100 km. The charging time of the batteries is from 2 to 4 hours. The road train operates as a part of the Scherm fleet, which specializes at the delivery of spare parts from the warehouse to the BMW plant. In comparison with the diesel analogue, when fuel consumption exceeds 50 thousand liters, the Terberg electric traction machine saves 35 thousand dollars.

At present, FSUE "NAMI", with the financial support of the Education and Science Ministry of the Russian Federation, together with PJSC "KAMAZ" as an industrial partner, is developing and piloting a prototype of the unmanned cargo vehicle with electric drive (UCVED) on the basis of KAMAZ chassis with a layout without a driver's cabin for logistics transportation.

Key principles in the development of promising UCVED are unmanned execution, electric drive application, platform volume increased due to the exclusion of the cabin from the chassis structure, driver exclusion from the management process, organization of traffic remote control by an operator, the creation of an electronic traffic control system, working with the use of new technologies for their construction [13, 14], usage of traffic control systems and technical vision.

The UCVED platform includes a cargo platform, an electric drive, a motion control, information processing and communication systems, and control actuators.

The layout solution of the UCVED special chassis is shown in Figure 4.

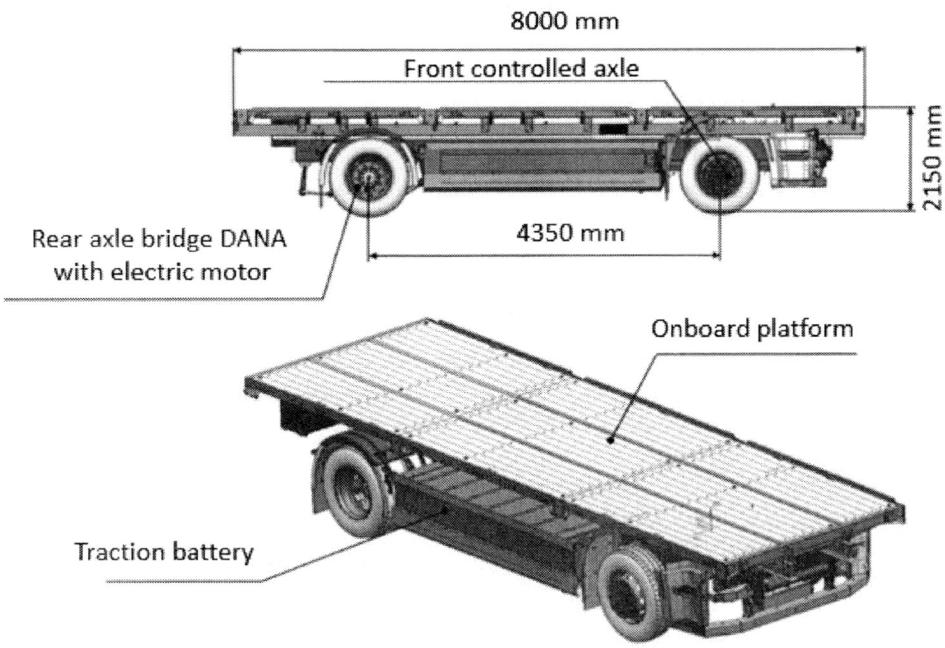

Figure 4. The layout solution of the UCVED special chassis.

The UCVED prototype control system includes a control unit with a microcontroller, a video processing unit, a secondary power source, an executive brake control system, a steering actuator, a navigation car terminal with receiving antennas, a free inertial navigation system, radars, a LIDAR, wheel speed sensors, video cameras. The architecture of the UCVED motion control module is shown in Figure 5.

IOP Conf. Series: Materials Science and Engineering **534** (2019) 012002 doi:10.1088/1757-899X/534/1/012002

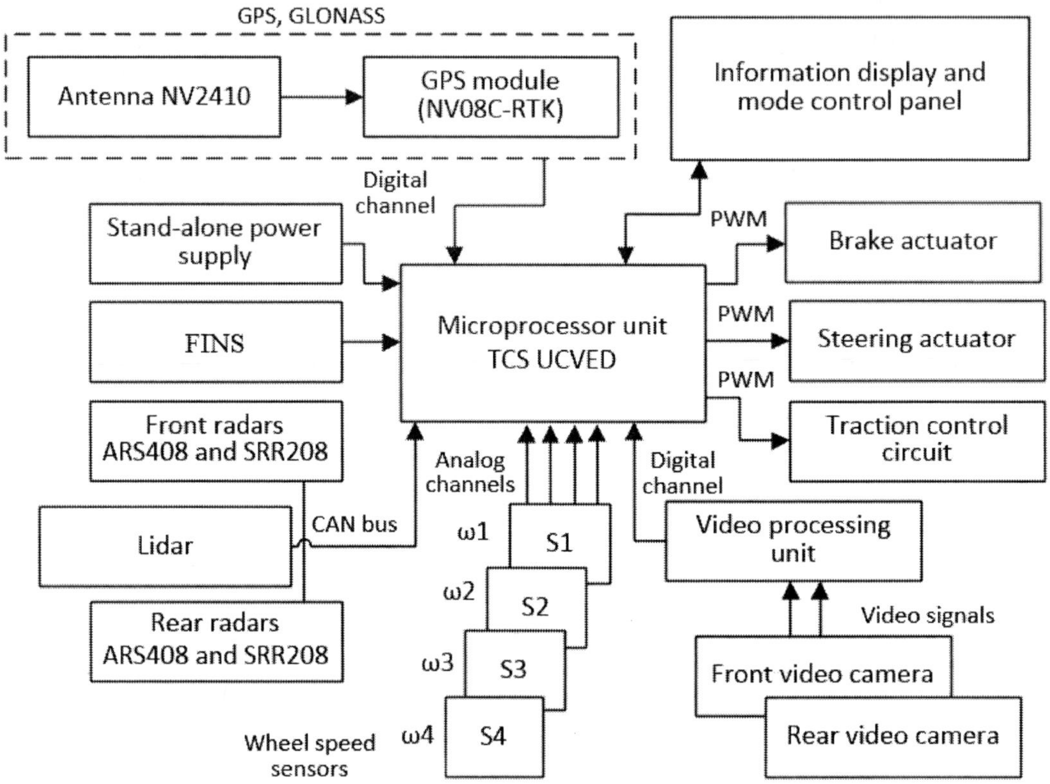

Figure 5. The architecture of the UCVED motion control module.

The main advantages of the project at its implementation are the time and costs reduction of goods and passenger transportation, improving transport and environmental safety, minimizing traffic accidents and the number of human victims in them, reduction of fuel consumption, emission of harmful substances into the atmosphere, more efficient use of road capacity.

At UV creation with an electric drive geographical and climatic features, long distances, low average temperatures for most territory of Russia, lack of proper infrastructure for the electric vehicles operation are taken into account. Application of foreign technical solutions developed for the movement control of unmanned trucks with electric drive in the conditions of their operation is problematic for the road and climatic conditions of the northern regions and the Arctic zone of the Russian Federation, which constitute more than 70% of the country's territory. The following features characterize road and climatic conditions of the northern regions and the Arctic zone of the Russian Federation:

- abnormally low temperatures (down to -60 °C), which significantly reduce the efficiency of battery cells and the functioning of electronic vision systems operating at minimum permissible temperatures not lower than -20 °C;
- long periods of low temperatures, snowfalls, fog, ice cover, etc., snow drifts that do not allow to recognize road markings and make it difficult for the UV systems to operate;
- difficult conditions of off-road driving, allowing the vehicle operation only in the summer and winter seasons;
- at off-road conditions in large areas of these regions, the UV cannot orient that makes it difficult to visualize the UV location on the route;

- limited visibility in the conditions of the polar night;
- unpredictable influence of the geomagnetic situation and the state of the troposphere and the ionosphere of the Earth in the polar latitudes on the uninterrupted satellite navigation signals receiving.

The negative impact on UV operation is also provided by unfavorable operating conditions as windshield contamination, poor illumination (insufficient illumination of traffic lanes), bad weather conditions (snowfall, icing, heavy fog and rain), absence or worn-out lanes marking, etc.

Summarizing the foregoing, it can be argued that the development of unmanned cargo transport with electric traction in the Russian Federation conditions should be carried out by developing additional control technologies, scientific and technical solutions that are comparable to the known ones. It should be noted that the motion control system and electric drive cannot have a universal performance for vehicles operated in various regions of the Russian Federation, for example, southern, northern, Arctic.

The developed cabinless construction of a cargo UV with an electric drive is promising for use in the vehicles platoon, the use of which allows to reduce fuel consumption and harmful substances emissions to the atmosphere by 10-20% at passenger and cargo transportation.

At the development of advanced cargo UVs with electric drive in terms of health safety ensuring of passengers and workers, it is also necessary to solve the problem of their protection against excess air pollution in the cabins and inhabited compartments of the UVs, which is currently not being given due attention.

3. Conclusion

Analysis of the accumulated experience, modern scientific, technical, regulatory, methodological literature on the development of UVs with an electric drive makes it possible to assert that solving this problem, taking into account operating conditions in the Russian Federation, requires the development of designs for advanced cargo UVs with electric drive and the implementation in their traffic control systems of additional, compared with similar European counterparts, functions as:

- identification of the road surface condition, reliable recognition of road markings and road signs;
- ensuring the operability, reliability and efficiency of the electric drive, motion control systems and their components in the conditions of low and ultralow temperatures of the ambient air;
- ensuring the reliable operation of orientation systems on the terrain, especially in the northern and Arctic regions of the country;
- prediction of the condition and destruction of tires;
- determination of suspension and steering dangerous malfunctions;
- recognition in the initial stages of detaching the wheels from the hubs and others.

The use of foreign stabilization systems is limited by unfavorable road conditions. The recognition of road markings by modern technical vision systems is difficult due to road surface contamination during the autumn-winter and spring periods of the year.

The motion control systems construction with physical sensors of information used in foreign analogues makes it necessary to use new materials, technologies and structures that are not available in the Russian Federation. Given the projected long-term isolation of the Russian Federation from the developed countries of the world, this path seems unrealizable.

In the current situation, the use of a new principle for the physical information sensors replacement by virtual technologies based on control technology implemented at FSUE "NAMI" on indirect measurements and integration of management functions is a new scientific and technical solution for creating an efficient and reliable traffic control system for unmanned cargo vehicles intended for operation in the conditions of the Russian Federation, and allows to reduce the dependence of ongoing developments on foreign technologies and components.

The creation of integrated intelligent traffic control systems based on new principles opens the possibility to implement the projects, starting with the operation of unmanned cargo vehicles with electric drive in limited areas and their subsequent introduction to public roads, including the northern regions and the Arctic zone of the Russian Federation. The conceptual core of such systems is mathematics and software, which requires concentration of the developers' efforts in this area.

Acknowledgments
The materials of the article are prepared based on the applied scientific research results and experimental development financially supported by the Russian Education and Science Ministry, the grant agreement No.14.624.21.0049 from 26.09.2017, unique identifier of the project is RFMEFI624X0049.

References

[1] SAE J3016 2014 Standard: Taxonomy and Definitions for Terms Related to On-Road Motor Vehicle Automated Driving Systems. Issued: 2014-01-16 – SAE International Inc.
[2] Information on https://venturebeat.com/2017/04/07/meet-t-pod-an-autonomous-electric-truck-that-can-also-be-controlled-remotely/
[3] Information on https://techcrunch.com/2017/04/06/einrides-electric-self-driving-t-pod-is-a-new-kind-of-freight-transport-vehicle/
[4] Information on http://www.ixbt.com/news/2017/07/05/einride-t-pod.html
[5] Information on https://www.trucksplanet.com/catalog/model.php?id=2275
[6] Information on https://hightech.fm/2017/10/09/driverless-fuel-cell-platform
[7] Information on https://www.popmech.ru/vehicles/news-390862-bespilotnaya-gruzovaya-platforma-na-toplivnyh-elementah/
[8] Information on https://media.ford.com/content/fordmedia/feu/en/news/2017/02/27/ford-employees_-self-driving-autolivery-concept-demonstrate-idea.html
[9] New Energy and Industrial Technology Development Organization, NEDO: URL: http://www.nedo.go.jp
[10] Nagaitsev M V and Eidinov A A 2014 *Electromobiles* (Moscow:NAMI) p 515
[11] Terenchenko A, Karpukhin K and Kurmaev R 2015 Operation features of electromobile transport in the conditions of Russia *Paper of EVS 28 International Electric Vehicle Symposium and Exibition* KINTEX Korea
[12] Shadrin S S, Ivanov A M and Karpukhin K E 2016 Using Data From Multiplex Networks on Vehicles in Road Tests, in Intelligent Transportation Systems, and in Self-Driving Cars *Russian Engineering Research* **36**(10) pp 811-814
[13] Saykin A, Bakhmutov S, Terenchenko A, Endachev D, Karpukhin K and Zarubkin V 2014 Tendency of driverless vehicles creation abroad Bioscience *Biotechnology Research Asia* **11** pp 241-246
[14] Saykin A M, Buznikov S E and Karpukhin K E 2016 The analysis of technical vision problems typical for driverless vehicles *Research journal of pharmaceutical, biological and chemical sciences* **7**(4) pp 2053-59

IOP Conf. Series: Materials Science and Engineering **534** (2019) 012003 doi:10.1088/1757-899X/534/1/012003

Cross-modal information flows in highly automated vehicles

V V Savchenko and S N Poddubko

State Scientific Institution "The Joint Institute of Mechanical Engineering of the National Academy of Sciences of Belarus", 12 Akademicheskaya str., 220072, Minsk, Republic of Belarus

E-mail: uus@tut.by

Abstract. The classification of cross-modal information flows in highly automated and autonomous vehicles is presented. The definition of cross-modal interaction in the highly automated vehicles is given and the goal of the analysis of heterogeneous information flows in onboard systems of vehicles is formulated. From the interdisciplinary perspective, on the basis of the system approach, the main provisions of the original method for analyzing the driver's potential to regain control of the highly automated vehicle are formulated. They are based on the driver's awareness of the situational circumstances along the route, monitoring the current functional state and individual driver's characteristics.

1. Introduction

The level of vehicles automation is incrementally growing. That not only makes it possible to solve the pre-existing problems at vehicles operation but also creates new ones to be explored and solved. With the increase in the level of automation, the volumes of heterogeneous information circulating in such metasystem as "driver – highly automated vehicle – road – intelligent transport system – information field" are growing at a significant rate. Obviously, in this case, the "information field" falls under the category of "Big data". Figure 1 shows the projected flow of data generated in autonomous vehicles and the main sources.

To implement algorithms of movement trajectory and correct them in compliance with the current terms, the autonomous vehicles, according to experts from Intel [1], need cameras that generate an information flow about 20-60 MB/s, radars and ultrasonic locators (each with 10-100 kB/s), GPS navigation systems – 50 kB/s (although we believe that it is preferable to use four-system navigation chipset GLONASS, GALILEO, BEIDOU, GPS – 50 kB/s for each GNSS, 200 kB/s, as having a functional that allows to determine the current coordinate with greater accuracy, higher noise immunity and reliability [2]). Lidar forms traffic of about 10-70 MB/s. According to approximate expert estimates, a vehicle using such technologies will generate about 4 TB of data per day [1]. To this the information is added generated by the "classic" active security systems and monitoring systems of major joints and units.

IOP Conf. Series: Materials Science and Engineering **534** (2019) 012003 doi:10.1088/1757-899X/534/1/012003

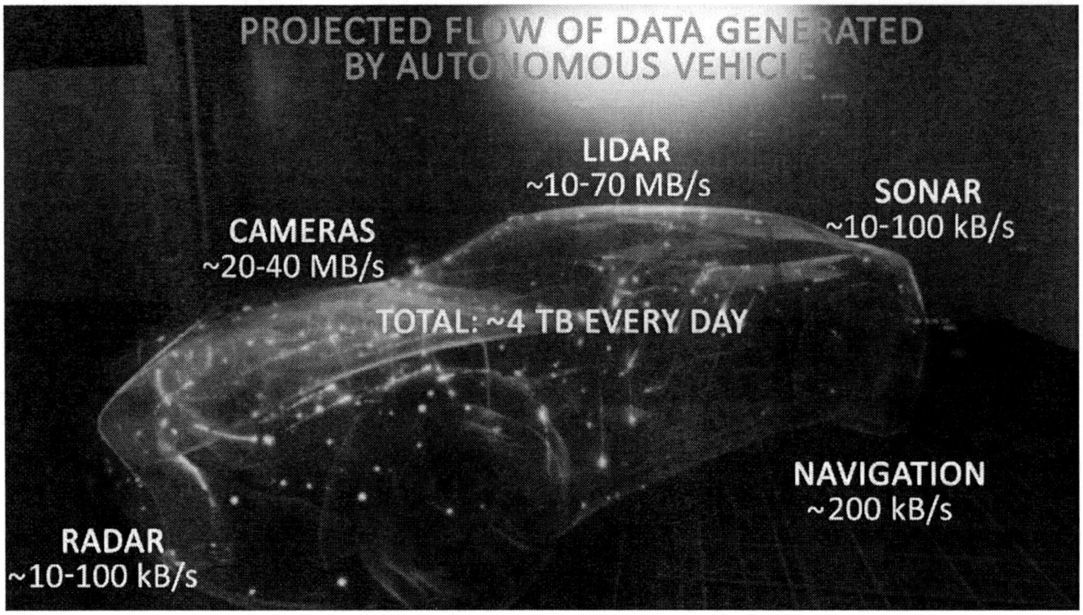

Figure 1. Projected flow of data generated by autonomous vehicle.

A distinctive feature of the multi-system navigation GNSS receiver is that it identifies more signals from different navigation satellite devices. The total number of satellites visible by the receiver can increase essentially (for two system satellites it increases on average by one and a half to two times). Due to this possibility, the multi-segment receiver selects the best geometry by a combination of accompanied satellites, which directly affects the positioning accuracy. Under the conditions of strong attenuation of received signals power due to wide choice possibilities, the receiver can choose the best signal-to-noise signals for the navigation solution, which significantly improves the reliability of navigation. For the same reasons, narrowband interference will not be able to block the signals reception from all systems of the multi-system receiver, which increases the accuracy of positioning in rough conditions of reception, i.e. increases the noise immunity of the satellite navigation receiver. In addition to positioning accuracy due to the multi-system for the same reasons, the time to obtain a fixed solution in RTK mode, which is widely used to obtain high-precision data, is significantly reduced. The multi-system satellite receivers have a higher level of security due to the implementation in the software of an algorithm for monitoring the integrity of data received from navigation satellites. The algorithm is designed to detect anomalies caused by failures of navigation satellites or ground control system. Such algorithms are implemented in single-system receivers as well, however, due to the fact that the multi-system receiver has more source data for this algorithm, the possibility of detecting anomalies is wider. And that allows to identify and exclude from the navigation solution not only individual satellites, but also the whole navigation systems, operating in abnormal modes intentionally or unintentionally.

2. Cross-modal information flows in highly automated and autonomous vehicles

In highly automated vehicles, information flows (IF) of various semantics obtained from intelligent transport systems (to inform the driver), telematics systems, cameras and sensors of monitoring the functional state of the driver are additionally added to these flows, including for solving the problem of automatic assessment of determining the time of possible control transfer to the driver, when the driver is potentially able to perform the required algorithms of activity, when switching from the automatic mode of vehicle movement control to the "manual" one. The autonomous and highly automated vehicles are also a significant consumer of information from the telematics systems (for example, high-precision

3D road maps (including the area at a distance of several meters from them), which will be constantly updated in an automatic mode using the resources of the vehicles, including high-precision navigation coordinates of road marking; the vehicle sends the received data to the "cloud", where they are converted into a high-precision digital copy of the route of the road, figure 2) and from the evolving communication platform Cellular Vehicle to Everything (C-V2X) for "connected" vehicles and the transition to the use of 5G communication technology. The functionality of C-V2X has significant prospects, as it uses both network (already existing coverage of cellular networks and distribution of smartphones, tablets and gadgets) and "direct" communication used for V2X services. The volume of data flows received by the highly automated and autonomous vehicle is commensurate with the amount of data generated by the autonomous vehicle.

Figure 2. Fragment of a high-precision 3D roads map [3].

2.1. Definition

The cross-modal interaction in highly automated vehicles is the transformation, synchronization and analysis of heterogeneous information flows in onboard vehicle systems in real time, with a single target function.

The purpose of the analysis of heterogeneous information flows in the vehicle onboard systems is to obtain an integrated assessment (synthesis of integrated evaluation criteria) and automatic development of control commands for efficient operation in highly automated driving (HAD) with support of ADAS and intelligent transport systems (ITS) in real time.

One of the priority areas is automatic monitoring and analysis of the fleet of connected vehicles, including the exchange of information. This will make it possible to identify and block emerging threats in a timely manner.

2.2. The main problems of the "human factor" arising in the operation of the highly automated vehicles

It is known that with the growth of automation in all classes of "human–machine" systems, there are more and more monotonous fragments in the algorithms of human-operator activity, which have a negative impact on the "regular" implementation of the activity algorithms, primarily reduce the current awareness of the integrated situational circumstances (which affects the adequacy of the assessment of the dynamics of its development) and increase the reaction time of the human-operator in case of the

IOP Conf. Series: Materials Science and Engineering **534** (2019) 012003 doi:10.1088/1757-899X/534/1/012003

need for emergency action, which directly affects the safety of the functioning of the "human–machine" systems. This also applies to the drivers of the vehicle with the HAD.

It is known that the use of ACC leads to a relatively small reduction in the workload of the driver, and the HAD leads to a significant reduction in the workload compared to manual driving. The paper [4] presents an overview of the experimental data obtained in twelve studies on the effectiveness of solving additional tasks by the driver when using the ACC and the HAD under load and on awareness of the driver's situation in comparison with manual driving. The analysis of the working load measured as efficiency at performance of the arbitrary task displayed on the onboard display is given as well. The average number of tasks completed was 100% for manual driving, 112% – when using the ACC, and 261% – when using the HAD. Stated differently, using the ACC, drivers can perform approximately 12% more tasks on the visual display than with manual driving. However, when using the HAD, drivers can perform more than 2.5 times more tasks than with manual driving. More than 4,000 literary sources were subjected to meta-analysis.

About twenty years ago, studies of the problem of "human factor" with the increasing level of automation in the vehicle, mainly focused on the interaction of the drivers with the ACC, were started, and even then it was found that an increase in the level of automation can lead to a decrease in drivers' awareness of the road situation, contributing to a decrease in the efficiency of the implementation of algorithms for handling the vehicle in case of emergency situations [5]. Subsequently, similar results were obtained in the study of the influence of the ADAS maintaining the lane on the driver at the levels of automation of the vehicle from the first to the third.

The presented results indicate the relevance of the task of monitoring the driver's functional state (the potential ability to perform the required algorithms) in real time, including when using the HAD, including for prevention of the onset of driver relaxation, the level of which has a negative impact on the readiness for emergency action. The activation time of functional driver systems is by times higher than the response time of onboard systems.

3. The main provisions of the analysis method of the driver's potential to regain control over the highly automated vehicle

Development of evaluation criteria and conditions for automatic transition of control to the driver on the move, when changing the paradigm of the vehicle control from automatic to manual, is one of the most urgent tasks today. The main task is to determine at what point in time the safe transition of control to the driver is possible, when the driver is already potentially able to perform the required algorithms of activity in accordance with the current functional state. Obviously, for different drivers this time will be different, because various internal and external factors having different weight coefficients exert influence. Basic ones are awareness of the situational circumstances along the route, the current functional state (largely due to the nature of the driver's activity in the previous 5-10 minutes) and the individual characteristics of the driver.

Situational circumstances along the route are the quality and variable value, depending on many factors (traffic intensity and potential difficulties on specific local road sections, emergency situations, incidents and accidents, weather conditions, time of day, etc.). The studies on the driver's awareness of the situational circumstances along the route are conducted in the world quite intensively [4,6,7, etc.]. According to various estimates, and depending on the complexity of the situation and the type of the driver's activity in the previous time, the driver takes from a few tens of seconds (often about 60 seconds) to 360 seconds to adequately assess the situation. The majority of research projects use methods of monitoring and evaluating the dynamics of the visual analyzer of the driver, using the appropriate experimental equipment, in relation to the restoration of driver's control over the motion trajectory of the vehicle, the situational circumstances and the implementation of algorithms of activity (or their fragments) in different variations and in real scope of time. Directly during the restoration of awareness of the situation, the dynamics of changes in psycho-physiological and physiological parameters of drivers is of particular importance.

Methods of monitoring and maintaining the current functional state of the driver on the basis of monitoring and analysis of electrodermal parameters of the driver in connection with the performed algorithms of activity, directly during the execution of the algorithms of activity, are developed and repeatedly tested, and the systems that implement them are produced in quantity. It is found that they ensure the maintenance of the driver in a state of readiness for the emergency action [8-11, etc.].

Individual features of the driver, in this context, is primarily the dynamics of psycho-physiological (physiological) parameters, the level and dynamics of its professionally important qualities (PIQ) for the controlling the vehicle. The PIQ of a specific driver, including individual quantitative values (for example, the time spent on the implementation of control actions, receiving signal information), some of which are in the databases of systems of professional selection and development of such qualities. For example, the paper [12] presents an approach that allows the operator (driver) to control the perception of semantically binary relevant information. The most widespread system of professional selection of drivers are hardware and software complex automobile "Universal psychodiagnostic complex" (UPDK-MK) [13] and "Psychological certification of drivers" (VIENNA testing system) [14], that use about 30-50 psychological techniques, psycho-physiological parameters and personality tests to assess the level of the PIQ of drivers in stationary conditions during the professional selection or inspection. For example, the UPDK-MK provides testing of psycho-physiological PIQ (perception of spatial relations and time, eye estimation, stability, switchability and distribution of attention, memory, psychomotor, emotional stability, dynamics of performance) and the properties and qualities of the driver's personality, which make it possible for him to drive safely (neuro-mental stability, temperament properties, risk tolerance, proneness to conflict, resistance to monotony). The UPDK-MK also includes methods of development of PIQ (training of selectivity and concentration, attention distribution, increase of emotional stability, flexibility of actions rate, resistance to monotony) and methods of the general development (training of associative processes, memory on images and symbols). All methods used in testing the PIQ are valid. Information from the databases of professional selection systems or inspection of a particular driver can be available for the ITS and using the information channels C-V2X, on the board of the highly automated vehicle with this driver, and can be used in the analysis algorithms of the potential of the driver to regain control over the highly automated vehicle, to improve their reliability. Some values of professionally important parameters can be updated in automatic mode, for example, all that is associated with the reaction times of the drivers during the "manual" control [12].

4. Conclusions

The classification of the cross-modal information flows in the highly automated and autonomous vehicles is presented. The definition of the cross-modal interaction in the highly automated vehicles is given and the goal of the analysis of the heterogeneous information flows in onboard systems of the vehicles is formulated.

From the interdisciplinary perspective, on the basis of the system approach, the main provisions of the original method for analyzing the driver's potential to regain control of the highly automated vehicle are formulated. They are based on the driver's awareness of the situational circumstances along the route, monitoring the current functional state and individual driver's characteristics.

The development of the ITS, ADAS, active safety systems (including high-precision navigation modules, working with multiple GNSS) and steadily increasing level of the vehicles automation make it possible to formulate and solve entirely new problems for the safe functioning of "human–machine" transport systems, and the automatic analysis of the heterogeneous information streams gives a synergistic effect. The cross-modal interaction in "human-machine" transport systems is the interaction of hierarchically ranked multi-functional onboard systems of monitoring and control (the technical components of the vehicle, characteristics of the movement trajectory in relation to a navigation coordinate, the results of the work of the ADAS, including the functional state of the driver) between them and external systems (primarily: the ITS, traffic control and emergency services) with objective function of increasing the efficiency of the "human-machine" systems. Analysis of the cross-modal information exchange in the "human-machine" transport system with a single methodological position

IOP Conf. Series: Materials Science and Engineering **534** (2019) 012003 doi:10.1088/1757-899X/534/1/012003

will make it possible, with precision acceptable for the solution of applied tasks, to form in the real time an integrated assessment of the status of basic components of the "human-machine" system and the dynamics of changes in the situational circumstances in the relationships, including external environment.

References

[1] Intel: autonomous vehicle to generate 4 TB of data per day URL: http://www.intel.ru/content/www/ru/ru/automotive/autonomous-vehicles.html

[2] Savchenko V V and Poddubko S N 2017 *Proc. Int. Automobile Scientific Forum "Intellectual transport systems" (Moscow)* R20171014

[3] The most detailed world maps to be necessary for vehicles, not for people URL: http://security-corp.org/programming/vebdev/39026-samye-podrobnye-karty-mira-budut-nuzhny-avtomobilyam-a-ne-lyudyam.html

[4] de Winter J C F, Happee R, Martens M H and Stanton N A URL: https://www.vdi-wissensforum.de/news/effects-of-adaptive-cruise-control-and-highly-automated-driving

[5] Endsley M and Kaber D 1999 *Ergonomics* **42**(3) pp 462–92

[6] Merat N 2012 *Human Factors* **54** pp 762–71

[7] Merat N, Jamson H, Lai F and Carsten O 2014 *Transportation Research Part F: Traffic Psychology and Behavior* **27** Part B pp 274–82

[8] Savchenko V V 2009 *J. of Automation and Information Sciences* **41**(1) pp 75–80

[9] Dementienko V V 2015 *Sci. J. of Transportation* Especial Issue **6** pp 110–14

[10] Savchenko V V, Svistun M S and Sikorskiy V V 2008 *Automotive industry* **1** pp 32-34

[11] Dementienko V V and Dorokhov V B 2013 *J. of Higher Nervous Activity* **63**(1) pp 24–32

[12] Savchenko V V 2005 *Proc. of the National Academy of Sciences of Belarus. Physico-technical series* **2** pp 29-37

[13] Kremez A S 2014 *Automation in industry* **2** pp 29–33

[14] Psychological certification of drivers. VIENNA test system URL: https://ru.asystems.as/produkty-uslugi/

IOP Conf. Series: Materials Science and Engineering **534** (2019) 012004 doi:10.1088/1757-899X/534/1/012004

Estimating operation modes for the individual wheel electric drive of the all-wheel drive vehicle with the use of the driving simulator

G O Kotiev[1], A V Miroshnichenko[2], A A Stadukhin[3] and B B Kositsyn[1,4]

[1] Department of the Wheeled Vehicles, Bauman Moscow State Technical University, 105005 Baumanskaya 2th st., 5, Moscow, Russian Federation
[2] LLC OKB Tekhnika, Novaya Zorya st., 6, Moscow, 115191 Russian Federation.
[3] Department of the Multi-purpose track vehicles and mobile robots, Bauman Moscow State Technical University, 105005 Baumanskaya 2th st., 5, Moscow, Russian Federation

E-mail: [1] kotievgo@yandex.ru; [2] avm-technika@mail.ru; [3] ant.m9@yandex.ru; [4] kositsyn_b@bmstu.ru.

Abstract. With the development of the electric drive systems and vehicle on-board power sources the electric transport is becoming more popular, in particular wheeled vehicles with individually driven wheels, while the question of choosing power plants for this type of transport remains unresolved. In the presented article, the authors illustrate an approach to determining the requirements for power plants of such vehicles by an example of a four-wheel drive car equipped with an individual electric wheel drive. The approach is based on the use of the real-time driving simulator allowing the operator to drive the vehicle along typical routes. The typical routes for the driving simulator are generated from the known statistical data on the excitation inputs during the vehicle motion. Thus, the developed driving simulator provides analysis of the vehicle traction motors operation in the conditions most close to the actual ones and collection of the necessary statistical data which can be used for statement of the requirements for new electric machines which would provide their most effective use on the vehicle.

Currently, the development of the wheeled vehicles with individual drive wheels is getting substantial boost. The increased interest in this type of drive can be explained by the fact that the use of the drive motor provides a number of advantages:

- increased cross-country ability of the wheeled vehicle;
- stability and control improvement;
- better dynamic properties of the wheeled vehicle;
- better comfort for the driver and passengers;
- simplified gearing system as compared to the mechanical drivetrain.

Nowadays, the main disadvantage of this transmission layout is the high weight and cost of the fuel and energy sources, which reduces its competitiveness.

For the high-speed wheeled vehicles the most difficult mode is the upslope motion. In this mode, the normal reactions in the contact patches of the front axle wheels reduce, thus reducing the maximum longitudinal tractive forces in the contact patch. To increase the average speeds it is necessary that the

IOP Conf. Series: Materials Science and Engineering **534** (2019) 012004 doi:10.1088/1757-899X/534/1/012004

drive motor of the more loaded wheels of the rear axles be able to realize higher torque and power in a wide range of rotation frequencies. Such requirements for the drive motor lead to a significant increase in their size and weight, which means, for the case of rectilinear motion of the transport vehicle, that the transmission developed on this principle will be oversized.

Since the transport wheeled vehicles work in the heavy duty modes only a small amount of their operating time, it is advisable to drive in such modes due to the operation of the drive motor in the short-term operation mode. Thus, in order to state the requirements for the electric machines and to ensure a minimum weight of the power drive it is necessary to collect statistical information about the short-term and long-term operation of the drive motor on the vehicle.

To solve this problem the authors have developed a driving simulator on the basis of the simulation mathematical model of the dynamics of a horizontal curvilinear and rectilinear upslope/downslope motion of a wheeled vehicle [1,2,3,4,5,6].

In the mathematical model [7], the authors consider the dynamics of the wheeled vehicle as the motion of a solid body in a horizontal plane on a planar undeformed ground surface. The dynamics of the wheeled vehicle also consists of the translational motion of the center of mass and the rotational motion around the center of mass (figure 1). The connection of the wheels with the vehicle frame in the vertical direction is considered to be rigid, that is the elastic properties of the suspension are neglected. At the same time, the model takes into account the redistribution of the wheel loads.

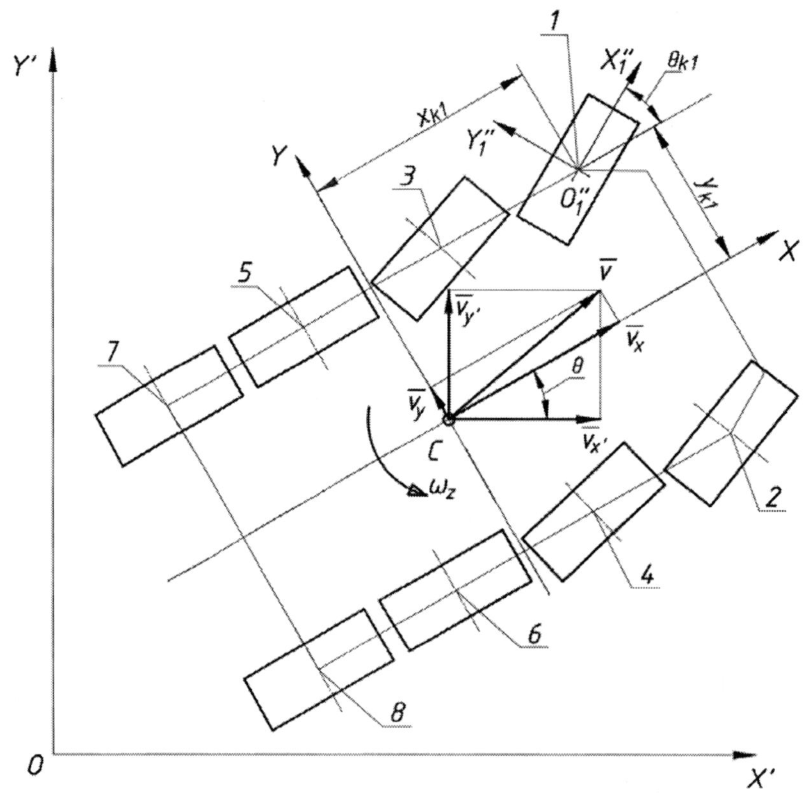

Figure 1. Planar motion of the wheeled vehicle.

According to the presented scheme of the vehicle motion (figure 1), the system of equations describing the planar motion of the wheeled vehicle has form (1). This system allows calculation of the current acceleration of the center of mass from the values of the effective forces and moments acting on the vehicle.

$$\begin{cases} a_x = \dfrac{dv_x}{dt} - \omega_z v_y = \dfrac{1}{m}\left(\sum_{i=1}^{n} R_{x_i} - P_w\right) \\ a_y = \dfrac{dv_y}{dt} + \omega_z v_x = \dfrac{1}{m}\left(\sum_{i=1}^{n} R_{y_i}\right) \\ J_z \dfrac{d\omega_z}{dt} = \sum_{i=1}^{n} M\left(R_{y_i}\right) - \sum_{i=1}^{n} M\left(R_{x_i}\right) - \sum_{i=1}^{n} M_{\text{сп}_i}, \\ v_{x'} = \dfrac{dx'}{dt} = v_x \cos(\theta) - v_y \sin(\theta) \\ v_{y'} = \dfrac{dy'}{dt} = v_x \sin(\theta) + v_y \cos(\theta) \\ \omega_z = \dfrac{d\theta}{dt} \end{cases} \quad (1)$$

where

m –vehicle mass; J_z is the vehicle mass moment of inertia relative to the z-axis passing through the center of mass; a_x, a_y – projections of the vehicle center of mass acceleration onto the x - and y-axes; v_x, v_y – projections of the vehicle center of mass velocity onto the x - and y-axes; $v_{x'}$, $v_{y'}$ – projections of the acceleration of the vehicle center of mass onto the x' - and y'-axes; $\frac{dv_x}{dt}, \frac{dv_y}{dt}$ - projections of the relative derivative of the velocity vector of the vehicle center of mass onto the x - and y-axes; ω_z – projection of the angular velocity vector of rotation of the vehicle onto the vertical z-axis; θ - the rotation angle, in the fixed coordinate system; x', y' – coordinates of the vehicle mass center in the coordinate system $x' - y'$; R_{x_i}, R_{y_i} – longitudinal and lateral components of the reaction acting at the contact patch of the *i*-th wheel; P_w – projection of the air resistance force onto the x -axis; $M_{\text{сп}_i}$ – moment of resistance to rotation of the contact patch of the *i*-th wheel around a vertical axis passing through the patch center; n – number of the vehicle wheels.

The force acting between the wheel and the ground in the plane of the road is determined with the use of the approach based on the friction ellipse, according to which the force of interaction of the tire contact patch with the ground is directed opposite to the slip velocity [8, 9, 10, 11, 12].

The steering wheel angle is connected with the angles of the steered wheels according to Ackerman geometry [10] and the steering gear ratio.

The following system of equations describes the connection of the drive motor torque of each wheel with the value of the control parameter and the rotor speed:

$$M_d(\omega_{id}, h) = \begin{cases} M_{id}^{max} \cdot h, \text{ если } \omega_{id} < \dfrac{N_{id}^{max}}{M_{id}^{max}} \text{ и } \omega_{id} \geq 0 \\ \dfrac{N_{id}^{max} \cdot h}{\omega_{id}}, \text{ если } \omega_{id} \geq \dfrac{N_{id}^{max}}{M_{id}^{max}} \text{ и } \omega_{id} < \omega_{id}^{max}, \\ 0 \quad, \text{если } \omega_{id} \geq \omega_{id}^{max} \end{cases} \quad (2)$$

where

M_{id}^{max} – maximum traction/braking torque generated by the motor; N_{id}^{max} – maximum traction/braking power of the motor; ω_{id}^{max} – limit rotation frequency of the rotor; ω_{id} – current rotation frequency of the rotor; h – the value of the control parameter ($h \in [-1,1]$).

The presented simulation mathematical model has been used for the development of the real-time driving simulator. The simulator software uses the 4-5 order Runge–Kutta–Fehlberg integrator [11] and can work under conventional MS Windows operating system providing synchronization of computation time and real time by waiting for synchronization [15].

The driving simulator includes:

- a computer which provides operation of the wheeled vehicle model in the real-time mode – determination of the coordinates of the vehicle on the route, calculation of the current speeds and acceleration of the vehicle under the influence of external forces;
- controls which transmit the driver inputs into the simulation model and provide feedback;
- a graphical visual display that allows the operator to estimate the curvature of the upcoming turn, the angle of the slope, driving conditions (the type of the road) in the same way as it is done by the driver of a real vehicle, as well as to receive information about the current parameters of the vehicle motion (speed, traveled distance, etc.).

Figure 2 shows the layout of the driving simulator.

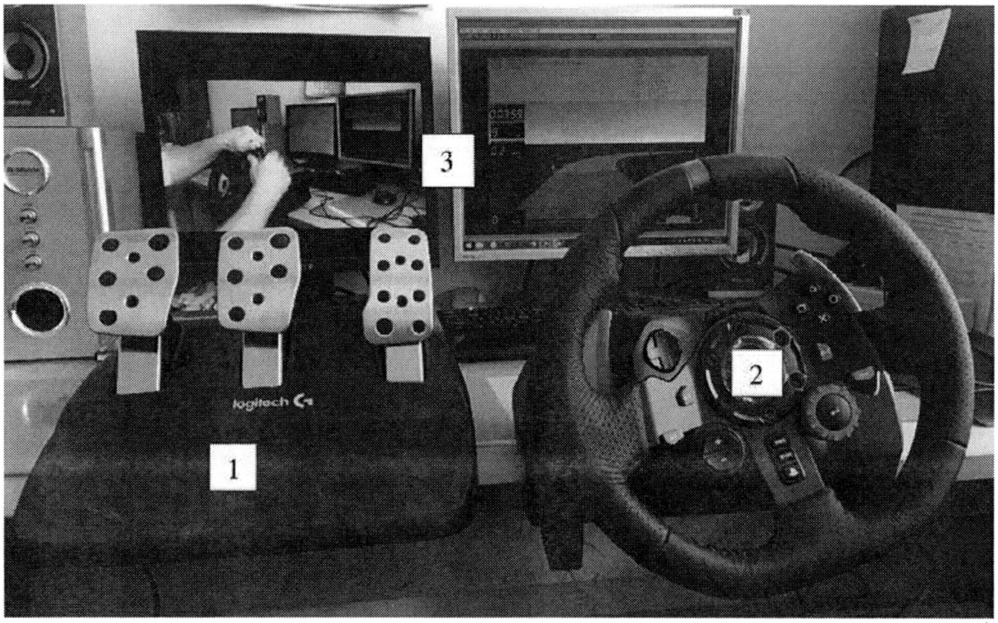

Figure 2. Driving simulator layout:

1,2 – controls; 3 – graphic information output station.

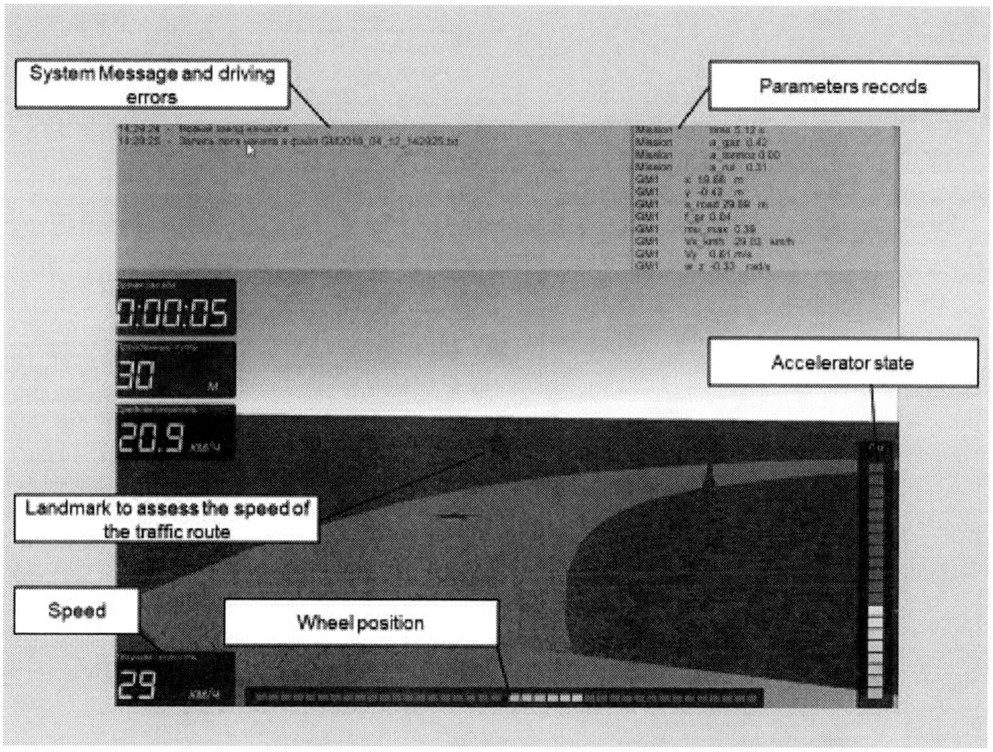

Figure 3. Graphical information output of the driving simulator.

The principle of operation of the presented driving simulator is as follows. The origin of the model fixed coordinate system is superposed with the start point of the route and the center of mass of the vehicle. The operator provides control signals which, by means of the sensors installed on the controls, are transmitted from the operator to the computer and used as the input parameters for the simulation model. During the simulation, according to the control actions and road conditions under each wheel of the vehicle, the model calculates the kinematic parameters of the vehicle (coordinates of the center of mass, wheel speeds, etc.) at a given step on the real-time scale by numerical integration. At the end of the time step the vehicle is transferred in the fixed coordinate system (relative to the start point of the route) to the position calculated during the integration process. The model parameters of interest to the operator (speed, traveled distance, etc.), and the position of the vehicle relative to the typical route are displayed on the simulator screen in a virtual three-dimensional space and on the dials of the virtual dashboard of the simulator. The driving simulator operator, visually receiving information from the screen and feeling the force on the controls, adjusts the control action based on the experience of operating real vehicles and the simulator again starts the process of calculating the kinematic parameters of the vehicle for the next time step.

Figure 4 shows the block diagram of the driving simulator elements interaction.

IOP Conf. Series: Materials Science and Engineering **534** (2019) 012004 doi:10.1088/1757-899X/534/1/012004

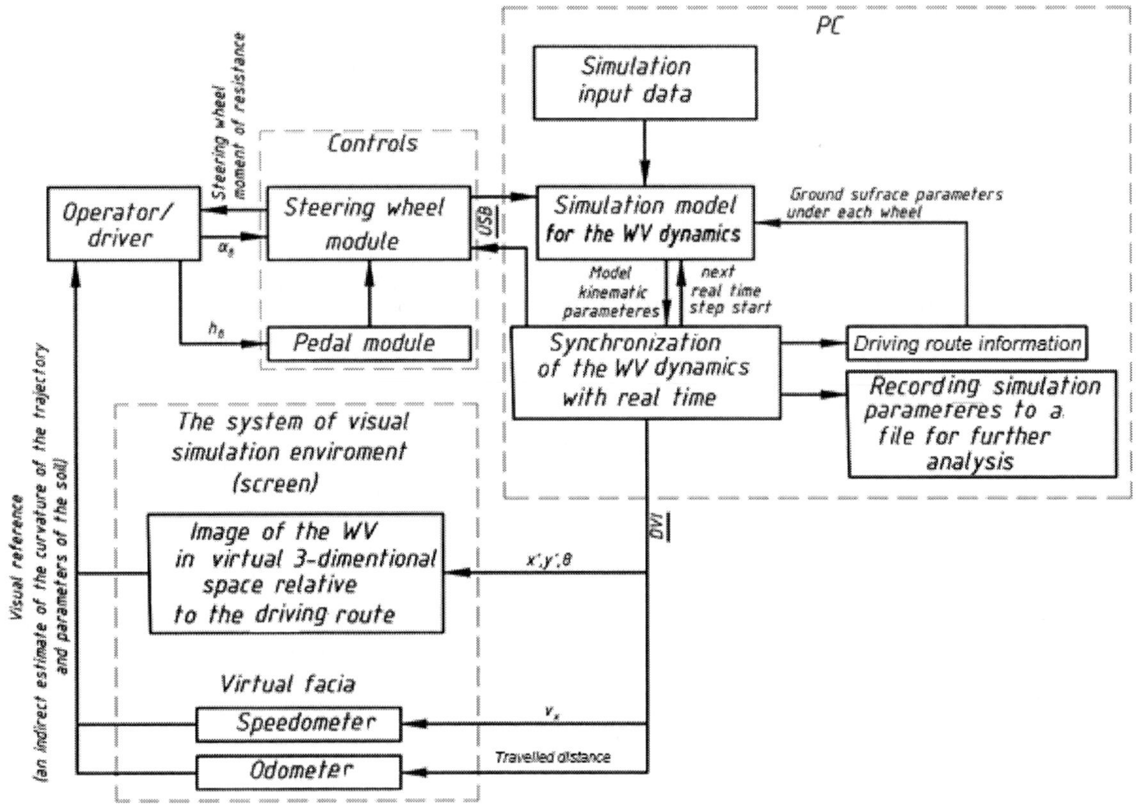

Figure 4. Block diagram of the driving simulator elements interaction (WV – wheeled vehicle).

Typical driving routes were generated with the use of the mathematical statistics methods and well-known probabilistic characteristics of excitation inputs affecting the dynamics of the transport wheeled vehicle. The values of the road curvature *k_r*, maximum road – tire friction coefficient *μ_s_max*, and total motion resistance coefficient *f_sum* were randomly distributed along the driving route with the use of the statistical data on these parameters by the non-canonical representations method [16, 17, 18, 19, 20]. Figure 5 shows an example of the driving route obtained from the realization of the random function of the road curvature $k_d(s)$.

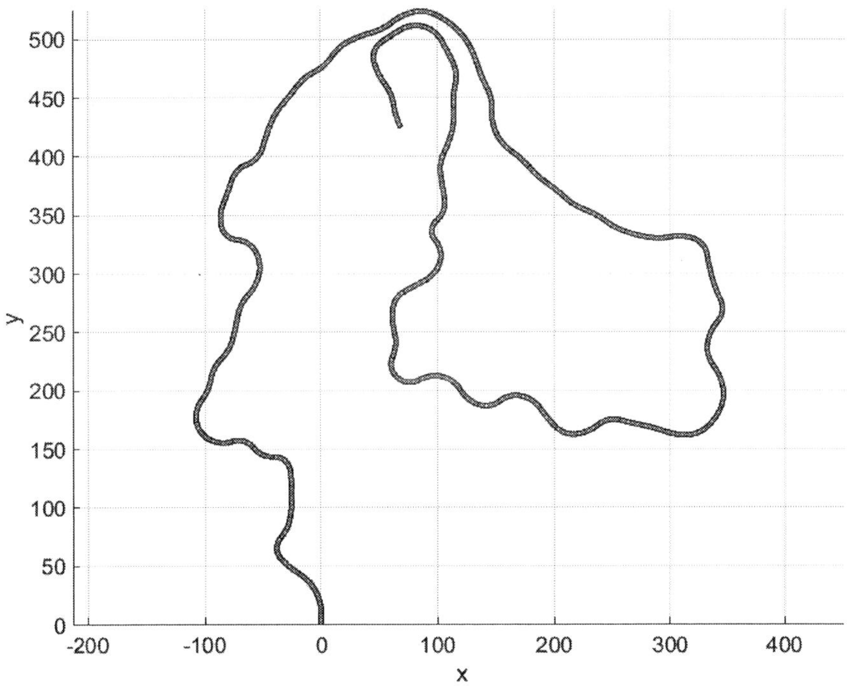

Figure 5. An example of the generated driving route.

To evaluate the required parameters of the drive motor cooling system in the vehicle operation conditions on the route, the corresponding mathematical model was developed (see Figure 6).

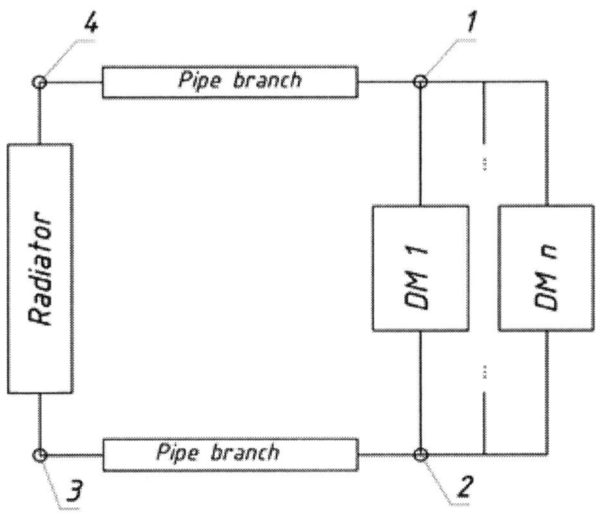

Figure 6. Vehicle cooling system schematic.

The following assumptions for the cooling system model have been made:

- Fluid flow in the cooling system is constant and does not depend on the rotor speed of the drive motor;

IOP Conf. Series: Materials Science and Engineering **534** (2019) 012004 doi:10.1088/1757-899X/534/1/012004

- In node 1, the coolant flow is distributed equally between the motors;
- Heat exchange between the cooling liquid and the external environment is carried out only by means of the radiator (there is no heat exchange through the pipe branches);
- The heat capacity and density of the cooling liquid are independent from the temperature;
- All the heat power generated by the drive motor is used to heat the cooling liquid.

The following system of equations describes the cooling system of the vehicle:

$$\begin{cases} c_j\rho_jV_{patr_{41}}\dfrac{dT_1}{dt} = \Phi_4 - \sum\limits_{i=1}^{n}\Phi_{1_i} \\ c_j\rho_jV_{DM}\dfrac{dT_{2_i}}{dt} = \Phi_{1_i} - \Phi_{2_i} + N_{Q_i} \\ c_j\rho_jV_{patr_{23}}\dfrac{dT_3}{dt} = \sum\limits_{i=1}^{n}\Phi_{2_i} - \Phi_3 \\ c_j\rho_jV_{rad}\dfrac{dT_4}{dt} = \Phi_3 - \Phi_4 - N_{rad} \end{cases}, \quad (3)$$

where

c_j – heat capacity of the cooling liquid; ρ_j – density of the cooling liquid; T_1, T_3, T_4– the temperature of the liquid in the respective nodes of the cooling system; T_{2_i} – the temperature the liquid coming from the *i*-th drive motor in the 2nd node of the cooling system; $V_{patr_{41}}$ – volume of the liquid in the pipe at the segment 4 to 1 of the cooling system; $V_{patr_{23}}$ – volume of the liquid in the pipe at the segment 2 to 3 of the cooling system; V_{DM} – liquid volume in the drive motor; V_{rad}–liquid volume in the radiator; Φ_3, Φ_4 – thermal energy flux in the 3rd and 4th node of the cooling system; Φ_{1_i} – flux of the thermal energy entering the segment of the cooling system containing the *i*-th drive motor from node 1; Φ_{2_i} – flux of the thermal energy entering the segment of the cooling system containing the *i*-th drive motor from node 2; N_{rad}– power of the heat exchange between the cooling liquid in the radiator and the environment; N_{Q_i} – thermal power transmitted to the coolant of the *i*-th drive motor; n – number of used drive motors.

Thus, riding along typical routes it is possible to register parameters of the drive motors and use these parameters for formulation of the requirements for their long-term and short-term operation modes and their cooling system in order to ensure effective use of the electric motors.

An example simulation has been performed for an 8x8 wheeled vehicle riding along a typical 2 km route (Figure 5) on the traffic-compacted soil road (*f_sum*=0,015…0,14, *μ_s_max*=0,31…0,85).

Table 1. Vehicle specifications.

Gross weight, kg	25000
Wheelbase, mm	4625
Track width, mm	2500
Height of the center of mass, mm	1400
Moment of inertia around the y-axis passing through the center of mass, kg. m2	57580
Tire radius, mm	630
Steering type	four-wheel steering on the first an the second axes
Tire width, mm	500
Distance from the rear axle to the center of mass, mm	2312
Distance from the rear axis to axis 1, mm	4625
Distance from the rear axis to axis 2, mm	3083
Distance from the rear axis to axis 3, mm	1542
Type of transmission layout	Individual electric drive of all wheels.
Steering gear ratio	10
Drive motor maximum power , kW	62,5 x 8
Drive motor maximum torque, Nm	1000
Drive motor speed at the maximum torque, rpm	4000

Figure 7 shows the time histories of the actual and average speed of the vehicle driving on the simulated route.

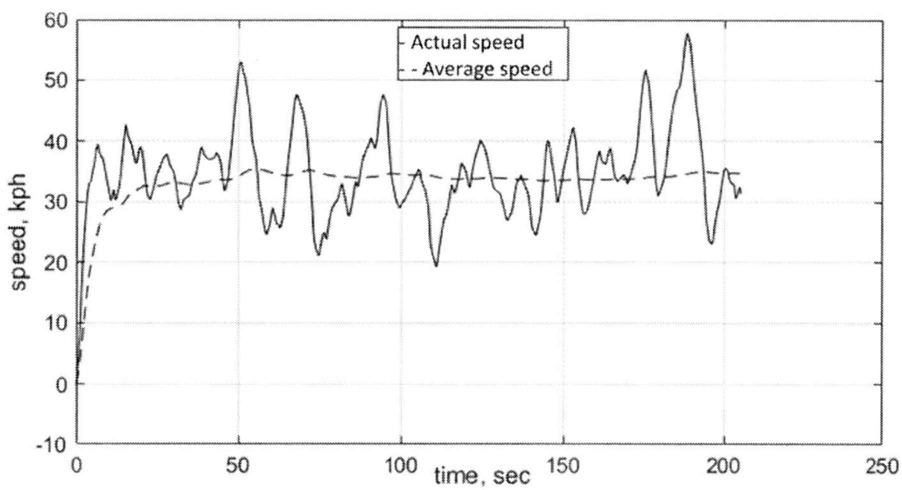

Figure 7. Actual and average speed of the vehicle on the simulated route.

It can be seen from the simulation results that the vehicle average speed for this type of the road is close to the steady-state value of 34.7 km/h. For this reason it can be concluded that the length of the route is sufficient and the simulation can be stopped since further statistics collection will not have an impact on the result.

Figures 8 and 9 show the time histories of the torques and rotor speeds of the drive motors for a typical route.

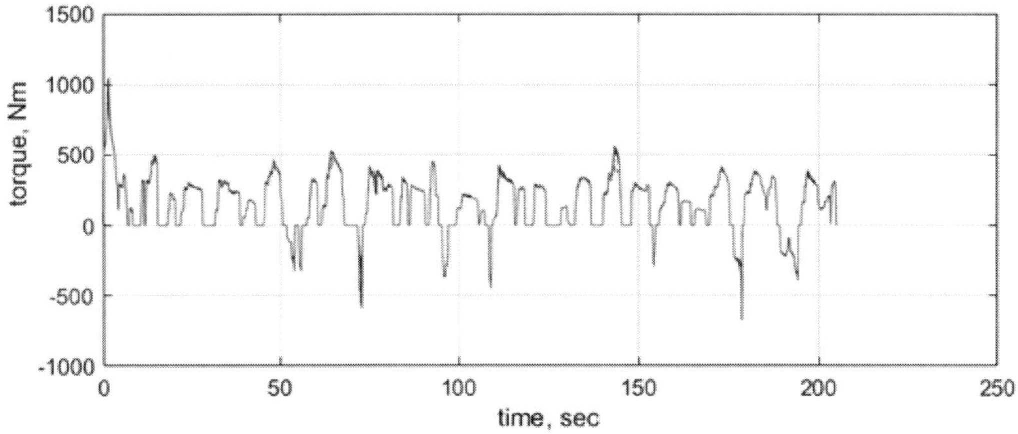

Figure 8. Torques of the drive motors on a typical route.

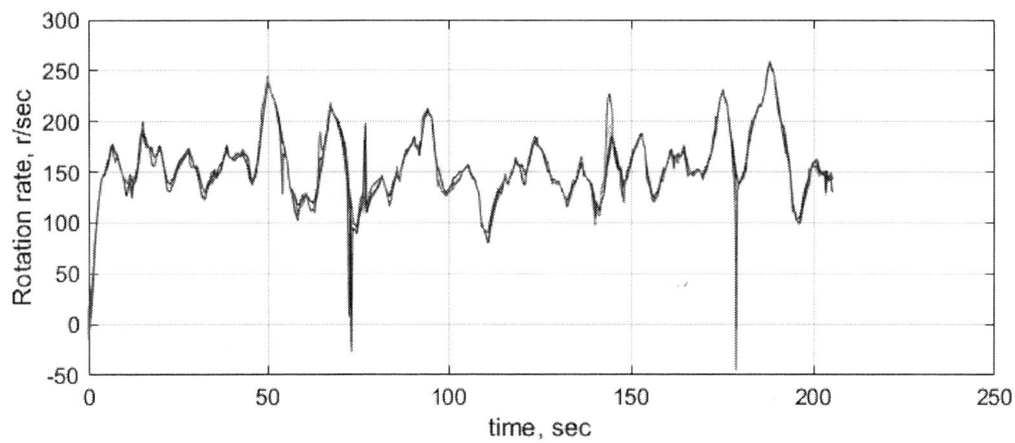

Figure 9. Rotor speeds of the drive motors on a typical route.

Figure 10 shows the time history of the coolant temperature in the cooling system nodes when driving on a typical route. The liquid flow rate in the system is taken to be 40 l/ min, the volume of liquid inside the drive motor $V_{DM} = 1$ l, the pipes $V_{{patr}_{41}} = 0{,}5$ l, $V_{{patr}_{23}} = 0{,}5$ l and the radiator $V_{\text{rad}} = 3$ l.

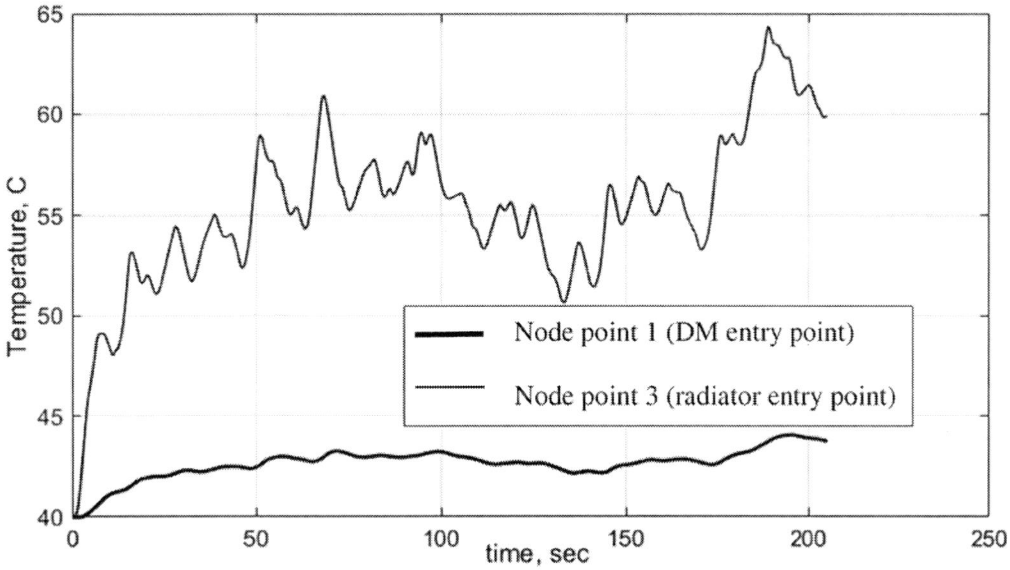

Figure 10. Coolant temperature for two node points of the cooling system.

The simulation results show that the drive motors used on the tested vehicle provide average speed 34.7 km/h on a typical route, and the cooling system provides the temperature of the cooling liquid exiting the drive motors no more than 65°C.

The proposed approach to the determination of the required parameters of the fuel and energy of transport vehicles with individual wheel drive allows to take into account the design features of the vehicle, road and ground conditions and qualification of the drivers. The data from the developed driving simulator make it possible to provide the desired torque – speed characteristics of the drive motor, rational gear ratio of the transmission, coolant consumption, the area of the radiator, etc at the early stages of the wheeled vehicle development.

The work has been carried out at Bauman Moscow State Technical University with the financial support of the Ministry of Education and Science of the Russian Federation under Agreement 14.574.21.0178 (Unique identifier of the work: RFMEF57417X0178).

References

[1] Kotiev G O, Butarovich D O and Kositsyn B B 2018 Energy efficient motion control of the electric bus on route *IOP Conference Series: Materials Science and Engineering* **315 issue 1** (DOI: 10.1088/1757-899X/315/1/012014)

[2] Zhileykin M M, Kotiev G O and Nagatsev M V 2018 Comparative analysis of the operation efficiency of the continuous and relay control systems of a multi-axle wheeled vehicle suspension *IOP Conference Series: Materials Science and Engineering* **315 issue 1** (DOI: 10.1088/1757-899X/315/1/012030)

[3] Zhileykin M M, Kotiev G O and Nagatsev M V 2018 Synthesis of the adaptive continuous system for the multi-axle wheeled vehicle body oscillation damping *IOP Conference Series: Materials Science and Engineering* **315 issue 1** (DOI: 10.1088/1757-899X/315/1/012031)

[4] Gorelov V A, Komissarov A I and Miroshnichenko A V 2015 8x8 wheeled vehicle modeling in a multibody dynamics simulation software *Procedia Engineering* **129** pp 300-307 (DOI: 10.1016/j.proeng.2015.12.066)

[5] Keller A V, Gorelov V A and Anchukov V V 2015 Modeling truck driveline dynamic loads at differential locking unit engagement *Procedia Engineering* **129** pp 280-287 (DOI: 10.1016/j.proeng.2015.12.063)

[6] Keller A V, Gorelov V A, Vdovin D S, et al. 2015 Mathematical model of all-terrain truck *Proceedings of the ECCOMAS Thematic Conference on Multibody Dynamics* pp 1285–1296
[7] Kotiev G O, Gorelov V A and Beketov A A 2008 Motion mathematical model off-road vehicle *AAE Journal* **1**(48) pp 50-54
[8] Larin V V 2010 *Theory of motion off-road wheel vehicle* (Moscow: BMSTU Press) p 391
[9] Keller A and Aliukov S 2015 Methodology of System Analysis of Power Distribution among Drive Wheels of an All-wheel-drive Truck *SAE Technical Paper* 2015-01-2788
[10] Dik A B 1988 Calculation of the stationary and non-stationary characteristics of the braking wheel during a drivable motion *PhD dissertation* (Omsk: SADI) p 224
[11] Keller A and Aliukov S 2016 Efficient Power Distribution in an All-Wheel Ground Vehicles *SAE Technical Paper* 2016-01-1105
[12] Rozhdestvenskij Yu L and Mashkov K Yu 1982 On the formation of reactions during the rolling of an elastic wheel along a non-deformable base *MVTU proceedings* **390** pp 56–64
[13] Rajmpel J 1987 Truck chassis. Steering control (Moscow: Mashinostroenie) p 232
[14] Belanger J, Venne P and Paquin J N 2009 The what, where and why of real-time simulation Transactions on Power Delivery **24**(2) pp 390-399
[15] Buzunov N V 2017 Method for developing control laws for the loader of a steering wheel in the absence of "hard" communication in the steering system for turning wheeled vehicles *PhD dissertation* p 186
[16] Shalygin A S and Palagin Yu I 1986 *Applied methods of statistical modeling* (Leningrad: Mashinostroenie) p 320
[17] Derzhanskij V B and Taratorkin I A 2012 *Stabilization of the movement of a high-speed tracked vehicle* (Kurgan: Kurgan State University publishing house) p 61
[18] Keller A and Aliukov S 2015 Effectiveness of Methods of Power Distribution in Transmissions of All-Wheel-Drive Trucks *SAE Technical Paper* 2015-01-2732
[19] Keller A and Aliukov S, Anchukov V, et al. 2016 Investigations of Power Distribution in Transmissions of Heavy Trucks *SAE Technical Paper* 2016-01-1100
[20] Savochkin V A and Dmitriev A A 1993 *Statistical dynamics of transport and traction caterpillar machines* (Moscow: Mashinostroenie) p 320

IOP Conf. Series: Materials Science and Engineering **534** (2019) 012005 doi:10.1088/1757-899X/534/1/012005

Creation concept and architecture principle for vehicle motion control competitive systems under road and climatic conditions of the Russian Federation

A M Saikin[1], S E Buznikov[2], A A Barashkov[3], D S Elkin[4], V O Strukov[4], N S Shabanov[4] and E V Lucenko[4]

[1] Special programs department, FSUE "NAMI", Moscow, Russian Federation
[2] Simulation department, FSUE "NAMI", Moscow, Russian Federation
[3] Vehicle safety department, Research center for testing and fine-tuning of automotive equipment, FSUE "NAMI", Moscow, Russian Federation
[4] Intelligent systems department, FSUE "NAMI", Moscow, Russian Federation

E-mail: sergey.buznikov@nami.ru

Abstract. The urgency of traffic safety improvement is determined by the global significance level of this problem. Current research provides the scientific substantiation of the creation concept and building principles for competitive motion control systems (MCSs) under road and climatic conditions of the Russian Federation. The study was carried out using system analysis methods and modern control theory. The safe vehicle control task is formulated as prevention of collisions with moving and stationary obstacles on the route taking into account the road and climatic conditions of the Russian Federation. To estimate the quality of the dynamic stabilization solution modified quadratic functional is introduced. The results of the study are presented as an analysis of properties of the modified quadratic control quality functional and associated consumer indicators. Based on the analysis of results of the minimization problem for control quality quadratic functional and the multicriteria problem could be solved by software and hardware modification of intelligent systems realized in the minimal configuration of hardware. The novelty of the work is the scientific substantiation of the concept for the competitive vehicle MCSs creation based on virtual information sensors, the use of new mathematical models and control algorithms to solve the problem of collision avoidance.

1. Introduction

The problem of creation of competitive products is traditionally the central problem of the developed countries with market economy. To solve it, intellectual, financial and labor resources of both private firms and state organizations are involved.

One of the most intensively developing industries in the developed world is the automobile manufacturing. The achievements of the modern science and technology are successfully used to increase the competitiveness level of the manufactured technical products and expand sales markets. Leading foreign companies and domestic organizations are conducting large-scale research in the field of vehicle active safety systems [1] and in the field of driverless vehicles creation [2] [3].

The aim of the research is the scientific substantiation of the concept and principles of creation of competitive motion control systems (MCSs) building under the road and climatic conditions of the Russian Federation.

The problem of collision prevention on the route is among the control tasks that should be solved with the help of MCSs. For driverless vehicles, the position stabilization is also necessary task [4].

To solve the first problem, various active safety systems are used, which are standard equipment of foreign cars.

To solve the second problem, navigation systems [5] and technical vision systems [6] are used, including radars, video cameras, lidars, ultrasonic sensors, etc.

The road and climatic conditions of the Russian Federation considerably differ from the conditions of car operation in Central Europe, North America and the countries of the Asian region.

For the Russian Federation, long periods of low temperatures, snowfalls, fog, ice slick are typical.

Unsatisfactory state of roads is added to the unfavorable climatic conditions: lack of visible marking, increased rutting, significant coating unevenness. These unfavorable conditions can affect the effectiveness of the active safety and technical vision systems and reduce it to unacceptable level, making the vehicles safe operation impossible.

The concept of creation of competitive systems for vehicles motion control under road and climatic conditions of the Russian Federation is a group of goals, the achievement of which is possible with the use of new technologies, models and principles of building of information control systems [7].

2. Task statement

Multicriterion optimization methods are used within hardware *(R_H)* and software *(R_S)* tools [8] for the scientific substantiation of the concept for competitive vehicle MCSs creation.

Summarizing the experience gained from work with the consumers of high-tech technical products in the field of automotive electronics allows us to determine the main consumer indicators of interest to the end user.

As general criteria for evaluating technical systems, the following consumer indicators can be considered:

- level of external design;
- completeness of implemented functions;
- level of observation and control errors;
- level of power consumption from external sources;
- degree of external factors influence;
- degree of influence on adjacent systems;
- level of fault tolerance;
- restrictions on installation;
- operating costs;
- price of the system.

The best solution to the multicriteria problem within software-hardware solutions $R = (R_H, R_S)^T$ is an R for which the system of inequalities is satisfied:

$$q_i(R) \leq q_{imin}, 0 \leq i \leq 9, \text{ where} \quad (1)$$

$q_{imin} = \min[q_i(R_1), \dots, q_i(R_n)]$

$R_1 \div R_n$ – known technical solutions.

The known technical solutions R satisfying to the inequalities (1) form a region of record decisions that correspond to the creation of systems with the highest level of the competitiveness concept.

3. Research results

In order to carry out a comparative analysis of possible technical solutions for the vehicle MCS, we use the properties of the quadratic control quality functional and consumer indicators.

At the conceptual level, the MCS should provide the solution of the following applied tasks:

- detect the approach and exceeding the dangerous states boundaries;
- stabilize the longitudinal and lateral slip of the wheels within safe limits;
- stabilize the safe speed and distance to the obstacle on the lane;
- perform automatic braking to a complete stop before a stationary obstacle on the lane;
- stabilize the course and position of the car on a given trajectory.

As a final goal, the MCS must provide the solution to the task of object state coordinate dynamic stabilization.

To evaluate the solution quality of the vehicle dynamic stabilization problem, to which the collision avoidance problem is reduced, a modified control quality functional $Q(t_2)$ is introduced to the final time interval of operation $t_1 \div t_2$:

$$Q(t_2) = Q_r(t_2) + Q_{\text{con}}(t_2) + Q_{exp}(t_2), \text{ where} \tag{2}$$

$$Q_r(t_2) = \int_{t_1}^{t_2} \sum_{i=1}^{N} C_{1i}[x_i(\tau) - x_{ib}^{U}]^2 d\tau + \int_{t_1}^{t_2} \sum_{i=1}^{N} C_{2i}[x_i(\tau) - x_{ib}^{L}]^2 d\tau;$$

$$Q_{con}(t_2) = \int_{t_1}^{t_2} \sum_{i=1}^{5} C_{3i} \cdot U_i^2(\tau) d\tau;$$

$$Q_{exp}(t_2) = \int_{t_1}^{t_2} \Delta C(R,\tau) d\tau + Q(t_1).$$

The summands of the functional $Q_r(t_2)$ characterize the solution quality of the dynamical stabilization problem of the state vector $X(t)$ within the upper and lower dynamic boundaries $X_b^U(t)$ and $X_b^L(t)$. If the state vector components $x_i(t)$ are within the upper and lower bounds, the penalty coefficients C_{1i} and C_{2i} are zeroed, otherwise they denote large positive values $C_{1i} \gg 0$ and $C_{2i} \gg 0$.

The risk functional $Q_r(t_2)$ is zero if control of the object U provides the satisfied constraints on the state vector. If the boundaries are violated, it means that the conditions for preventing one or more typical collisions are not meet for the control object and prerequisites for an accident arise. The more time the object is in a dangerous state, the more accumulated risk becomes.

The summand of the control cost functional $Q_{con}(t_2)$ corresponds to the equivalent energy of the driver spent for control. The control vector components $U = (U_1, U_2, U_3, U_4, U_5)^T$ correspond to the control actions on the gearbox (U_1), the throttle (U_2), the braking system (U_3), the steering wheel (U_4) and the clutch pedal (U_5).

The choosing and justifying problem of the MCS optimal version at the content level is reduced to modified functional $Q(t_2)$ minimization due to the controls U, realized in the hardware and software environment R:

$$Q(t_2) \to min \tag{3}$$

at $U \in U_{add}, R \in R_{add}$

The best or optimal solution meets the criteria (3) is the control $U(t) = F(X, X_b^U, X_b^L)$ realized by the driver and control system using data generated by the software and hardware environment R, providing solution of the dynamic stabilization problem and characterized by minimal acquisition and operation costs.

The summand of the cost functional for the software and hardware acquisition and operation: $Q_{exp}(t_2, R) = \int_{t_1}^{t_2} \Delta C(R,\tau) d\tau + Q(t_1, R)$ is determined by the number $N_c = N + N_m + N_D$ used the information sensors on object state variable – N, driver – N_D, actuators for control – N_m and software:

$$Q(t_1, R) = \sum_{i=1}^{N_c} C_{Hi} + C_s \tag{4}$$

where $C_{Hi}, 1 \le i \le N_c$ – the cost of information sensors and actuators purchased or produced by the manufacturer;

C_s – cost of special software replication for the developer, incommensurable with C_{Hi}, $C_{Hi} \gg C_s$.

IOP Conf. Series: Materials Science and Engineering **534** (2019) 012005 doi:10.1088/1757-899X/534/1/012005

The costs intensity for operational state maintenance is also determined by the number of serviced primary information sensors:

$$\Delta C(R,\tau) = \sum_{i=1}^{N_c} \Delta C_i(\tau). \quad (5)$$

To reduce the cost $Q_{exp}(R, t_2)$ in the information processing and control systems virtual sensors that replace physical ones are used. Their operation principle is based on the use of mathematical models and algorithms of indirect measurements, which allow using the minimum number of physical information sensors and actuators $N_{min} \ll N_c$.

In this case, for all $R \in R_{add}$ the following inequalities are observed:

$$\begin{cases} Q_{\exp}(R_{min}, t_2) < Q_{exp}(R, t_2); \\ Q(R_{min}, t_2) < Q(R, t_2), \end{cases} \quad (6)$$

where R_{min} is a technical solution with a minimum set of sensors for primary information and actuators N_{min} and virtual sensors for measuring N_c components of the object and the driver.

For consumer indicators, this means that $q_i(R_{min}) < q_i(R), 0 \leq i \leq 9$ and the developed MCS acquires competitiveness properties in a certain segment of applied technical systems.

Thus, for the motion control of a driverless vehicles platoon with a leading controlled vehicle, it becomes possible to significantly reduce the influence of external factors such as illumination, fog, precipitation in the form of rain and snow, the lack of recognizable markings.

The use of data from the wheeled navigation system for distance estimation between objects in a platoon equipped with means for exchanging information about the traveled distance and the angle of steered wheels rotation makes it possible to almost completely eliminate the influence of external factors on the results of stabilization problem solution. The use of video cameras to determine the deviation of the car axis from the center of the leader vehicle tailgate and the distance to it makes it possible to exclude the need for recognition of marking and compensate the accumulation of distance measurement errors by the wheeled navigation system.

The use of new anti-lock system algorithms based on identification of the wheel slip friction coefficients top values allows avoiding the intermittent braking mode and using only one actuator to act on the brake pedal.

To implement the stabilization functions the same brake actuator is used, which is controlled in accordance with the original algorithm.

The principle of replacing of physical information sensors by virtual ones involves the use of indirect measurements of physical variables. For its implementation, mathematical models and algorithms are used, which allow obtaining reliable information about object state parameters without additional equipment.

4. Conclusion

The concept of creation of competitive system for vehicles motion control involves the achievement of the following objectives:

- implementation of the proposed MCS in the minimum configuration of equipment by using algorithms for indirect measurements (virtual sensors) and knowledge about the control object in the form of formalized descriptions of its properties;
- increase of efficiency and safety of cargo and passenger transportation;
- reduction of maintenance costs for vehicles due to the timely detection of critical failures;
- ensuring better connectivity of the Russian Federation territory through the creation of intelligent transport and telecommunication systems, as well as engaging and retaining leadership positions in the creation of international transport and logistics systems;

- increasing the capacity of the road network, reducing the amount of harmful emissions to the environment by increasing the average speed of traffic flow due to decrease in the number of accidents.

The analysis of the accumulated experience, modern scientific, technical, normative and methodological literature on the problem of MCSs allows us to state that the application of mathematical models and algorithms of indirect measurements, along with more sophisticated control algorithms are the basis for creation of efficient technologies and applied control systems in this area.

The implementation of the proposed MCS in the minimal possible configuration of hardware, by using algorithms and mathematical models of indirect measurements (virtual sensors) is accompanied by an abrupt improvement in almost all consumer indicators, including reliability, operation costs, energy consumption and total cost.

Acknowledgments

This scientific paper was prepared based on the applied scientific research results conducted with the financial support of the Education and Science Ministry of the Russian Federation, Agreement No. 14.625.21.0043. The unique identifier of the project is RFMEFI62517X0043.

References

[1] Jeong E and Oh C 2017 Evaluating the effectiveness of active vehicle safety systems *Accident Analysis & Prevention* **100** pp 85–96 (DOI: 10.1016/j.aap.2017.01.015)

[2] Katrakazas C, Quddus M, Chen W and Deka L 2015 Real-time motion planning methods for autonomous on-road driving: State-of-the-art and future research directions *Transportation Research Part C: Emerging Technologies* **60** pp 416-442 (DOI: 10.1016/j.trc.2015.09.011)

[3] Ni L, Gupta A, Falcone P and Johannesson L 2016 Vehicle Lateral Motion Control with Performance and Safety Guarantees *IFAC Proceedings Volumes* **49**(11) pp 285–290 (DOI: 10.1016/j.ifacol.2016.08.043)

[4] Buznikov S E, Endachev D V, Elkin D S and Strukov V O 2018 The problem of the driverless vehicle specified path stability control *IOP Conf. Series: Materials Science and Engineering* **315** (DOI: 10.1088/1757-899X/315/1/012006)

[5] Kiss D and Papp D 2017 Effective navigation in narrow areas: a planning method for autonomous cars *IEEE 15th Int. Symp. on Applied Machine Intelligence and Informatics* (DOI: 10.1109/SAMI.2017.7880346)

[6] Saikin, A M, Buznikov S E and Karpukhin K E 2016 The Analysis of technical vision problems typical for driverless vehicles *Research Journal of Pharmaceutical, Biological and Chemical Sciences* **7**(4) pp 2053-59

[7] Shadrin S, Ivanov A and Karpukhin K 2016 Using Data From Multiplex Networks on Vehicles in Road Tests, in Intelligent Transportation Systems, and in Self-Driving Cars *Russian Engineering Research* **36**(10) pp 811-814

[8] Buznikov S E 2018 System analysis of vehicle active safety problem *IOP Conf. Series: Materials Science and Engineering* **315** (DOI: 10.1088/1757-899X/315/1/012005)

IOP Conf. Series: Materials Science and Engineering **534** (2019) 012006 doi:10.1088/1757-899X/534/1/012006

Autonomous tracked vehicles effectiveness estimation

V N Naumov, K Y Mashkov and K E Byakov

Bauman Moscow State Technical University, 105005 Baumanskaya 2th st., 5, Moscow, Russian Federation

Abstract. Effectiveness of application of autonomous tracked vehicles depends on many factors. For the manned car, the driver decides how to choose the optimal speed. The speed of autonomous vehicles, especially over highly rough terrain, is significantly lower and is caused by the operation of an autonomous control system. In tracked chassis, one of the agility components is the turnability, which characterizes the control ability of the vehicle under the given conditions. Controllability and ride comfort define how the power-weight ratio influence the average vehicle speed. At the same time, the behavior of an autonomous vehicle during rotation depends not only on the power-weight ratio, but also on the dimensions of the tracks support surfaces; pressure on the ground; steering radius provided by steering drive; traction properties, as well as on road conditions (path curve, traction and steering resistance coefficients). The report presents the analysis of the dependence of tracks slipping on the speed of the tracked autonomous vehicle chassis. The parameters that determine the state of the chassis of an autonomous tracked vehicle are listed. The results of mathematical simulation of the tracked mobile robot movement for different types of soils and with different speeds are presented.

The mobility requirements of both manned and unmanned transport systems are significantly stiffened with the development of technologies. In general, the autonomous wheeled vehicles are mainly considered. However, lack of attention is paid to the effectiveness of autonomous tracked transport systems.

Traditionally, the speed of tracked vehicles increased due to the increase in power-weight ratio and the ride comfort and due to the steering drive improving [1,2]. At the same time, the degree of influence of power-weight ratio on the average speed depends on such properties of the chassis as controllability and ride comfort, and with the increase of power-weight ratio the influence of controllability on the tracked vehicles speed is increased too.

In the figure 1 the graphic representation of the main chassis parameters influence on the maximum tracked vehicle speed (*Vmax*) and average tracked vehicle speed on dry ground roads (*V*) is given.

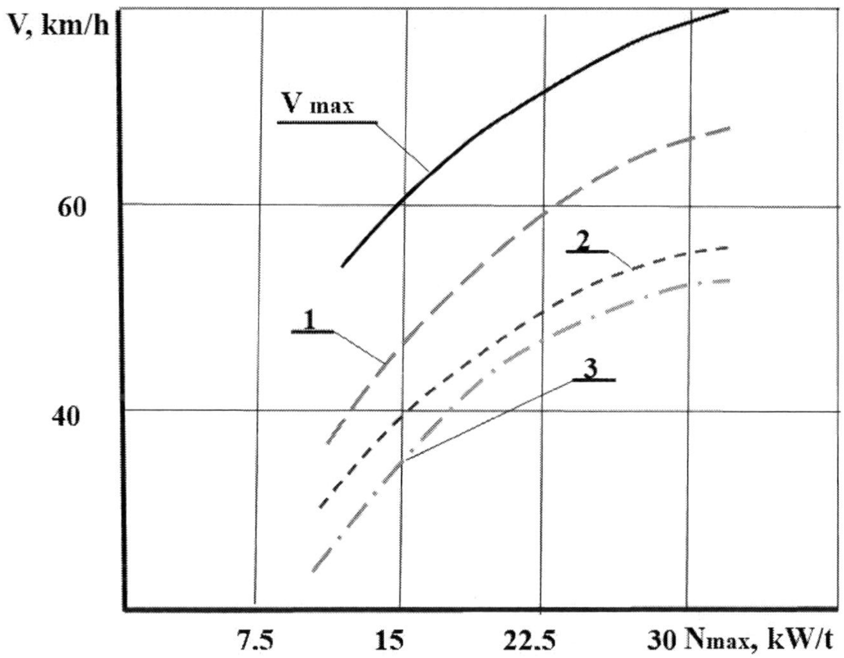

Figure 1. Influence of limiting factors on the maximum *Vmax* and average velocity *V* at various levels of power-weight ratio *Nmax*: 1– factor of external conditions; 2 – controllability; 3 – ride comfort.

From the plots in the figure 1 follows: the dependence of the average speed on the power-weight ratio is essentially nonlinear in the 10-15 kW/t power-weight ratio range, and 20% power-weight ratio increase leads to an 15 % average speed increase. In tracked chassis, one of the agility components is the turnability, which characterizes the control ability of the vehicle under the given conditions. At the same time, the behavior of an vehicle during steering depends not only on the power-weight ratio, but also on the other technical parameters (dimensions of the tracks support surfaces; pressure on the ground; steering radius provided by steering drive; traction properties, tracks stability in the caterpillar contour and others) as well as on road conditions (path curve, traction and steering resistance coefficients).

In the 25-30 kW/t range, the same increase of *Nmax* by 20% corresponds to increase of speed *V* by 6% only.

In addition, the speed of the tracked vehicle during the steering is influenced by factors such as the turning road radius and the steering drive radius mismatch, the requirement to define angle steering, the trajectory uncertainty with partial movement of the steering control, the ratio between the external dimensions of the tracked vehicle and the roadway width etc.

All this intensifies differences between the tracked vehicle speed technical capabilities and the degree of their operation implementation. In practice, the dynamics of tracked vehicle is realized through the driver's skill [3, 4].

Modern design technologies of the engine, drivetrain and the chassis allow the movement of tracked robotic systems with speeds not lower than the existing tracked vehicles speeds on roads with a hard surface (60-70 km/h) and on rough terrain (25-35 km/h).

It should be emphasized that such speeds of movement are limited by the possibility of the control system and the potential growth of average and maximum speeds will be carried out as the development of robotics technologies and autonomy increasing.

During the unmanned driving, the most difficult for the control system of tracked vehicle is to provide a curvilinear motion with the drift and traction limit steering speeds [5, 6, 7]. The signs of the

IOP Conf. Series: Materials Science and Engineering **534** (2019) 012006 doi:10.1088/1757-899X/534/1/012006

sidewise drift appearance are not known by the control system, and it should reduce the movement speed in order to avoid loss of control or emergency situation.

The critical drift speed is determined experimentally [8] during the passage of the «snake»-type track shown in the figure 2.

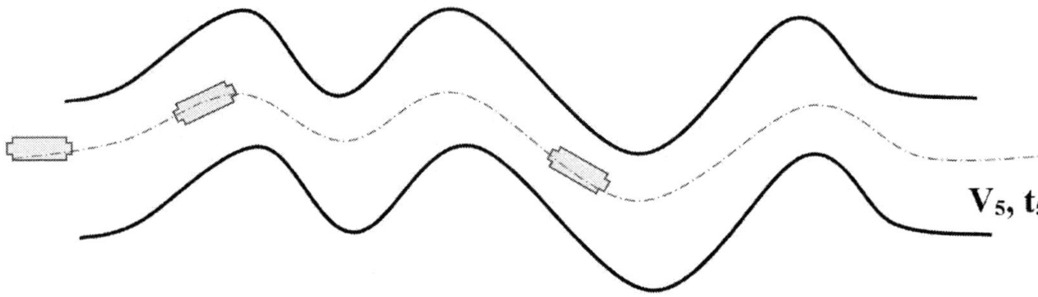

Figure 2. Experimental route.

The dependence of the relative displacement of the instantaneous steering center of vehicle on the relative speed is shown in the figure 3:

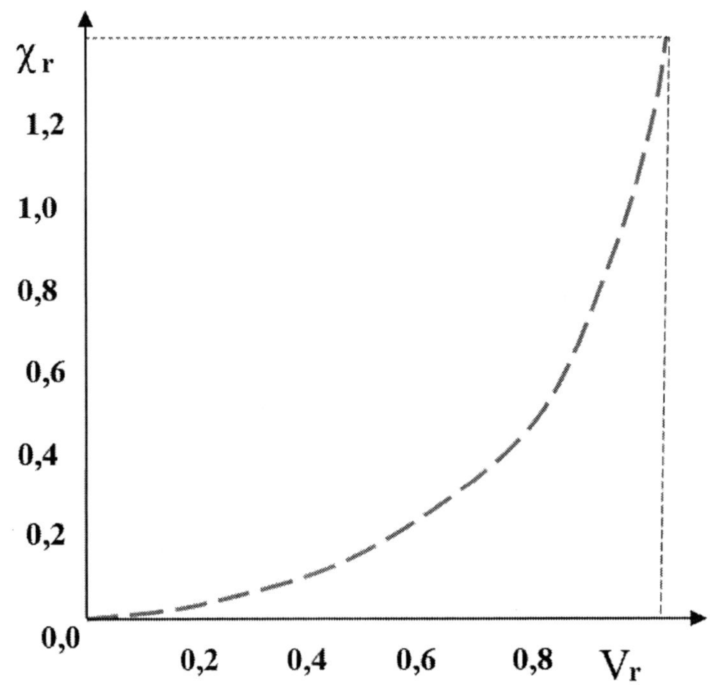

Figure 3. Plot of the dependence of the relative displacement of the vehicle's instantaneous steering center on the relative speed [8].

The main controllability evaluation criterion of robotic tracked vehicle is the average speed on the «snake»-type track. It characterizes the ability of vehicle to realize its potential under the controllability restrictions.

Auxiliary indicators are the critical drift speed and the acceleration characteristics.

The «snake»-type track with a variable curvature makes it possible to evaluate the controllability under the most critical driving regimes - during the change of vehicles movement direction to the opposite.

During the controlled movement, the center of mass of tracked vehicle will move tangentially to the trajectory of movement. The longitudinal axis of vehicle will rotate around the center of mass at an additional angle relative to the tangent to the trajectory of movement [5].

The longitudinal axis rotation of the robotic vehicle chassis relative to the tangent to the trajectory occurs due to the skidding and slipping of the tracks. The lagging caterpillar track switches from the skidding mode to the slipping at the approach to the critical controllability speed. The slipping speed of the outrun caterpillar is increased and during the drift both caterpillar tracks have comparable slipping index (figure 4).

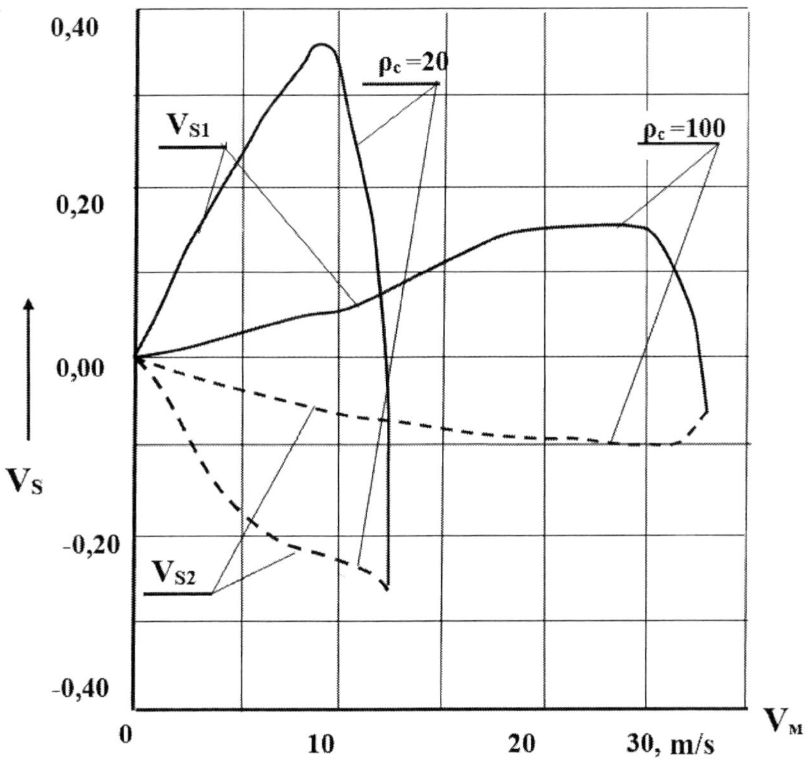

Figure 4. The dependence of the slippage speed of tracks *Vs* on the speed of the robotic tracked vehicle chassis *Vm* for different ρ values.

In this process there is a condition that there is the driving mode of the vehicle, when the slipping speeds of the caterpillar tracks will be the same $V_{S2} = V_{S1}$. In this case, the boundary conditions of the controlled movement of the robotic tracked vehicle chassis are characterized by the following expression $\omega_t = \omega$. That is the equality of the real angular speed of the robotic chassis (given the external conditions) to the angular velocity determined by the kinematics of its steering drive (as the speed of the chassis moves to the controllability critical speed) [8].

The steering is carried out from a certain angular velocity ω, due to the characteristic of the steering drive and the movement mode of the tracked chassis, taking into account that the boundary condition of the controlled motion $\omega_t = \omega$ characterizes accurately the moment of the uncontrolled movement beginning.

On the dependencies $\omega_t / \omega = f(V / V_{kr})$ (figure 5) it can be seen that the established controlled movement boundary condition characterizes accurately the moment of the uncontrolled movement beginning. Here *V* – is the current speed of the center of mass of the vehicle chassis, *Vkr* – is the critical value of the chassis speed for this steering radius.

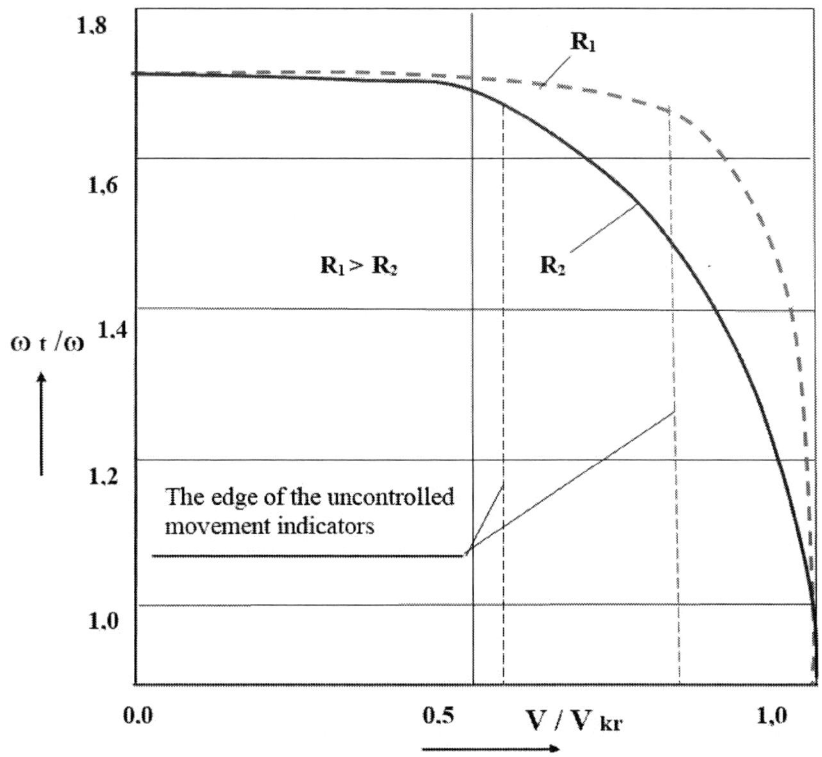

Figure 5. Dependence of the steering relative angular speeds on the relative speed of the chassis for different steering radii.

From the qualitative and quantitative analysis of the controlled curvilinear movement of the robotic chassis, shown in figure 5, it is obvious that during the steering automated control system creating, the relative angular velocity can be used to predict the uncontrolled movement.

The advantage of this parameter is that it does not require the complex design solutions. It will allow to create a reliable parameter measurement system, which characterizes the reliability of the entire automated movement control system of the robotic tracked vehicle chassis.

The parameters that determine the control status of the robotic chassis are the following:

- angular velocity of the chassis;
- rotation frequency of the chassis driving wheels.

The critical control algorithm of the tracked robotic chassis uncontrolled movement preventing (drift overcoming) can be represented as follows: when the robotic vehicle is moving, the angular velocities of the right and left boards are determined. The angular velocities enter the valuator unit for calculating the theoretical angular velocity of the chassis steering. Next, go to the divider unit. The actual angular velocity from the sensors is fed into the divider unit. Next - the comparison with the reference signal and the command to the actuator.

The structural diagram of the system is shown in the figure 6.

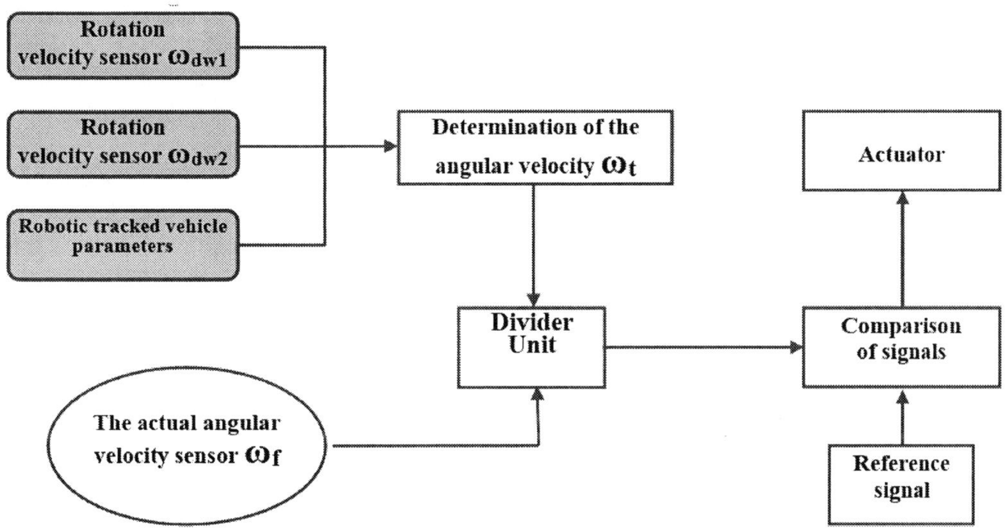

Figure 6. Scheme of the robotic tracked vehicle uncontrolled movement prevention system.

Simulation of the tracked vehicle movement along the arc of a circle was performed in the software packages «Universal Mechanism» and «Simulink MATLAB» (figure 12). The simulation was performed at different speeds with a steering radius of 25 meters. As a model of the soil, a linear elastic model was chosen. The simulation results are presented in the figures 7-9.

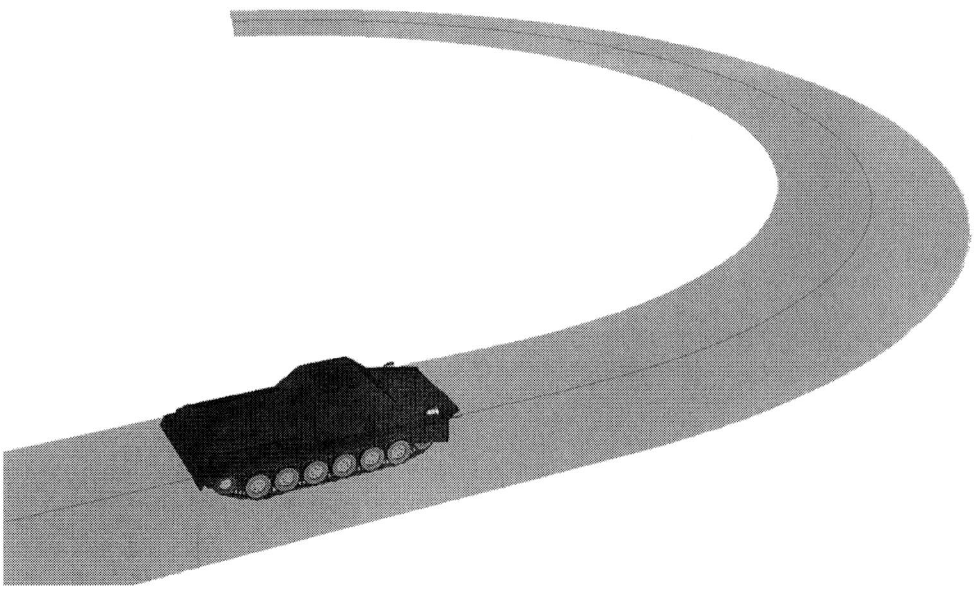

Figure 7. The tracked vehicle simulation model in the «Universal mechanism» software.

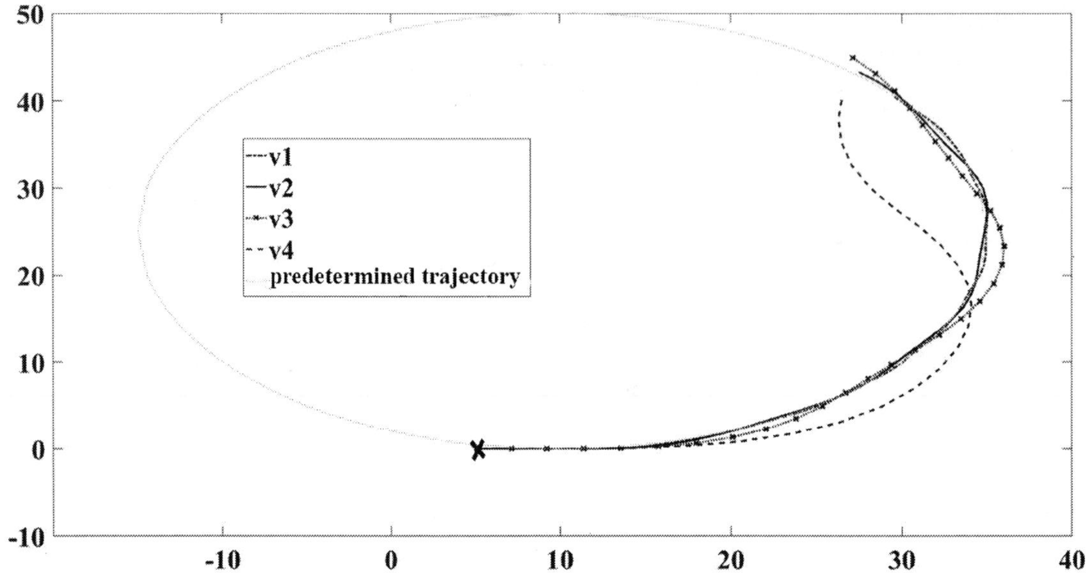

Figure 8. The steering of the tracked vehicle simulation results.

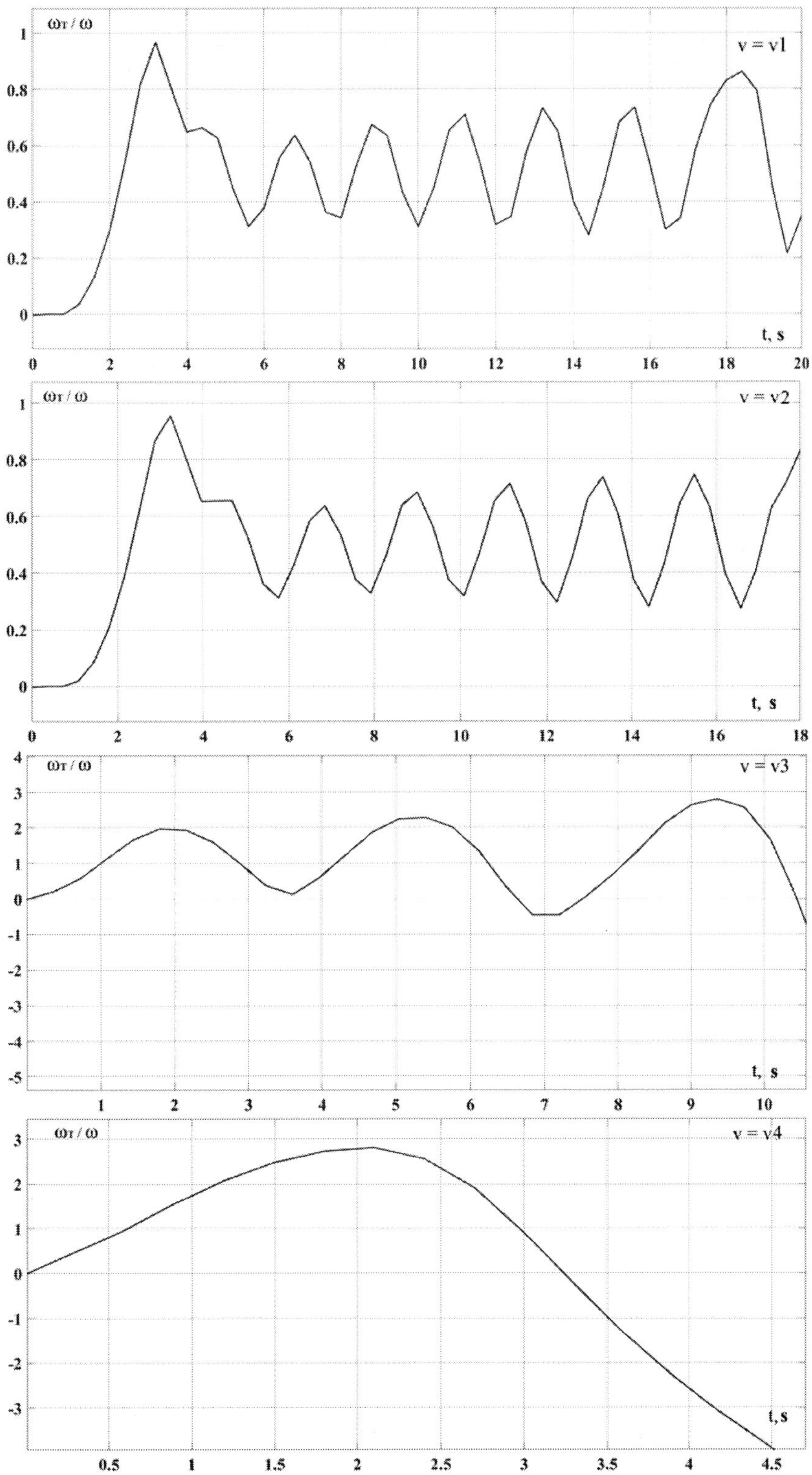

Figure 9. Relative angular velocity ω_t/ω for different speeds.

IOP Conf. Series: Materials Science and Engineering **534** (2019) 012006 doi:10.1088/1757-899X/534/1/012006

Conclusion

The uncontrolled movement preventing algorithm for tracked robotic vehicles is presented. A mathematical simulation of the vehicle movement for various types of soils has been carried out. Analysis of the results showed that the value $\omega_t/\omega = 1$ is a boundary condition for the signs of uncontrolled movement. It coincides with the theoretical calculations presented above.

References

[1] Brilev O N et al 1977 M*obility of armored vehicles assessment* (Moscow: VA BTV) p 107
[2] Brilev O N 2015 *Tanks* (Moscow: Planet Press) p 564
[3] Belyakov V V et al 2004 *All-terrain transport and technological vehicles* (N Novgorod: Talam Press) p 960
[4] Nikitin A O and Sergeev L V 1962 *The theory of tank* (Moscow: ABTV Press) p 583
[5] Vasil'chenkov V F et al 1976 *Military vehicles* (Moscow: ABTV Press) p 659
[6] Savochkin V A and Dmitriev A A 1993 *Statistical dynamics of transport tracked vehicles* (Moscow: Mechanical engineering) p 320
[7] Shapkin A N 2012 Methods of the tracked vehicles controllability assessing *77th Int. Sc. and Techn. Conf, of the Auto-Tr. Eng, in Russia* pp 243–252
[8] Krasnen'kov V I and Haritonov S A 1984 *J. Bauman MVTU* **339** pp 367–369

Mivar models of reconstruction and expertise of emergency events of road accidents

D A Chuvikov[1,2]**, O O Varlamov**[1,3]**, D V Aladin**[1,3]**, V M Chernenkiy**[3] **and A V Baldin**[3]

[1] MIVAR, 72 Oktyabrskaya str., Moscow 127521, Russian Federation
[2] Department of Automated control systems, Moscow Automobile and Road Construction State Technical University (MADI), Leningradskiy Pr. 64, Moscow 125319, Russian Federation
[3] Department of Information Processing and Control Systems, Bauman Moscow State Technical University, 2-ya Baumanskaya Street 5/1, Moscow 105005, Russian Federation

E-mail: d.chuvikov@mivar.ru

Abstract. The article deals with the problem of creating expert systems for the reconstruction and analysis of motor vehicle accidents (MVA). The need for development of the special expert system of the analysis and examination of road accident for the solution of this problem was formulated. Also, in the article the compatibility models of the Mivar Expert System and the abstract system of simulation modeling were described. In this article, the Mivar expert system "Analysis MVA" is considered as a promising solution for the reconstruction and analysis of traffic accidents. Much attention is drawn to the principles of operation of the mivar systems in the expert analysis of accidents.

1. Introduction

The number of cars in the world has been growing steadily over the years. If this trend continues, in Russia by 2020 the number of road motor vehicles per 1,000 inhabitants will reach 550 units. The stable increasing number of vehicles on the roads is one factor that raises the risk of accidents. It is estimated that every year more than a million people die from road traffic injuries around the world. An important aspect in the analysis of the accidents is the question of establishing the objective circumstances and causes of the accident with the assessment of the resulting traffic situation and the actions of the participants from a technical point of view. Such analysis allows creating a technical basis for identification of the perpetrators. For each case, it is necessary to consider all possible factors that could have an impact on the occurrence of the road transport situation. Therefore, there is a need to develop models of reconstruction and examination of road accidents [1–4]. It is also worth noting that for the reconstruction of the car crash there is a need in a simulation model of the accident. It takes a lot of time to create a high-quality simulation model, as it is important to consider a huge number of parameters, their combinations, variations and interactions.

This article proposes to solve the problem of creating high-quality simulation models by using a specialized Mivar Expert System (MES) of analysis and expertise of accidents [5]. The proposed system

IOP Conf. Series: Materials Science and Engineering **534** (2019) 012007 doi:10.1088/1757-899X/534/1/012007

will determine the optimal parameter values for the simulation. In other words, MES will allow experts to obtain the most reliable set of input parameters for a simulation modeling system (SMS) [6].

2. Models of compatibility of Mivar Expert and Simulation Systems

Let us consider models of compatibility of Mivar Expert and Simulation Systems to explain principles of work of MES [3–4]. $P_{\exp} = \{P_{\exp 1}, P_{\exp 2}, ..., P_{\exp n}\}$ denotes a set of parameters of MES. The set $P_{\exp}$ consists of a set of input parameters $P^{in}_{\exp}$ $(P^{in}_{\exp} \subseteq P_{\exp})$ and a set of output parameters $P^{out}_{\exp}$ $(P^{out}_{\exp} \subseteq P_{\exp})$. $P_{sim} = \{P_{sim1}, P_{sim2}, ..., P_{sim\ m}\}$ – a set of parameters of SMS. P^{in}_{sim} and P^{out}_{sim} are sets of input and output parameters of SMS ($P^{in}_{sim} \subseteq P_{sim}$ .and $P^{out}_{sim} \subseteq P_{sim}$). There are three types of models of compatibility of MES and SMS: MES is a dominant – a forward model: $P^{out}_{\exp} \rightarrow P^{in}_{sim}$; SMS is a dominant – a reverse model: $P^{out}_{sim} \rightarrow P^{in}_{\exp}$; an equivalent relation between MES and SMS – an equivalent model: $P_{\exp} \leftrightarrow P_{sim}$. We denote a set of forward and inverse models by $M_{\exp \rightarrow sim}$ and $M_{\exp \leftarrow sim}$ respectively. Then we get $[P_{\exp} \leftrightarrow P_{sim}] \in M_{\exp \rightarrow sim} \cap M_{\exp \leftarrow sim}$.

3. Structure of Mivar Expert System "Analysis MVA"

The Mivar model of MES "Analysis MVA" [7–8] consists of 16 classes, 144 parameters, 93 relations, 352 rules which include 109 restrictions. The structure of all classes, parameters, relations and restrictions is shown in Figure 1.

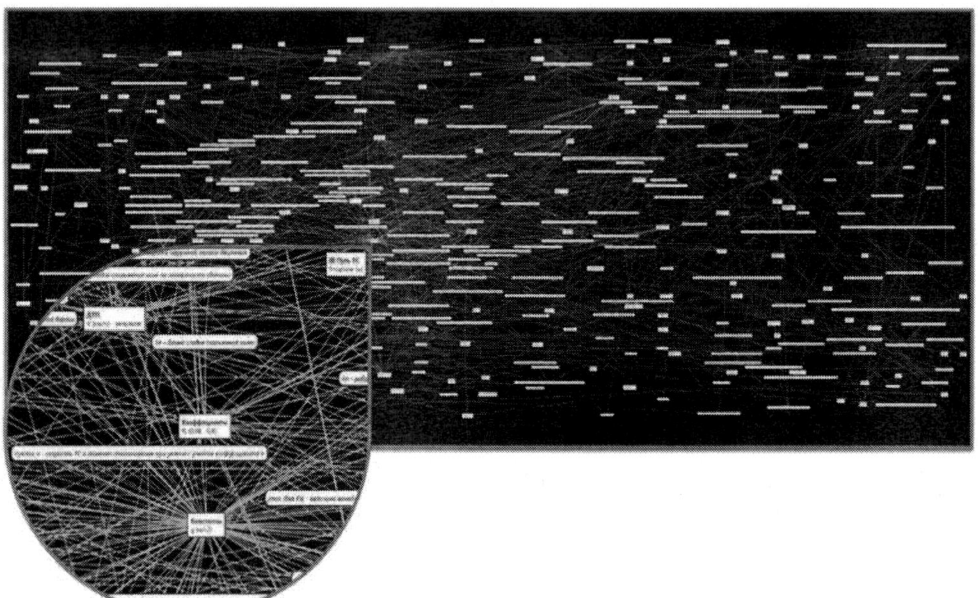

Figure 1. Structure of the Mivar model of the system.

The model parameters [7] are a set $P_{\exp} = \{P_1, P_2, ..., P_{144}\}: P_1$ – value of the deceleration parameter $j_{\max}$; P_2 – gravitational acceleration *g*; P_3 – coefficient of the tyre adhesion to the road surface *f*; P_4 – correction factor for the braking performance K_3; P_5 – vehicle speed before the start of braking V_0; … P_{144} – braking distance S_4.

IOP Conf. Series: Materials Science and Engineering **534** (2019) 012007 doi:10.1088/1757-899X/534/1/012007

The rules of the model are a set $R_{\exp} = \{R_1, R_2, ..., R_{352}\}$:

$$R_1(P_1, P_2): P_3 = \frac{P_1}{P_2} \Leftrightarrow f = \frac{j_{\max}}{g}$$

$$R_4(P_2, P_3, P_4): P_1 = \frac{P_3 \cdot P_2}{P_4} \Leftrightarrow j_{\max} = \frac{f \cdot g}{K э};$$

$$R_7(P_2, P_3, P_{144}) \in P_5 = \sqrt{2 \cdot P_2 \cdot P_3 \cdot P_{144}} \Leftrightarrow V_0 = \sqrt{2 \cdot g \cdot f \cdot S_4};$$

$$\ldots$$

$$R_{352}(P_1, P_5): P_{144} = \frac{P_5^2}{2 \cdot P_1} \Leftrightarrow S_4 = \frac{V_0^2}{2 \cdot j_{\max}}.$$

$C_{\exp} = \{C_1, C_2, ..., C_{109}\}$ is a set of restrictions of the model, where $C_{\exp} \subseteq R_{\exp}$:

$$C_1(P_3): 0{,}08 \le P_3 \le 0{,}8 \Leftrightarrow 0{,}08 \le f \le 0{,}8;$$

$$C_2(P_4): 1 \le P_4 \le 1{,}6 \Leftrightarrow 1 \le K_Э \le 1{,}6;$$

$$C_3(P_1): P_1 \ge 0 \Leftrightarrow j_{\max} \ge 0;$$

$$\ldots$$

$$C_{109}(P_{144}): P_{144} \ge 0 \Leftrightarrow S_4 \ge 0.$$

A set of parameters and rules $\{P_{\exp}, R_{\exp}\}$ is used to build a Mivar network [9–13]. The network consists of the following elements of a bipartite oriented graph: vertex sets of first type (the parameters P_{exp}), vertex sets of second type (the rules R_{exp}) and edges of the graph that connect vertices from the first and second set (Figure 2) [14].

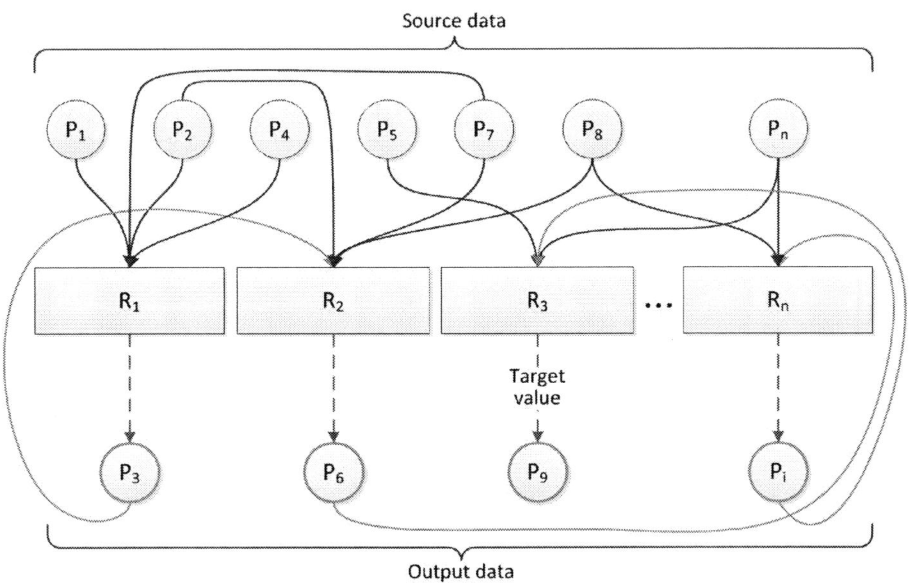

Figure 2. Algorithm of work MES "Analysis MVA".

MES "Analysis MVA" is adaptive and represents a "white box" model. In another words, the expert has an access to change the knowledge base, add new rules (formulas) and parameters. It is also possible to change the existing parameters for individualization of the system. Therefore, the system is very flexible and allows the expert to create new scenarios for simulation of MVA.

4. Results of experimental researches

For our experiments we used the simulation system Virtual CRASH 3.0. The articles [1–7] discusses in detail the principles of collaboration of MES "Analysis MVA" and Virtual CRASH 3.0. As an example, let us consider the results of the experimental research of a road accident with a pedestrian. In this research, the name of the street was changed and a vehicle registration plate of the car was hidden in order to preserve anonymity.

Description of the motor vehicle accident: The car Toyota Yaris hit a pedestrian who crossed Kruglaya street on the right side. Kruglaya street was a two-way street. It was found that the driver was under the influence of alcohol. The reaction time of the driver was 1.5 seconds; the time delay actuation of the actuator was 0.3 seconds; the braking time was 0.2 seconds. The pedestrian's speed was 5 km/h. The pedestrian went along the lane of the car distance equal 1.4 meters. According to witnesses, the speed of the car Toyota Yaris at the time of collision was 50 km/h.

The road had the asphalt concrete pavement. The weather at the time of the accident was clear. The permitted maximum speed on this section of the road was 60 km/h. According to the results of measurements, the length of the brake track of the vehicle was 32 meters. The main force of the hit fell on the right front of the bumper and the hood of the car. The distance from the right edge of the road to the car was 2 meters. The overall width of the car is 1.66 meters.

The purpose of the research is to answer the question: "Did the driver have the technical abilities to avoid hitting the pedestrian?".

Let us define the parameters of the road situation as follows: $V = 13.89\ m/s$; $V_p = 1{,}39\ m/s$; $t_1 = 1{,}5\ s$; $t_2 = 0{,}3\ s$; $t_3 = 0{,}2\ s$; $S_4 = 32\ m$; $S_p = 2\ m$; $\Delta Y = 2\ m$; $B = 1{,}66\ m$; $f = 0{,}75$; $ly = 0{,}7\ m$; $SpeedLimit = 60\ km/h$.

For ethical reasons, the photo materials of the accident were replaced with photos of the crash test of the event in question. The photos are presented in Figures 3, 4, 5 and 6.

Figure 3. Moment when the driver of the vehicle Toyota Yaris could see the pedestrian.

Figure 4. Moment of the first contact.

Figure 5. Braking of the vehicle.

Figure 6. Car damage.

IOP Conf. Series: Materials Science and Engineering **534** (2019) 012007 doi:10.1088/1757-899X/534/1/012007

The source data were entered into MES "Analysis MVA". The initial data and the calculation results are shown in Figure 7.

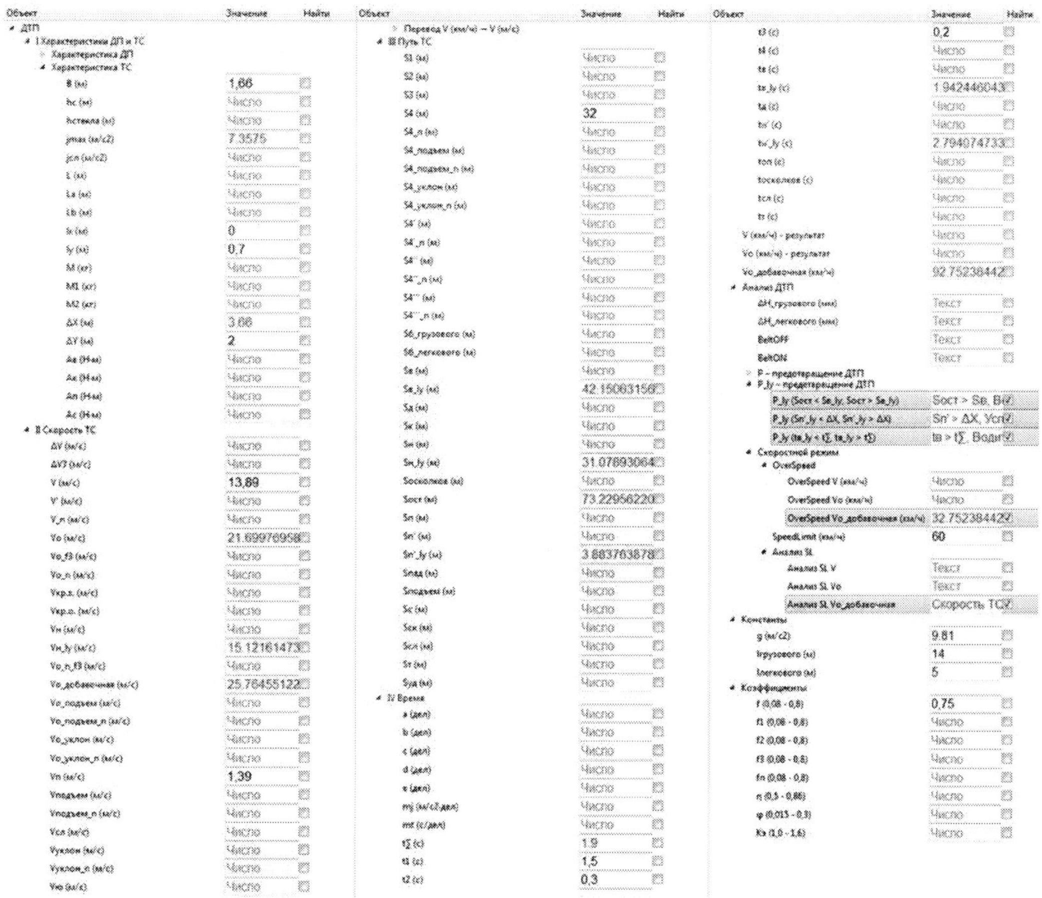

Figure 7. Screenshot of MES "Analysis MVA": source data and calculation results.

The solution search model is shown in Figure 8 as a graph.

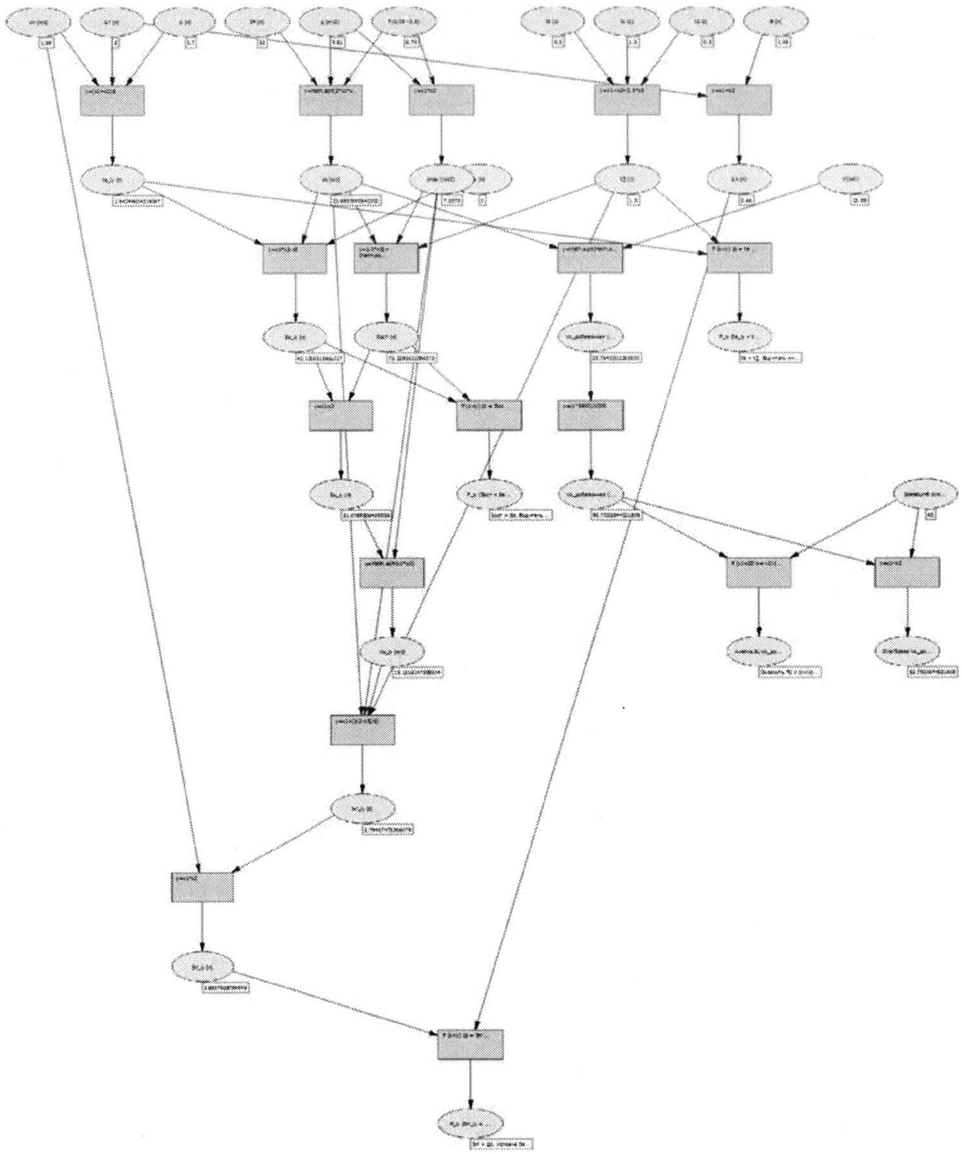

Figure 8. Solution search model.

The following is a listing of the solution search algorithm. The algorithm consists of 18 steps.

Step # 0
Relation: tv_ly – pedestrian visibility time in the driver's field of vision [depends on ly]
Rule: tv_ly – pedestrian visibility time in the driver's field of vision [depends on ly]
Input parameters:
Vp (m/s)=1.39;
ΔY (m)=2;
ly (m)=0.7;
Formula:
y=(x1+x2)/z
Result: tv_ly (s)=1.94244604316547;

IOP Conf. Series: Materials Science and Engineering **534** (2019) 012007 doi:10.1088/1757-899X/534/1/012007

Step # 1
Relation: Vo – vehicle speed before braking
Rule: Vo – vehicle speed before braking
Input parameters:
S4 (m)=32;
g (m/s2)=9.81;
f (0,08 - 0,8)=0.75;
Formula:
y=Math.sqrt(2*x1*x2*x3)
Result: Vo (m/s)=21.6997695840302;

Step # 2
Relation: Sv, Sv_ly
Rule: Sv_ly – pedestrian visibility distance (distance from the car to the collision) [depends on ly]
Input parameters:
lx (m)=0;
Vo (m/s)=21.6997695840302;
tv_ly (s)=1.94244604316547;
Formula:
y=x1*x2-x3
Result: Sv_ly (m)=42.1506315661017;

Step # 3
Relation: t∑ - total time
Rule: t∑ - total time
Input parameters:
t3 (s)=0.2;
t1 (s)=1.5;
t2 (s)=0.3;
Formula:
y=x1+x2+0.5*x3
Result: t∑ (s)=1.9;

Step # 4
Relation: t1, t2, t3, t4, jmax, jmax (without Kbe), S1, S2, Sp', Sp'_ly, Sdr, Sim
Rule: jmax (without Kbe) - deceleration value without Kbe – brake efficiency correction factor
Input parameters:
g (m/s2)=9.81;
f (0,08 - 0,8)=0.75;
Formula:
y=x1*x2
Result: jmax (m/s2)=7.3575;

Step # 5
Relation: Ssw – length of the stopping distance
Rule: Ssw – length of the stopping distance

IOP Conf. Series: Materials Science and Engineering **534** (2019) 012007 doi:10.1088/1757-899X/534/1/012007

Input parameters:
jmax (m/s2)=7.3575;
t∑ (s)=1.9;
Vo (m/s)=21.6997695840302;
Formula:
y=(x1*x2)+(Math.pow(x2,2)/(2*z))
Result: Ssw (m)=73.2295622096573;

Step # 6
Relation: Sn, Sn_ly, La, Lb, M1, M2, ΔV, ΔV3, Vu, ΔV3 [depends on jsl], Vu [depends on jsl], OverSpeed V, OverSpeed Vo, OverSpeed Vo_additional
Rule: Sn_ly – distance that the car will move after crossing the line of the trajectory of the pedestrian if the driver starts to brake in a timely manner [depends on ly]
Input parameters:
Ssw (m)=73.2295622096573;
Sv_ly (m)=42.1506315661017;
Formula:
y=x1-x2
Result: Sn_ly (m)=31.0789306435556;

Step # 7
Relation: Vn, Vn_ly
Rule: Vn_ly – speed of the car at the time of crossing the line of the trajectory of the pedestrian if the driver starts to brake in a timely manner [depends on ly]
Input parameters:
Sn_ly (m)=31.0789306435556;
jmax (m/s2)=7.3575;
Formula:
y=Math.sqrt(x1*x2)
Result: Vn_ly (m/s)=15.1216147355354;

Step # 8
Relation: tn', tn'_ly
Rule: tn'_ly – time of movement of the car from the moment of occurrence of the dangerous situation to crossing the line of the trajectory of the pedestrian (if the driver starts to brake in a timely manner) [depends on ly]
Input parameters:
Vn_ly (m/s)=15.1216147355354;
jmax (m/s2)=7.3575;
t∑ (s)=1.9;
Vo (m/s)=21.6997695840302;
Formula:
y=x1+((x2-x3)/z)
Result: tn'_ly (s)=2.79407473306079;

Step # 9
Relation: t1, t2, t3, t4, jmax, jmax (without Kbe), S1, S2, Sp', Sp'_ly, Sdr, Sim

IOP Conf. Series: Materials Science and Engineering **534** (2019) 012007 doi:10.1088/1757-899X/534/1/012007

Rule: Sp'_ly – movement of the pedestrian during tn'_ly
Input parameters:
tn'_ly (s)=2.79407473306079;
Vp (m/s)=1.39;
Formula:
y=x1*x2
Result: Sp'_ly (m)=3.8837638789545;

Step # 10
Relation: Sls_truck, Sls_auto, ΔX, L, M, Sfragments, Vo, tоп, Vсл
Rule: ΔX – sum of the distances from the lateral surface of the vehicle to the boundary of the danger zone and the width of the track of the vehicle
Input parameters:
ΔY (m)=2;
B (m)=1.66;
Formula:
y=x1+x2
Result: ΔX (m)=3.66;

Step # 11
Relation: P (Sp' < ΔX, Sp' > ΔX), P_ly (Sp'_ly < ΔX, Sp'_ly > ΔX) – traffic accident prevention
Rule: Sp'_ly < ΔX, Sp'_ly > ΔX
Input parameters:
Sp'_ly (m)=3.8837638789545;
ΔX (m)=3.66;
Formula:
if (x<y) {z = 'Sp' <ΔX, condition of unsafe crossing of the vehicle trajectory';} else {z = 'Sp' > ΔX, condition of safe crossing of the vehicle trajectory";}
Result: P_ly (Sp'_ly < ΔX, Sp'_ly > ΔX)=Sp' > ΔX, condition of safe crossing of the vehicle trajectory;

Step # 12
Relation: Vo_additional – vehicle speed before braking, considering the additional speed V at the time of collision of the vehicle
Rule: Vo_additional – vehicle speed before braking, considering the additional speed V at the time of collision of the vehicle
Input parameters:
V (m/s)=13.89;
Vo (m/s)=21.6997695840302;
Formula:
y=Math.sqrt(Math.pow(x1,2)+Math.pow(x2,2))
Result: Vo_additional (m/s)=25.7645512283835;

Step # 13
Relation: Conversion V (m/s) → V (km/h)
Rule: Vo_additional (m/s) → Vo_additional (km/h)
Input parameters:

Vo_additional (m/s)=25.7645512283835;
Formula:
y=(x*3600)/1000
Result: Vo_additional (km/h)=92.7523844221808;

Step # 14
Relation: SpeedLimit - maximum speed limit
Rule: Analysis SL Vo_additional – parameter that determines the driver's violation of the speed limit, considering Vo_additional
Input parameters:
SpeedLimit (km/h)=60;
Vo_additional (km/h)=92.7523844221808;
Formula:
if (x1 >= x2) {y = 'Vehicle speed V (km/h) ≤ SpeedLimit, The driver observed the speed limit';}
else {y = ' Vehicle speed V (km/h) > SpeedLimit, The driver did not observe the speed limit';}
Result: Analysis SL Vo_additional=Vehicle speed V (km/h) > The driver did not observe the speed limit;

Step # 15
Relation: P (tv < t∑, tv > t∑), P_ly (tv_ly < t∑, tv_ly > t∑) – traffic accident prevention
Rule: tv_ly < t∑, tv_ly > t∑
Input parameters:
tv_ly (s)=1.94244604316547;
t∑ (s)=1.9;
Formula:
if (x<y) {z = ' tv <t∑, The driver did not have the technical ability to prevent an accident';} else {z = 'tv > t∑, The driver had the technical ability to prevent an accident';}
Result: P_ly (tv_ly < t∑, tv_ly > t∑)=tv > t∑, The driver had the technical ability to prevent an accident;

Step # 16
Relation: Sn, Sn_ly, La, Lb, M1, M2, ΔV, ΔV3, Vu, ΔV3 [depends on jsl], Vu [depends on jsl], OverSpeed V, OverSpeed Vo, OverSpeed Vo_additional
Rule: OverSpeed Vo_additional – exceeding the maximum speed Vo_additional
Input parameters:
Vo_additional (km/h)=92.7523844221808;
SpeedLimit (km/h)=60;
Formula:
y=x1-x2
Result: OverSpeed Vo_additional (km/h)=32.7523844221808;

Step # 17
Relation: P (Ssw < Sv, Ssw > Sv), P_ly (Ssw < Sv_ly, Ssw > Sv_ly), P (Ssw < Sp, Ssw > Sp) – traffic accident prevention
Rule: Ssw < Sv_ly, Ssw > Sv_ly
Input parameters:
Ssw (m)=73.2295622096573;

Sv_ly (m)=42.1506315661017;
Formula:
if (x<y) {z = 'Ssw <Sv, The driver had abilities to prevent the accident by timely braking';} else {z = 'Ssw > Sv, The driver did not have abilities to prevent the accident by timely braking. The vehicle was not able to stop in front of the pedestrian's trajectory';}
Result: P_ly (Ssw < Sv_ly, Ssw > Sv_ly)=Ssw > Sv, The driver did not have abilities to prevent the accident by timely braking. The vehicle was not able to stop in front of the pedestrian's trajectory;

Using MES "Analysis MVA" we get the following results:

1) The driver had technical abilities to prevent the accident ($t_v > t_\Sigma = 1{,}94\ s > 1{,}90\ s$).

2) Since the condition $S_{sw} > S_v = 73{,}22\ m > 42{,}15\ m$ is met, the vehicle did not stop in front of the pedestrian's trajectory. As a result, the pedestrian was hit by the vehicle.

3) The pedestrian could have left the danger zone if he had changed his speed and direction of the movement ($S_p' > \Delta X = 3{,}88\ m > 3{,}66\ m$).

4) The driver did not comply with the speed limit and exceeded the maximum speed limit by 32.75 km/h ($V_o > SpeedLimit = 92{,}75\ km/h > 60\ km/h$).

In order to get a complete picture of the situation, we used the results obtained from MES "Analysis MVA" for the reconstruction of the accident in Virtual CRASH 3.0. Figure 9 shows the reconstruction [15] of the accident events.

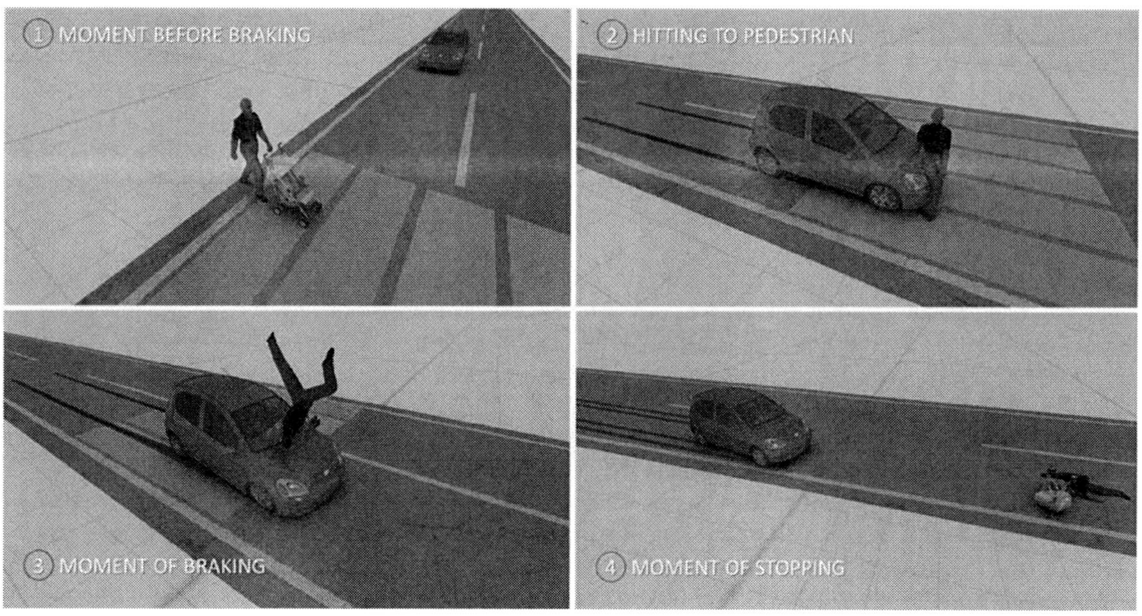

Figure 9. Reconstruction of the accident.

The conclusion is thus that the driver acted technically wrong and he is completely guilty of the accident.

5. Conclusion

The Mivar Expert System "Analysis MVA" [8] helps to reduce the complexity of the process of analysis and reconstruction of disputed motor vehicle accidents [16–17], to minimize the occurrence of errors in the calculation, to save human resources on experts and to improve the speed in decision-making [3–4].

In the future, the developed mivar mathematical models [18–19] of reconstruction and expertise of road accidents can be used in autonomous transport [20–23]. This will allow law enforcement agencies to automatically determine the cause and culprits of the accident and to form the conclusion with detailed information about the technical prerequisites of the accident.

6. References

[1] Chuvikov D A 2017 The use of expert system "Analysis MVA" and simulation system virtual crash 3.0 for solving tasks of criminal expertise centre *Industrial automatic control systems and controllers* **5** pp 23–34

[2] Chuvikov D A 2017 About the methodology of combining the expert system constructed on the basis of mivar approach, and simulation system for solving the class of problems of reconstruction and expertise of mva *Proceedings of the congress on intelligent systems and information technologies. "IS&IT'17"* **1** pp 313–321

[3] Chuvikov D A 2017 *Models and algorithms of reconstruction and examination of emergency events of road accidents* Abstract of PhD Thesis Moscow p 24

[4] Chuvikov D A 2017 *Models and algorithms of reconstruction and examination of emergency events of road accidents* PhD Thesis Moscow p 318

[5] Chuvikov D A 2017 Simulation as a tool to provide additional information for human decisions *Industrial automatic control systems and controllers* **4** pp 44–53

[6] Chuvikov D A 2017 Application of expert modeling in new knowledge obtained by man *Radio industry* **2** pp 72–80

[7] Chuvikov D A 2017 About the expert system "Analysis MVA", based on the concept of the Mivar approach *Problems of artificial intelligence* **2** pp 78–88

[8] Chuvikov D A 2017 *Expert system "Analysis MVA"* (Russian Federation Patent No.2017660355)

[9] Belousova A I, Varlamov O O, Ostroukh A V and Krasnyanskiy M N 2011 The approach to the development of multilevel model of multi-agent system using mivars *Science prospects* **5** pp 57–61

[10] Varlamov O O 2012 Mivar technologies: transition from productions to bipartite graphs mivar nets and realization of automated constructor of algorithms handling more than three million production rules *Artificial intelligence* **4** pp 11–33

[11] Varlamov O O, Adamova L E, Eliseev D V, Maiboroda Y I, Antonov P D, Sergushin G S and Chibirova M O 2013 Comprehensive computer modeling of understanding of texts, images and speech based on mivar *Artificial intelligence* **4 pp** 15–27

[12] Varlamov O O 2004 Systems of processing of information and interaction between groups of mobile robots on the basis of mivar information space *Artificial intelligence* **4** pp 695–700

[13] Sergushin G S, Varlamov O O, Chibirova M O, Eliseev D V and Muravyova E A 2013 Information modeling of complex automated process control systems, based on mivar technology *Artificial intelligence* **3 pp** 126–138

[14] Chuvikov D A 2017 Models of reconstruction and examination of MVA in a formalism of the knowledge base of the bipartite focused mivar networks *Proceedings of the congress on intelligent systems and information technologies. "IS&IT'17"* **1** pp 308–313

[15] Chuvikov D A, Kazakova N A, Varlamov O O and Golovizin A V 2014 3D modeling and 3D objects creation technology analysis for various intelligent systems *International Journal of Advanced Studies* **4** pp 16–22

[16] Chuvikov D A 2017 The universal algorithm of interaction between expert system and simulation system *T-Comm* **11** pp 34–40

[17] Chuvikov D A 2017 Methodology of combining expert system and simulation system *Industrial automatic control systems and controllers* **3** pp 11–18

[18] Chuvikov D A and Nazarov K V 2016 Designing algorithms for solving physics problems on the basis of mivar approach *International Journal of Advanced Studies* **3** pp 31–50

[19] Panferov A A, Zhdanovich E A, Yufimychev K A and Chuvikov D A 2016 Designing algorithms for service robots on the basis of mivar approach *International Journal of Advanced Studies* **3** pp 72–86

[20] Shadrin S S, Varlamov O O and Ivanov A M **2017** Experimental autonomous road vehicle with logical artificial intelligence *Journal of advanced transportation* ID 2492765
[21] Chuvikov D A, Teplov E V, Saraev D V, Varlamov O O and Punam J 2016 Methods of automation of supervisory control system based on the urban passenger traffic software expertise system *Radio industry* **4** pp 89–95
[22] Varlamov O O and Aladin D V 2018 About the creation of mivar control systems for monitoring the keeping of traffic rules on the basis of logical kernel (razumator) and expert systems *Radio industry* **2** pp 25–35
[23] Varlamov O O 2018 Wi!Mi Expert System Shell – A Novel Tool To Build Knowledge-Based Systems With Linear Computational Complexity *International Review of Automatic Control* **6**

Improving traction and active safety of the wheeled vehicle by the distribution of the driving torque between its axles

O I Chudakov, V A Gorelov and E B Sarach

Bauman Moscow State Technical University (BMSTU), 105005, Moscow, Russian Federation, 2nd Baumanskaya St., 5

Abstract. One method to improve dynamic qualities, stability and cross-country ability of wheeled vehicles is to use Active Torque Transfer Systems (ATTS). An operating algorithm for the ATTS that ensures improved traction and dynamic properties, road-holding and trajectory stability of a two-axle wheeled vehicle has been suggested. Efficiency and feasibility of the suggested algorithm have been proved through simulation modeling. Lines of future research have been defined.

1. Introduction

There has been observed a trend in the modern automobile industry to improve power-to-weight ratio of wheeled vehicles (WV), which is followed by issues related to efficient implementation of increased tractive capacity. In bad weather conditions and, first of all, in vehicles with high-performance engines, significant wheel sliding can appear due to low friction coefficient combined with high tractive capacity. As a result, not only the created tractive capacity is decreased, but the response of vehicle as well. Solution of this problem requires new design concepts of the powertrain, which would allow the full use of the tractive potential and improvement of the powertrain characteristics by the optimization of its control laws and algorithms.

Currently, there are many known all-wheel drive systems providing redistribution of the driving torques between both the axles and the individual wheels of the vehicle.

One of the most popular options to implement four-wheel drive on two-axle passenger WVs is connection of the second driving-axle (front or rear axle depending on the vehicle layout) using a multi-plate frictional clutch [1-5].

The following companies produce multi-plate clutches for WV: Borg Warner (used on Honda, Hyundai, Porsche), Magna Steyr (BMW, Mercedes-Benz), GKN Driveline (Nissan, Renault), Haldex Traction (Volkswagen, Volvo). Clutch plates pack compression may be performed using mechanical (BMW xDrive system), hydraulic (Haldex) and electromagnetic forces (Porsche PTM system).

The use of the Part-time 4WD systems have the following primary objectives:

1) increasing tractive capabilities of the vehicle;

2) improving stability of the vehicle driving in curves;

A possible criterion of the meeting of the first objective is the longitudinal acceleration value and the capability to drive over the low friction or non-homogenous surfaces, including the situation of driving up-slope. The second objective achievement can be assessed by the speed of the vehicle performing the test without loosing yaw and trajectory control. Efficiency of the Part-time 4WD systems mostly depends on their multi-disk friction clutch control algorithms and can be assessed by means of the mathematical simulation [6-14]. Operating algorithm of Part-time 4WD systems made by foreign manufactures is a commercial secret of particular manufacturing companies and it can not be found in public domain.

Operating algorithm of existing designs depends on many parameters, including temperature conditions of a clutch and its interaction with other electronic active safety systems (Anti-lock Braking System, Electronic Breakforce Distribution, Dynamic Stability Control, etc.). On the initial stage of development, we can evaluate the operating algorithm regarding its impact on tractive and dynamic properties, as well as WV stability [15-16].

The objective of this paper is to increase tractive and dynamic properties, as well as stability of two-axle WVs with a part-time front axle through torque biasing.

2. Theory

The authors developed an algorithm of connecting the first axle of the two-axle vehicle. As the initial data for the calculation of the torque that should be transmitted by the friction clutch, the algorithm uses the following sensors signals: crankshaft speed of the internal combustion engine (ICE), current gear in the gearbox; longitudinal and lateral accelerations of the vehicle center of mass; rotational rate of the vehicle; steer angle; rotational speeds of the wheels. The placement of the sensors is shown in figure 1.

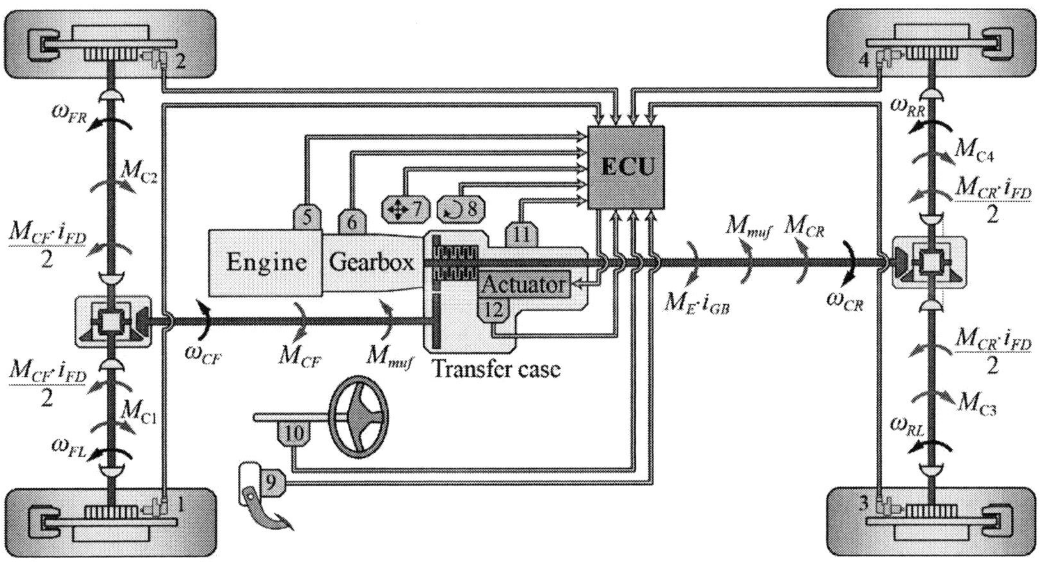

Figure 1. Scheme of transmission and placement of sensors: 1, 2, 3, 4 – wheel speed sensors; 5 – ICE speed sensor; 6 – sensor of the current gear in the gearbox; 7 – linear acceleration sensor; 8 – rotation rate sensor; 9 – accelerator pedal position sensor; 10 – steering angle sensor; 11 – temperature sensor; 12 – actuator position sensor.

In case of linear motion (it is generally thought that a linear motion is motion at steering-wheel angle of $\theta_{SW} \leq 15°$), the system operates in a mode preventing rear axle skidding. We should calculate the relative difference between angular velocities of the front and the rear axles:

$$\Delta\omega_{rel} = 1 - \frac{\omega_{FL} + \omega_{FR}}{\omega_{RL} + \omega_{RR}}, \quad (1)$$

where ω_{FL}, ω_{FR} – angular velocities of the front left and right wheels respectively, s^{-1}; ω_{RL}, ω_{RR} – angular velocities of the rear left and right wheels respectively, s^{-1}.

Apart from the relative difference of angular velocities, torque ratio transmitted to the front axle (h_M) depends on the accelerator pedal position (h_A). Value h_M is calculated according to the curve in figure 2.

IOP Conf. Series: Materials Science and Engineering **534** (2019) 012008 doi:10.1088/1757-899X/534/1/012008

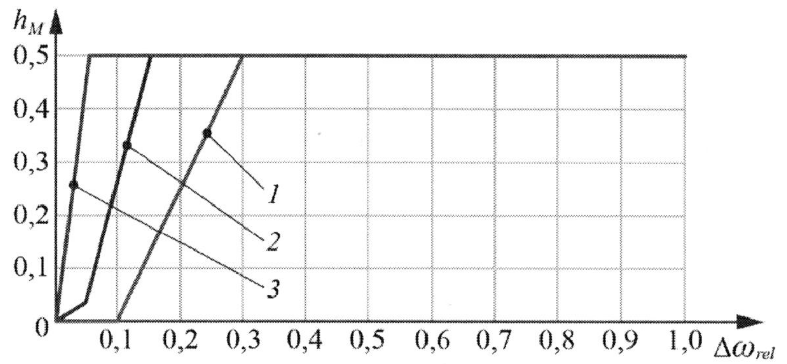

Figure 2. Dependence $h_M(\Delta\omega_{rel})$: 1 – h_A is less than 30%; 2 – h_A is 60%; 3 – h_A is more than 90%.

During nonlinear motion ($\theta_{SW} > 15°$), all-wheel-drive system operates in Dynamic Stability Control mode. One of the main problems in developing Dynamic Stability Control (DSC) operation algorithm is to receive accurate information on quantitative values of WV movement parameters, which would allow to consider how such parameters go in lines with the ones chosen by a driver, predict when an emergency situation occurs and diagnose such situations (for instance, front or rear axle drift, a risk of roll-over, etc.).

In this paper we use the following definitions from Global Technical Regulation No. 8. Electronic Stability Control Systems:

Ackerman Steer Angle means the angle whose tangent is the wheelbase divided by the radius of the turn at a very low speed;

Oversteer means a condition in which the vehicle's yaw rate is greater than the yaw rate that would occur at the vehicle's speed as result of the Ackerman Steer Angle;

Understeer means a condition in which the vehicle's yaw rate is less than the yaw rate that would occur at the vehicle's speed as result of the Ackerman Steer Angle;

Yaw rate means the rate of change of the vehicle's heading angle measured in degrees/second of rotation about a vertical axis through the vehicle's centre of gravity.

Considering the definitions above, WV motion state is diagnosed, i.e. based on sensor values it should be determined to which of three possible states belongs the current data set:

– standard situation, no correction required
– understeer in progress, correction required
– oversteer in progress, correction required

Expressions to calculate theoretical values for WV wheels angular velocities for a WV steering at the Ackerman angle:

$$\begin{cases} R_t = L / \mathrm{tg}|\theta_{SW}|; \\ R_{RL} = R_t - 0{,}5 \cdot B \cdot \mathrm{sgn}\,\theta_{SW}\ ; \\ R_{FL} = \sqrt{R_{RL}^2 + L^2}\ ; \\ R_{RR} = R_t + 0{,}5 \cdot B \cdot \mathrm{sgn}\,\theta_{SW}\ ; \\ R_{FR} = \sqrt{R_{RR}^2 + L^2}\ ; \\ \omega_{FRt} = [\omega_{FR} + (\omega_{FL} / R_{FL} + \omega_{RR} / R_{RR} + \omega_{RL} / R_{RL}) \cdot R_{FR}] / 4\ ; \\ \omega_{FLt} = (\omega_{FRt} / R_{FR}) \cdot R_{FL}\ ; \\ \omega_{RRt} = (\omega_{FRt} / R_{FR}) \cdot R_{RR}\ ; \\ \omega_{RLt} = (\omega_{FRt} / R_{FR}) \cdot R_{RL}\ , \end{cases} \tag{2}$$

where R_t – theoretical turning radius WV, m; L – wheelbase, m; B – gauge, m; R_{FL}, R_{FR}, R_{RL}, R_{RR} – radiuses of trajectories of wheels, m; ω_{FL}, ω_{FR}, ω_{RL}, ω_{RR} – actual values of angular velocities of the wheels, s^{-1}; ω_{FLt}, ω_{FRt}, ω_{RLt}, ω_{RRt} – theoretical values of angular velocities of the wheels, s^{-1}.

Based on the wheels angular velocities theoretical values, a WV linear velocity is calculated:

$$V_{ar} = 0.25 \cdot (\omega_{FLt} + \omega_{FRt} + \omega_{RLt} + \omega_{RRt}) \cdot r_{st}, \tag{3}$$

where r_{st} – static wheel radius, m.

Then, estimated value of WV yawing velocity is calculated:

$$\omega_{zt} = \frac{a_y}{V_{ar}}, \tag{4}$$

where a_y – lateral acceleration of the WV, m/s^2.

Relative difference between the actual and calculated WV yawing velocities is considered to be taken as an indicator to diagnose oversteer or understeer occurrence:

$$\Delta\omega_z = \frac{|\omega_z| - |\omega_{zt}|}{|\omega_z|}. \tag{5}$$

If $\Delta\omega_z$ is less than 0, a understeer is diagnosed, if $\Delta\omega_z$ is greater than 0, a WV oversteer is diagnosed. Value h_M is calculated according to the curve in figure 3.

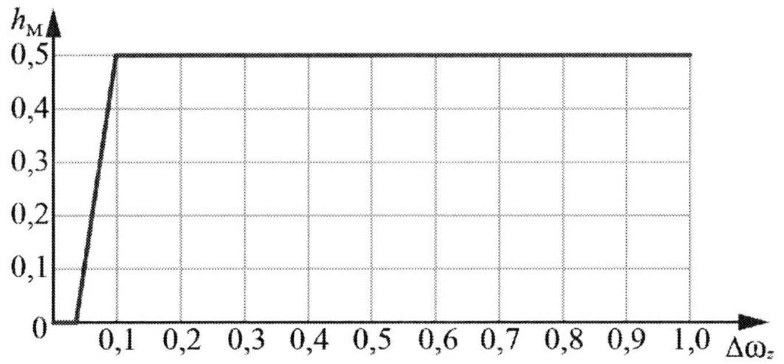

Figure 3. Dependence $h_M(\Delta\omega_z)$.

Calculation of the torque value transmitted by a multi-plate clutch is shown in [9]. It may be interpreted to be used within a numeric model as follows:

1. Using readings from the ICE crankshaft rotation velocity sensor and the ration of depressing of accelerator, as well as partial characteristics of the engine (figure 4), the current engine torque is defined.

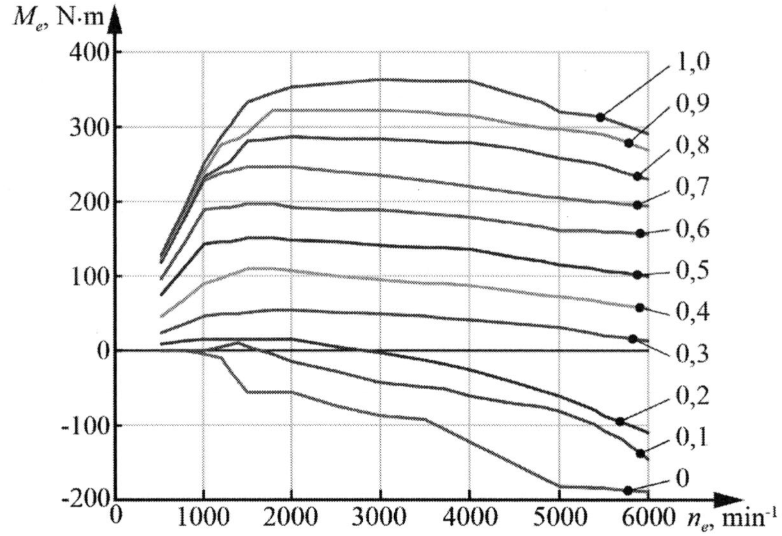

Figure 4. External and partial speed characteristics of the engine.

2. Input torque on the transfer case is calculated:

$$M_{TC} = M_e \cdot i_{GB} \cdot \eta_{GB} \ , \tag{6}$$

where i_{GB} – ratio of the current gear in the gearbox; η_{GB} – efficiency of the gear.

3. Using the relations above the front axle torque ratio (h_M) is calculated, and then the torque transmitted by the multi-plate clutch:

$$M_{Muf} = M_{TC} \cdot h_M \ . \tag{7}$$

The developed algorithm is integrated within a numeric model of non-linear motion of a WV, which is described in detail in papers [10]. Equations of transmission gear motion of a two-axle WV with a part-time front axle (figure 1):

$$\begin{cases} J_k \cdot \dot{\omega}_{FL} = 0.5 \cdot i_{FD} \cdot M_{CF} - M_1; \\ J_k \cdot \dot{\omega}_{FR} = 0.5 \cdot i_{FD} \cdot M_{CF} - M_2; \\ J_k \cdot \dot{\omega}_{RL} = 0.5 \cdot i_{FD} \cdot M_{CR} - M_3; \\ J_k \cdot \dot{\omega}_{RR} = 0.5 \cdot i_{FD} \cdot M_{CR} - M_4; \\ J_{CF} \cdot \dot{\omega}_{CF} = M_{muf} - M_{CF}; \\ J_{CR} \cdot \dot{\omega}_{CR} = M_e \cdot i_{GB} - M_{muf} - M_{CR}; \\ \dot{\omega}_{CF} = 0.5 \cdot i_{FD} \cdot (\dot{\omega}_{FL} + \dot{\omega}_{FR}); \\ \dot{\omega}_{CR} = 0.5 \cdot i_{FD} \cdot (\dot{\omega}_{RL} + \dot{\omega}_{RR}), \end{cases} \tag{8}$$

where J_k – moment of inertia of the wheel, kg·m^2; i_{FD} – final drive ratio; M_i – moment of resistance on the i-th wheel, N·m; $\dot{\omega}_{CF}$ – angular acceleration of the front cardan shaft, s^{-2}; $\dot{\omega}_{CR}$ – angular acceleration of the rear cardan shaft, s^{-2}; J_{CF}, J_{CR} – moments of inertia of the front and rear cardan shafts, respectively, kg·m^2; M_{CF}, M_{CR} – torque on the front and rear cardan shafts, respectively, N·m.

3. Research results

The mathematical model of the two-axle rear-drive vehicle with connectible first axle and the algorithm of the part-time all-wheel drive are implemented in the Matlab Simulink.

The results of numeric modeling of a two-axle WV motion with the total weight of 2,100 kg with a part-time front axle are provided. To study how the algorithm influences tractive and dynamic properties of a WV, an acceleration on a bearing surface with non-homogenous traction properties has been simulated (the interaction ratio of the mover with the bearing surface at full skidding under the rear axle was $\mu_{s100\%} = 0.3$, under the front axle – $\mu_{s100\%} = 0.75$). The modeling results are provided in figure 5.

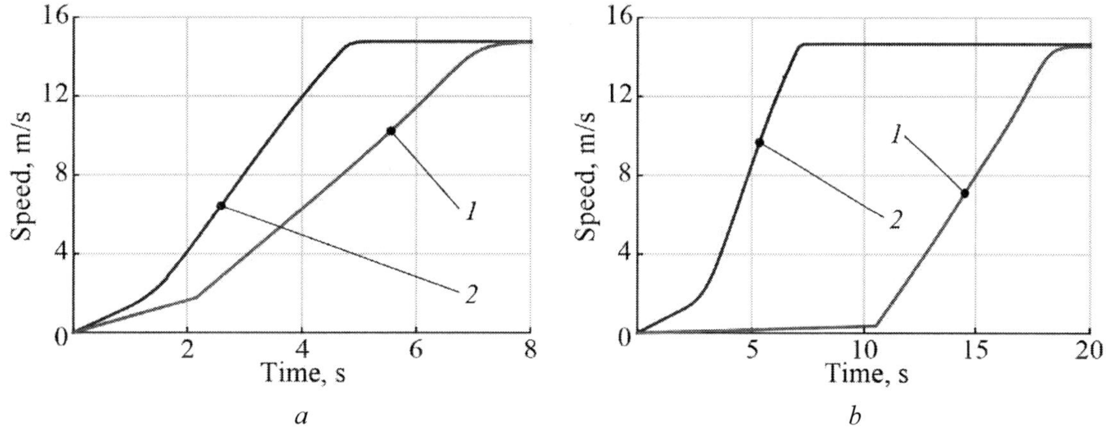

Figure 5. Change in linear velocity on a level track (*a*) and a 7.5° elevating track (*b*): *1* – a rear-wheel drive WV; *2* – a WV with a part-time front axle.

Applying algorithm of four-wheel drive connection while accelerating on a level track allowed a 3 seconds decrease in time of acceleration to reach maximum speed in 1st gear in the WV (figure 5, a). Maximum angle of the elevation being overcome for the two-wheel drive vehicle equals 7.5° (figure 5, b), and the acceleration time to the maximum speed in 1st gear reached 19 seconds (11 seconds greater than for the 4-wheel drive one). For the WV with a part-time front axle, the maximum angle of the elevation being overcome increased up to 13°.

The impact of the algorithm performance in a WV stability has been evaluated based on the modeling results of two maneuvers: going into corner and turning movement with a fixed position of the steering wheel; changing a traffic line (lane change). Movement along a surface "wet asphalt" was studied (interaction ration of the mover with the surface at full skidding was $\mu_{s100\%} = 0.4$). The results of the turning maneuver R = 35 m are shown in figure 6 below.

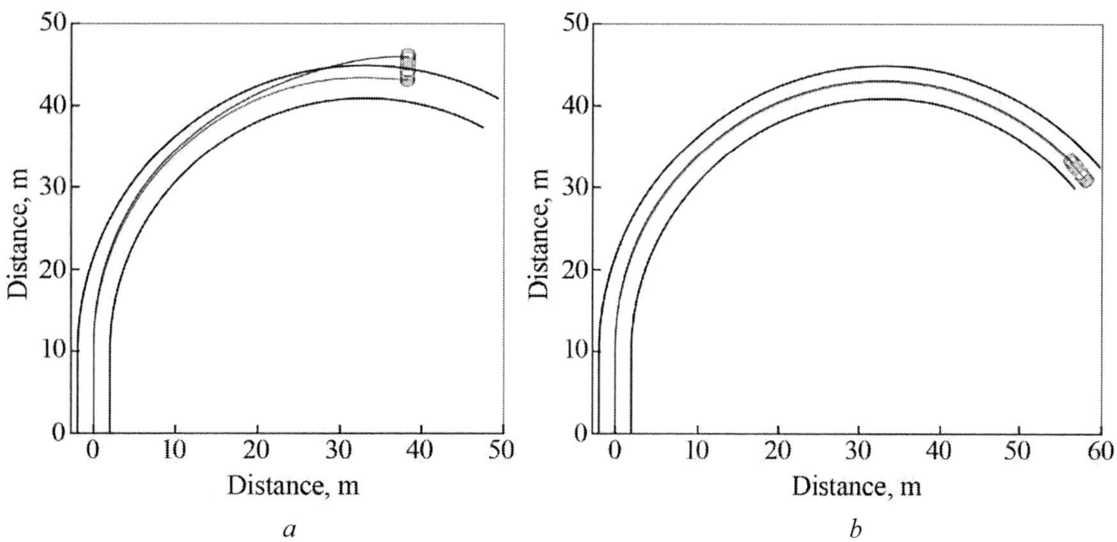

Figure 6. The trajectory of the movement when performing turning maneuver R = 35 m: *a* – a rear-wheel drive WV; *b* – a WV with a part-time front axle.

When the 4-wheel drive system is isolated, a WV loses its road-holding ability at the speed of 38 km/h (figure 6, *a*). With the 4-wheel drive system activated, a WV maneuvers with the speed of 40.7 km/h (figure 6, *b*), which meet the sweep critical speed at a given interaction ration. Further acceleration of a WV will lead to breaking the marked tunnel limits, however, no loss of road-holding ability is observed.

The results of the lane change maneuver S = 20 m are shown in figure 7. When the 4-wheel drive system is isolated, a WV loses its road-holding ability at the speed of 49 km/h (figure 7, *a*). With the 4-wheel drive system activated, a WV is able to maneuver with the speed of 54 km/h (figure 7, b). Further acceleration of a WV will lead to breaking the marked tunnel limits, however, no loss of road-holding ability is observed.

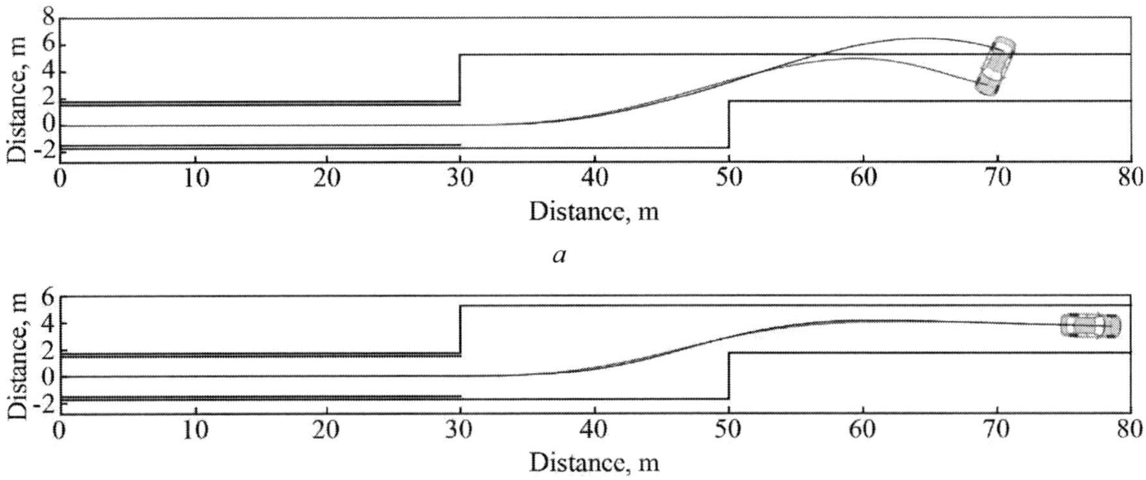

b

Figure 7. The trajectory of the movement when performing lane change maneuver S = 20 m:
a – a rear-wheel drive WV; *b* – a WV with a part-time front axle.

Simulation results showed the following: During acceleration on the road with non-uniform friction, the algorithm eliminates the wheel spin, decreases the time needed for the acceleration up to the given speed, and increases gradeability. When driving in curves, the developed algorithm considerably increases the speed of the vehicle during the tests in comparison to the single-axle drive variant. If the vehicle speed is close to the critical speed, the cornering ability of the vehicle changes from oversteer to understeer which means improvement in the yaw stability and results in better safety. Thus, the mathematical simulation proved efficiency of the developed algorithm of the part-time all-wheel drive.

4. Conclusions

1. We have suggested an algorithm of the Part-time 4WD to increase tractive and dynamic properties and stability of two-axle wheeled vehicles with a part-time front axle.
2. Efficiency and feasibility of the developed algorithm have been proved through simulation modeling of movement of such two-axle wheeled vehicles.
3. The next line of the study is the algorithm improvement oriented at limitations imposed by real-life constructions (temperatures, delay, etc.), as well as better interaction with other active safety systems.

References

[1] Bosch R 2014 *Automotive Handbook* 9th Edition (Robert Bosch GmbH) p 1550
[2] Keller A, Aliukov S 2015 Effectiveness of Methods of Power Distribution in Transmissions of All-Wheel-Drive Trucks *SAE Technical Paper* 2015-01-2732

[3] Keller A, Aliukov S, Anchukov V, et al. 2016 Investigations of Power Distribution in Transmissions of Heavy Trucks *SAE Technical Paper* 2016-01-1100
[4] Rajamani R 2012 *Vehicle Dynamics and Control* (Springer) p 498
[5] Chudakov O I, Volskaja N S and Shiryaeva E Y 20101 *Constructions of drives and distributive mechanisms of modern cars* (Moscow Moscow State Industrial University Press) p 32
[6] Keller A V, Gorelov V A, Vdovin D S et al. 2015 Mathematical model of all-terrain truck *Proceedings of the ECCOMAS Thematic Conference on Multibody Dynamics* (Spain Universitat Politecnica de Catalunya) pp 1285–1296
[7] Gladov G I, Popov S D, Sarkisov P I et al. 2015 Physical tire model based on understanding of separate components of tire deformation *The Dynamics of Vehicles on Roads and Tracks – Proceedings of the 24th Symposium of the International Association for Vehicle System Dynamics* (Austria IAVSD) pp 749–758
[8] Keller A and Aliukov S 2015 Methodology of System Analysis of Power Distribution among Drive Wheels of an All-wheel-drive Truck *SAE Technical Paper* 2015-01-2788
[9] Gorelov V A and Komissarov A I 2016 Mathematical model of the straight-line rolling tire-rigid terrain irregularities interaction *Procedia Engineering* **150** (Elsevier) pp 1322–1328
[10] Keller A, Aliukov S. 2015 Efficient power distribution in an all-wheel drive truck *Lecture Notes in Engineering and Computer Science* **2218** pp 1201-1206
[11] Gorelov V A, Komissarov A I and Miroshnichenko A V 2015 8×8 wheeled vehicle modeling in a multibody dynamics simulation software *Procedia Engineering* **129** (Elsevier) pp 300–307
[12] Keller A V, Gorelov V A and Anchukov V V 2015 Modeling truck driveline dynamic loads at differential locking unit engagement *Procedia Engineering* **129** (Elsevier) pp 280–287
[13] Cring B 2016 *Control strategy for active torque control* (Patent no. US 7125364B2)
[14] Keller A and Alyukov S 2016 Power Distribution in Transmissions of Multi-Wheeled Vehicles *SAE Technical Paper* 2016-01-1103
[15] Keller A and Aliukov S 2016 Efficient Power Distribution in an All-Wheel Ground Vehicles *SAE Technical Paper* 2016-01-1105
[16] Kotiev G O, Gorelov V A and Beketov A A 2008 Mathematical model of movement of a cross-country vehicle *Journal of Automotive Engineers* **1** (Moscow: AAE Press) pp 50–54

Performance analysis of the vehicle electronic stability control in emergency maneuvers at low-adhesion surfaces

I Kulikov[1] and J Bickel[2]

[1] Department of Powertrains, Federal State Unitary Enterprise "NAMI", Moscow, Russian Federation
[2] Faculty of Engineering and Computer Science, University of Applied Sciences "Hochschule Osnabrück", Osnabrück, Germany

Email: i.kulikov.mami@gmail.com

Abstract
Essential principles of the vehicle electronic stability control (ESC) operation are analyzed. Mathematical model employed for the performance investigation of vehicles equipped with ESC is described. Modeling results are presented, simulating the actual road test, in which ESC-equipped vehicle performed an emergency maneuver at the surface covered with packed snow. Modeling allowed identifying ESC interventions occurred in the road test. Also, modeling results are presented, simulating the same road test with disabled ESC and with replacement of the original ESC by an experimental yaw stability controller. Conclusions are drawn about performance of both original ESC and the experimental controller. Analysis of the simulation results allowed disclosing changes of the vehicle steering behavior during the maneuver, as well as stability loss factors emerging at particular points of the maneuver. Variation of controller's parameters allowed defining conditions in which ESC should intervene to prevent the vehicle from a skid.

1. Introduction

Vehicle emergency handling constitutes an example of transient maneuvers with a substantial risk of losing directional stability. This may take place during a sharp lane changing in order to avoid collision with a sudden obstacle. At low-adhesion surfaces, such maneuvers become especially unsafe.

At present, performance of vehicle electronic stability control systems at low adhesion surfaces can be considered as an insufficiently investigated issue. Global Technical Regulation No. 8 (GTR 8) is entirely dedicated to the ESC and includes a substantial analytical section; however, while stating the importance of said issue, this Regulation nevertheless offers neither analysis nor testing procedures for ESC operation at low adhesion surfaces. Among known literature on this problem, one can notice the works [1, 2], which describe experimental investigations of ESC performance at icy surfaces on vehicle equipped with studded tires. The experiments included "sine with dwell" maneuver specified by GTR 8. At the same time, for getting a deeper insight into vehicle dynamics at low adhesion surfaces with ESC interventions, theoretical investigations are also required. However, their number is small as well. One of the rare examples is [3] describing problem analysis and tire modeling taking into account ESC operation at slippery roads. Therefore, elimination of the research gap in this field can be considered an urgent task, which is addressed in the work described by this paper. To allow the reader better grab approaches of this investigation, as well as its results, first sections of the paper provide descriptions of the essential principles of operation and control of ESC systems.

2. ESC basic operation principles

During transient directional motion, vehicle steering behavior changes both quantitatively and qualitatively, i.e. showing transitions between oversteering and understeering. Transient steering behavior is defined by the yawing moment. The latter is built by the sum of tire force moments exerted by individual wheels. Difference between front and rear tire forces makes that sum deviate from zero when transient directional motion takes place. On the one hand, the yawing moment constitutes an immanent feature of the transient turn. On the other hand, this moment makes rising vehicle deviation from the neutral steering behavior. With certain vehicle velocity, steering wheel angle and steering dynamics, transient process does not converge to a stable state bringing vehicle into a skid. In other cases and conditions, transient process may lead to a stable state, which nevertheless prevents vehicle from following the trajectory specified by the driver, which is called drift.

The ESC system is intended for preventing skids and drifts through altering of the vehicle yawing moment [4, 5]. By its operation principle, ESC can be called a yaw controller of extremal type, because it only intervenes into vehicle motion if there is a risk of stability loss. ESC corrects vehicle steering behavior by regulation of yawing moments created by the individual wheels. This is achieved by altering tire force magnitude, or tire force arm, or both. These corrections are implemented through the known tire-surface adhesion property, i.e. association between longitudinal and lateral adhesion coefficients. The simplest model of this property is the Kamm's circle. It shows the relation between the tire force vector direction and the tire slip in both longitudinal and lateral directions: changes in these slip components cause rotation of the tire force vector. In figure 1, the tire force vector R_{12} at the front left wheel rotates from the initial position ($R_{12,i}$) into the final position ($R_{12,f}$). This rotation changes the tire force arm from $h_{12,i}$ to $h_{12,f}$. Assuming the vector magnitude constant, the change in yawing moment reads $\Delta M_z = R_{12} \cdot (h_{12,i} - (-h_{12,f}))$. Figure 1 shows a vehicle performing right turn, in which the front left tire creates the largest yawing moment increasing oversteering and, possibly, leading into a skid. Rotating this tire force vector clockwise, as it is shown in figure 1, diminishes oversteering behavior, thus preventing vehicle from a skid.

Since tire sideslip depends on directional motion dynamics, which is defined by the vehicle velocity, the steering wheel angle, and steering dynamics (in other words, defined by the driver and driving conditions), it cannot be employed for the controllable change of the tire force vector. This change can only be performed by the longitudinal tire slip control, which is achieved through the regulation of the wheel torque (T_w). The latter is provided by the engine (altered by the transmission) and brakes. Automatic control of these components, allowing regulation of torques at individual wheels, implements the slip control, and thus provides tire force rotation feature. This mechanism is essential for the ESC operation.

Backward rotation of the tire force vector at the front outer wheel (with respect to the center of the turn) constitutes the main operation performed by the ESC. This operation is relatively simple to implement – by means of a brake mechanism – and creates unambiguous effect: both decreasing of R_y component and increasing of $-R_x$ component diminish the yawing moment, which leads into a skid. Note that decreasing of the R_y requires a substantial longitudinal slip to be attained, which implies bringing tire beyond the maximum point of its adhesion characteristic (since this is the only way to rotate the tire force vector at a large enough angle). Additionally, ESC can operate at the rear outer wheel. It also should be braked, however, braking torque should have the magnitude that allows increasing $-R_x$ component, which acts against skid, without decreasing R_y component, which also counteracts skid. To meet this requirement, tire should be operated below its adhesion maximum, preventing the lateral force from falling. ESC operation at inner wheels is a more contradictory and complicated option, because: it requires creating non-symmetrical traction torques (which is technically more difficult than braking), produces rather ambiguous effect (increasing the vehicle speed and, as a result, the lateral acceleration), and is less effective in the sense of yawing moment correction (because the tire force is relatively small due to decreased weight acting on inner wheels).

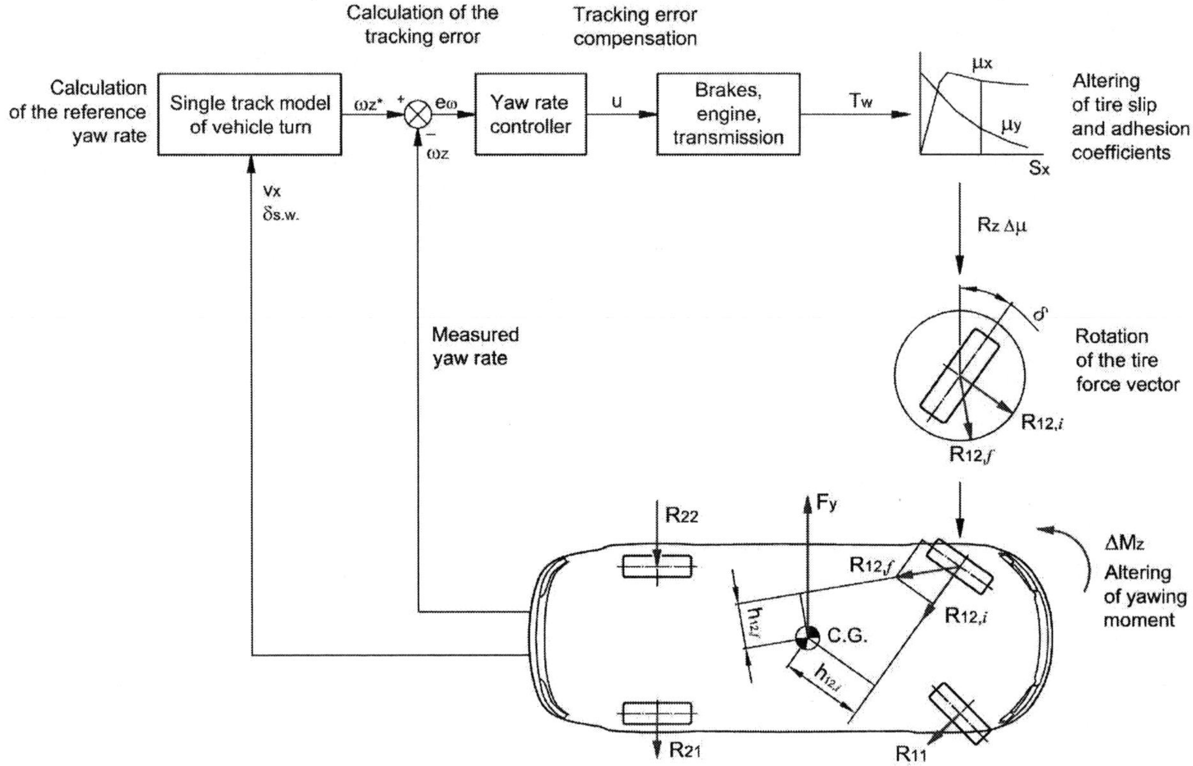

Figure 1. Vehicle electronic stability system operation and control loop.

3. ESC control

The ESC is a feedback yawing moment controller (see figure 1 for the control loop). The main components of such a system are the regulator generating control signal, the command signal for controlled variable, and the feedback signal. In the ESC, controlled variable is the yaw rate. This variable is involved as the basic parameter in a number of vehicle handling characteristics [6], being a sufficient quantitative measure of the vehicle steering behavior. Reference steering behavior of the vehicle can be defined by calculation of the corresponding yaw rate. The simplest way for calculating this reference yaw rate is to employ the single-track kinematic model of the vehicle turn (figure 2).

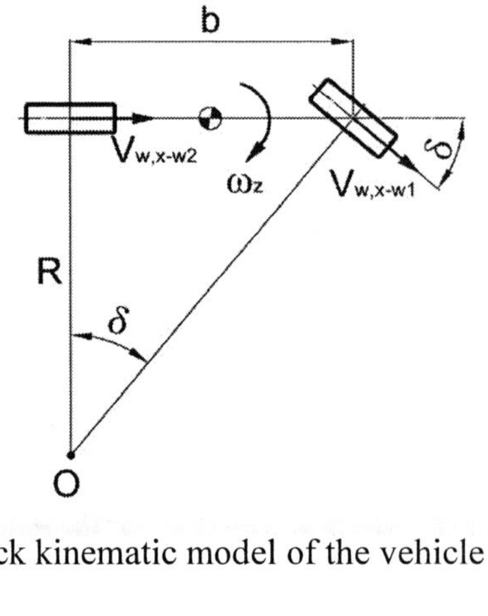

Figure 2. Single-track kinematic model of the vehicle turn on rigid wheels.

From this figure, an expression for the yaw rate can be derived, which depends on small number of parameters (see nomenclature of variables and parameters in the appendix):

$$\omega_z = \frac{V_{w-x,w2} \cdot \delta_{s.w.} \cdot u_{steer}}{b} \tag{1}$$

All these parameters can be determined in the ESC controller with sufficient precision, including vehicle velocity, although usually it does not coincide with the wheel linear velocity due to slip, which implies some sort of observer to be utilized to calculate it.

Yaw rate calculated by equation (1) corresponds to the Ackermann's kinematics, since there are no sideslip angles in this model. To make this model providing desired characteristics of understeering or oversteering, it is augmented with certain modifications [5]. One of them adds the cornering stiffness of tires. It must be noted that cornering stiffness coefficients depend on tire-surface adhesion, which should be identified by ESC controller in real-time. Second modification adds a dead zone at the input of the yawing moment regulator to adjust its sensitivity to the deviation between the actual yaw rate and the required one. Having that deviation within limits of the dead zone will prevent the ESC regulator from undesirable interventions into vehicle motion. Yet another modification is introduced to address difference between the steady state turn calculated by the model (1) and actual dynamics of the vehicle directional motion. This modification implies passing the calculated yaw rate signal through a filter having time constant, which corresponds to vehicle inertial properties. All mentioned modifications allow defining thresholds between save maneuvering, intervention to which is not necessary and negatively perceived by drivers, and critical maneuvering, which requires ESC interventions.

In addition, calculated yaw rate should be limited by the maximum lateral tire adhesion [5], which is defined for the steady-state single-track model as follows:

$$a_y = \frac{V_{w,x-w2}^2}{R} = \omega_z \cdot V_{w,x-w2} \leq \mu_{y,max} \cdot g \rightarrow \omega_z \leq \frac{\mu_{y,max} \cdot g}{V_{w,x-w2}}$$

Maximum lateral adhesion coefficient should be identified by indirect methods. Accuracy of this identification influences ESC performance at a given road surface.

The feedback signal of the ESC regulator is the actual yaw rate measured onboard by a gyroscopic sensor. Difference between required and measured yaw rate is fed into the regulator input. In the sophisticated production ESC systems, the regulator outputs correcting yawing moment command, which is distributed between separate wheels by means of weighting function [5]. In turn, yawing moment assigned to each wheel is converted into the longitudinal slip, which is required for corresponding rotation of the tire force vector. This conversion is performed by means of the Kamm's circle and the following variables calculated by the controller: wheel normal force, maximum tire-surface adhesion coefficient, and sideslip angle. The slip command is compared to the actual slip, which is also obtained by means of calculations. Resulting slip difference is fed into low-level controllers, which implement required slip by providing appropriate torques by brakes, engine, and transmission.

Production ESC systems employ yaw rate controllers in the form of PID regulators with variable gains [5], which depend on the vehicle velocity and the tire-road adhesion. Additionally, the proportional gain constitutes a non-linear function of the calculated vehicle sideslip angle.

Performance of an ESC system having control signals defined by indirect calculations of unmeasured quantities (so called virtual sensors) depends on accuracy of employed models, which is obviously limited and can deteriorate in complicated operating conditions like unsteady road adhesion, tire wearing, and so on. For compensation of inaccuracy or/and disturbances, ESC can use some methods of adaptive control. In the literature, different ways of implementing such systems are proposed – from employing of auxiliary relay regulators [7], to fuzzy logic [8], and artificial neural networks [9].

One can notice that existing literature on ESC systems mostly focuses on control systems and methods [7-10] paying much less attention to the controlled object, i.e. the vehicle. In many works, vehicle is treated as a "black box" modeled by library-based software such as ADAMS Car or CarSim. Such works show an approach involving analysis of two or three basic quantities (lateral acceleration,

IOP Conf. Series: Materials Science and Engineering **534** (2019) 012009 doi:10.1088/1757-899X/534/1/012009

yaw rate, and body roll angle) of the vehicle directional motion, which only allows stating that ESC has prevented or has not prevented the loss of stability; however this approach does not answer the question "why did that happen"?

Research and development of the ESC should be conducted along with thorough analysis of vehicle handling properties. It is necessary to study conditions and factors of stability loss, as well as mechanisms of its prevention by the ESC. In particular, it should be investigated at which stage of a handling maneuver the ESC has to intervene to maintain vehicle stability. This paper describes results of the research, which was aimed at addressing these issues. The main tools of this research were the mathematical model described in the next section and the results of road tests performed on ESC-equipped vehicle. For implementation of the vehicle stability control feature, a relatively simple yaw rate regulator was elaborated, which only outputs torque command for the front outer wheel, which yawing moment is the major factor of skid prevention. During the investigation, regulator adjustments were varied in order to disclose their effect on the vehicle handling ability.

4. Model of vehicle dynamics

In this work, vehicle was modelled as a multibody system comprising six masses: vehicle itself performing a planar motion, the sprung mass moving in lateral plane (body roll), and four rotating masses of the wheels. Wheel normal forces were calculated based on force equilibrium in both longitudinal and lateral planes. Neglecting a small angular speed of the sprung mass around the lateral axis allowed excluding association between angular speeds and accelerations from the Euler dynamic equations. Products of inertia, also assumed negligible, were excluded from the equations of vehicle motion and body roll.

Figure 3 shows schematics used to derive the equations of vehicle dynamics.

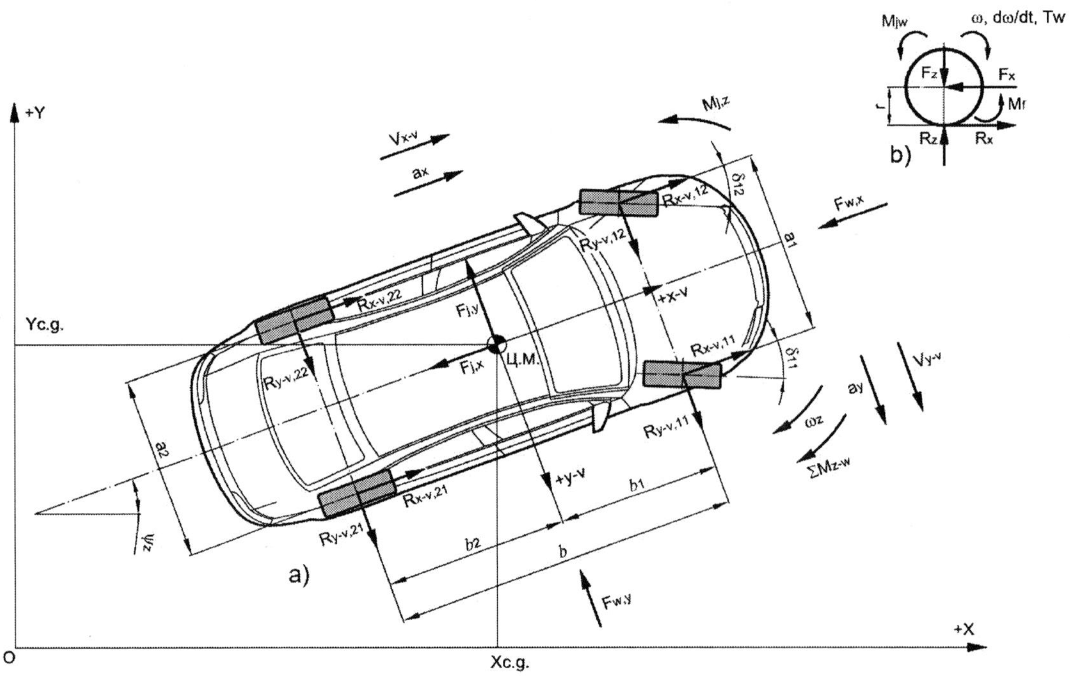

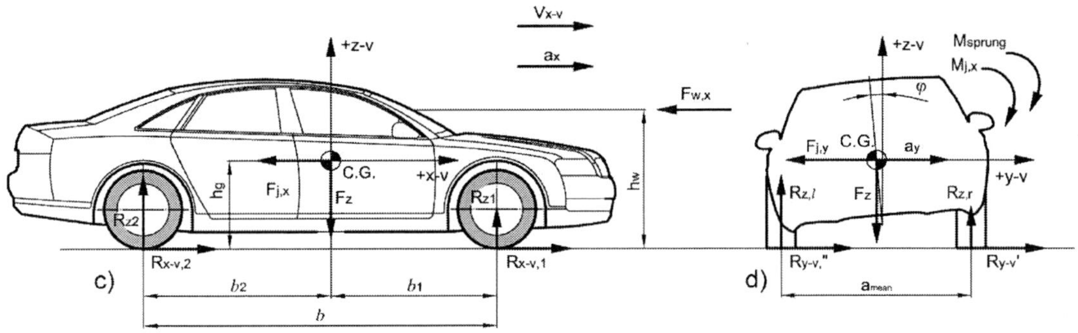

Figure 3. Schematics for the vehicle dynamics model. a – vehicle planar motion; b – wheel rotational dynamics; c – calculation of the tire normal forces in longitudinal plane; d – body roll and tire normal forces in lateral plane.

The following coordinate systems were implemented within the model: associated with the road plane, associated with the vehicle (corresponding variables are denoted by "v" index), and associated with each wheel (corresponding variables are denoted by "w" index). Also, designations of variables and parameters associated with the wheels include indices *ij*, where *i* denotes axis number, and *j* denotes wheel number associated with *i*-th axis.

The equation system derived from these schematics and assumptions, reads as follows (see nomenclature of variables and parameters in the end of the paper):

$$\begin{cases} m \cdot a_{x,c.g.} = m \cdot \left(\dot{V}_{c.g.,x-v} - \omega_z \cdot V_{c.g.,y-v}\right) = \sum_{i=1}^{n}\sum_{j=1}^{k} R_{x-v.ij} - F_{w,x} \\ m \cdot a_{y,c.g.} = m \cdot \left(\dot{V}_{c.g.,y-v} + \omega_z \cdot V_{c.g.,x-v}\right) = \sum_{i=1}^{n}\sum_{j=1}^{k} R_{y-v.ij} - F_{w,y} \\ \mathcal{J}_z \cdot \ddot{\psi}_z = \mathcal{J}_z \cdot \dot{\omega}_z = \sum_{i=1}^{n}\sum_{j=1}^{k} M_{z-v.ij} \\ \mathcal{J}_{x,sprung} \cdot \ddot{\varphi} = \sum_{i=1}^{n}\sum_{j=1}^{k}\left(R_{y-v.ij} \cdot h_{c.g.}\right) - \varphi \sum_{i=1}^{n} c_{\varphi i} - \dot{\varphi} \sum_{i=1}^{n} d_{\varphi i} \\ \mathcal{J}_{w,ij} \cdot \dot{\omega}_{w,ij} = T_{w,ij} - \left(R_{z,ij} \cdot f_{ij} + R_{x-w.ij}\right) \cdot r_{ij} \end{cases} \tag{2}$$

5. Model of vehicle kinematics

Tire force vector components included in the system (2) constitute functions of tire longitudinal slip and sideslip angles. For calculation of these parameters, a kinematic model of the vehicle was employed considering vehicle as a body performing a planar motion. With this assumption, the velocity graph of vehicle was drawn; figure 4 shows a part of this graph corresponding to the front left wheel.

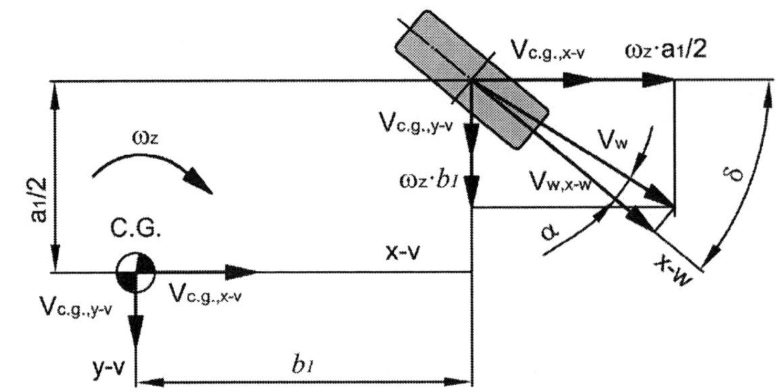

Figure 4. Velocity graph for the vehicle kinematic model.

Baseline point of this graph is the vehicle center of gravity (C.G.), which linear velocity components ($V_{c.g.,x-v}$, $V_{c.g.,y-v}$), as well as the vehicle yaw rate, are calculated by the model of vehicle dynamics (2). With these quantities, projections of the wheel linear velocity on the vehicle axes are calculated as follows:

$$V_{w,x-v,ij} = V_{c.g,x-v} \pm \omega_z \cdot \frac{a_i}{2}$$

$$V_{w,y-v,ij} = V_{c.g,y-v} \pm \omega_z \cdot b_i + \frac{\partial y_{w-v,ij}}{\partial \varphi} \dot{\varphi}$$

For "x-v" axis projection, the second term is taken with "+" for the right wheels. For "x-y" axis projection, the second term is taken with "+" for the front wheels. Lateral velocity component $V_{w,y-v,ij}$ is calculated taking into account lateral displacement of the wheel resulting from the body roll. This is accomplished by employing derivative $\frac{\partial y_{w-v,ij}}{\partial \varphi}$, which constitutes a kinematic characteristic of the suspension [6, 11].

Sideslip angle is calculated as follows:

$$\alpha_{ij} = arctan\left(\frac{V_{w,y-v,ij}}{V_{w,x-v,ij}}\right) - \left(\delta_{ij} + \frac{\partial \delta_{ij}}{\partial \varphi} \varphi\right).$$

Besides the wheel steer angle δ_{ij} resulted from the steering wheel rotation, this equation takes into account additional angle resulted from the body roll and calculated by another kinematic characteristic of the suspension $\frac{\partial \delta_{ij}}{\partial \varphi}$, which introduces relation between the body roll and the toe-in angle of the wheel [6, 11].

The full vector of the wheel center linear velocity is defined by its components:

$$V_{w,ij} = \sqrt{V_{w,x-v,ij}^2 + V_{w,y-v,ij}^2}$$

Projecting the wheel center velocity $V_{w,ij}$ on the wheel longitudinal plane and dividing it by the wheel angular velocity yields the wheel rolling radius:

$$r_{e,ij} = \frac{V_{w,ij} \cdot cos\alpha_{ij}}{\omega_{w,ij}}.$$

IOP Conf. Series: Materials Science and Engineering **534** (2019) 012009 doi:10.1088/1757-899X/534/1/012009

For calculation of the tire longitudinal slip, one of the following expressions is utilized constituting ratios between the current wheel rolling radius and the effective rolling radius $r_{e0,ij}$ for traction and braking modes of the wheel:

$$S_{x,tr.} = 1 - \frac{r_{e,ij}}{r_{e0,ij}};\ S_{x,br.} = -\left(1 - \frac{r_{e0,ij}}{r_{e,ij}}\right).$$

Effective rolling radius $r_{e0,ij}$ is determined experimentally under free rolling conditions when the vehicle is driving along a straight line.

6. Tire model

In this work, an empirical model called the Magic Formula (MF) devised by H. Pacejka [12] is employed for calculation of tire-surface adhesion characteristics. The MF model approximates both longitudinal and lateral components of the tire force vector with unified trigonometric functions:

$$R_{x-w,0} = D_x \cdot \sin\left(C_x \cdot arctan\left(B_x \cdot S_x - E_x\left(B_x \cdot S_x - arctan(B_x \cdot S_x)\right)\right)\right),$$

$$R_{y-w,0} = D_y \cdot \sin\left(C_y \cdot arctan\left(B_y \cdot \acute{\alpha} - E_y\left(B_y \cdot \acute{\alpha} - arctan(B_y \cdot \acute{\alpha})\right)\right)\right),$$

where $\acute{\alpha} = -\alpha$ (the lateral tire force acts in opposite direction to the sideslip angle). Index "0" denotes that these equations calculate tire forces for the cases of pure longitudinal slip ($\alpha \approx 0$) and pure sideslip ($S_x \approx 0$). Combined slip is taken into account by the correcting functions described below. B, C and E coefficients are identified by means of experimental data acquired from vehicle or tire test results. D coefficient constitutes a product of the maximum longitudinal/lateral adhesion coefficient and the tire normal force: $\mu_{x/y,max} \cdot R_{z,\cdot}$.

Tire aligning moment is calculated by the following function:

$$M_{z-w} = -R_{y-w,0} \cdot D_t \cdot \cos\left(C_t \cdot arctan\left(B_t \cdot tan(\acute{\alpha}) - E_t\left(B_t \cdot tan(\acute{\alpha}) - arctg(B_t \cdot tg\acute{\alpha})\right)\right)\right).$$

Equation of the longitudinal tire force for the combined slip reads:

$$R_{x-w} = R_{x-w,0} \cdot G_{xy},$$

where $G_{xy} = \cos\left(C_{xy} \cdot arctan\left(B_{xy} \cdot tan(\acute{\alpha}) - E_{xy}\left(B_{xy} \cdot tan(\acute{\alpha}) - arctan(B_{xy} \cdot tg\acute{\alpha})\right)\right)\right)$ is the correction function taking into account sideslip. A similar formula is applied for the correction of the lateral tire force, with the longitudinal slip S_x as an argument of the correction function.

Tire forces calculated by the MF model are projected from the wheel axes onto the vehicle axes in accordance with schematics of figure 5.

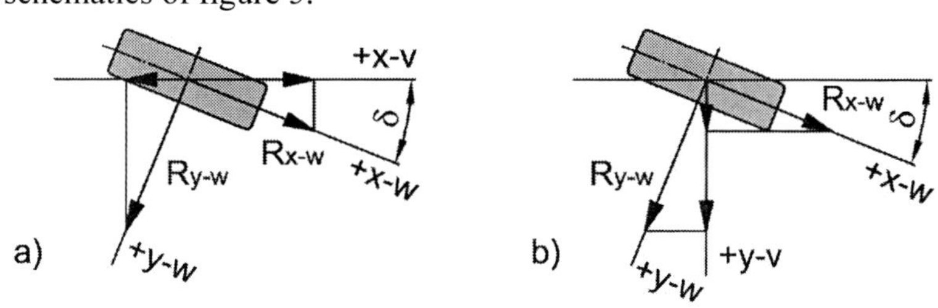

Figure 5. Projection of the tire force vector from the wheel axes onto the vehicle axes. a – on the longitudinal axis of the vehicle; b – on the lateral axis of the vehicle.

From these schematics, the following equations of longitudinal and lateral components of the tire force projected on the vehicle axes were derived:

$$\begin{cases} R_{x-v,ij} = R_{x-w,ij} \cdot cos\delta_{ij} - R_{y-w,ij} \cdot sin\delta_{ij} \\ R_{y-v,ij} = R_{x-w,ij} \cdot sin\delta_{ij} + R_{y-w,ij} \cdot cos\delta_{ij} \end{cases}$$

Using these projections and schematics shown in figure 3a allows defining yawing moment exerted by the *ij*-wheel:

$$M_{z-v,ij} = \pm R_{x-v,ij} \cdot \frac{a_i}{2} \pm R_{y-v,ij} \cdot b_i + M_{z-w,ij}.$$

According to the accepted sign convention, the moment of the longitudinal tire force is taken with "+" for the left wheels, and with "–" for the right wheels; the moment of the lateral tire force is taken with "+" for the front wheels, and with "–" for the rear wheels.

7. Experimental research

The studied vehicle was tested at the NAMI Testing Center by the Vehicle Safety Department. Physical and design parameters of the vehicle, which are essential for modeling, were determined by means of laboratory tests. In particular, vehicle weight parameters were measured, and center of gravity coordinates determined. Also, measurements were conducted, which allowed calculation of the steering mechanism characteristics. The road tests were performed at surfaces covered with packed snow and ice. The testing program consisted of both ESC-on and ESC-off emergency maneuvers (figure 6).

Figure 6. Performing of a "lane change" road test.

The employed test equipment allowed measuring the following signals: steering wheel angle, vehicle longitudinal velocity, wheel rpm, linear accelerations and angular velocities at certain points of the vehicle body. Description of the conducted tests as well as the measuring equipment can be found in [13].

8. Tire model parameters identification

Since no separate tire tests were conducted, adhesion characteristics have been identified by means of modeling. Identification implies varying MF coefficients with assessment how these affect certain variables of the vehicle dynamics (longitudinal velocity, linear accelerations, and yaw rate). Coefficient values, which provide modeling with minimum relative mean square errors (MSE) for those dynamic variables are considered acceptable. The identification procedure was organized as follows. First to be determined were characteristics of the pure tire slip. Pure longitudinal slip parameters can be determined from tests where maximum longitudinal acceleration/deceleration is achieved. This takes place in emergency braking tests. Pure sideslip takes place in tests where vehicle is steered at constant speed achieving maximum lateral acceleration. An example of such test is emergency lane changing. Having the pure slip characteristics defined, identification of the combined slip correction functions can be performed using experiments with ESC interventions. It should be stressed that the correction functions are crucial for ESC modeling, since the operation principle of ESC is based on tire combined slip properties.

Figure 7 shows an example of identified adhesion characteristics for the surface covered with packed snow. Note that the tested vehicle was equipped with studded tires.

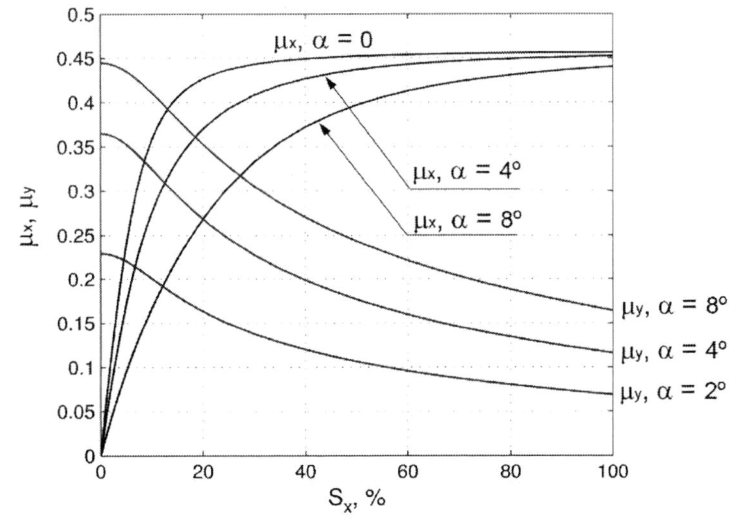

Figure 7. Identified tire adhesion characteristics for the packed snow.

9. Identification of the wheel torques

Modeling of active safety systems implies determination of the wheel torques. Since in this work these were not measured during the road tests, in the model they were calculated by means of the torque observers [13, 14]. These were implemented by the feedback control systems regulating wheel torques (figure 8). For such torque controller, the command signal is the wheel angular velocity measured during a test, and the feedback signal is the angular velocity of the same wheel calculated by the vehicle model. In order to compensate difference between the command value and the feedback value, the regulator calculates the wheel torque.

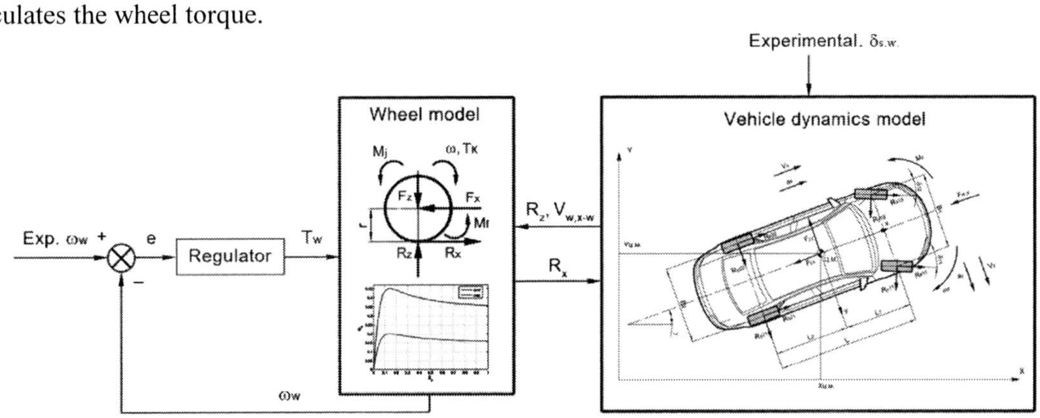

Figure 8. Identification of unmeasured wheel torques (torque observer).

Accuracy of identification depends on overall validity of the vehicle model and is evaluated indirectly, i.e. by relative mean square errors of vehicle velocity and acceleration. Another criterion is

IOP Conf. Series: Materials Science and Engineering **534** (2019) 012009 doi:10.1088/1757-899X/534/1/012009

the qualitative correspondence of the identification results to operation principles of the considered active safety system. For example, in maneuvers with an ESC system enabled, one can expect pronounced correction torques to be identified at individual wheels.

10. Simulation results and analysis thereof

To get an insight into the ESC operation and evaluate its performance, a modeling procedure was elaborated comprising three basic steps:

1) Simulation of a road test with ESC intervention.

2) Simulation of the same road test with the ESC disabled (if simulation results show vehicle skid, the ESC intervention is considered appropriate and efficient).

3) Simulations of the same test with the experimental yawing moment controller enabled.

The third step implies a number of simulations with different settings of the yawing moment controller. Basic parameters to be adjusted are gains P, I and D, which determine "stiffness", accuracy, and response of the regulator. In addition, regulator "dead zone" is adjusted, which specifies controller sensitivity.

Described below is the investigation of a "lane change" maneuver at packed snow, which was performed during the vehicle road testing. Figure 9 shows results of simulations conducted in accordance with steps 1 and 2 of the above-described procedure. Plot colors are explained in legend only briefly, implying the following clarifications:

– "Experiment with ESC" denotes data that were acquired by means of measuring equipment during the road test;

– "Model with ESC" means modeling results simulating the road test (i.e. the first step of the procedure);

– "Model without ESC" corresponds to the modeling with deactivation of the ESC (i.e. the second step of the procedure);

– "wheel11", "wheel12", "wheel21", and "wheel22" denote model variables associated with vehicle wheels (see using ij indices described above).

The plots show following variables (top-down):

– Longitudinal velocity at the vehicle center of gravity ($V_{c.g.,x-v}$): measured during the road test, resulted from the road test simulation, and resulted from the simulation of ESC deactivation.

– Steering wheel angle ($\delta_{s.w.}$) measured during the road test.

– Wheel torques (T_w) identified in simulation of the road test with ESC intervention.

– Vehicle yawing moment (ΣM_{z-v}): resulted from the road test simulation, and from the simulation of ESC deactivation.

– Yaw rate (ω_z): measured during the road test, resulted from the road test simulation, resulted from the simulation of ESC deactivation, and calculated by the single-track vehicle kinematic model (denoted as "Kinem. model").

– Lateral acceleration of the vehicle center of gravity ($a_{y,c.g.}$): measured during the road test, resulted from the road test simulation, and resulted from the simulation of ESC deactivation.

– Wheel sideslip angles (α) calculated during simulation of the road test with ESC intervention.

Model accuracy was estimated by MSEs for three variables, namely, longitudinal velocity of the vehicle, yaw rate, and lateral acceleration at the vehicle center of gravity. Resulted MSEs are shown next to corresponding plots.

From the graphs of identified wheel torques, one can see that ESC exerts the primary (i.e. the largest) correction torque at the front outer wheel, and the secondary torque – at the rear outer wheel. The first ESC intervention is seen in the time interval 6.5…7.2 s (i.e. duration of the intervention was 0.7 s). The major intervention, aimed at prevention of the right skid of the vehicle, took place between 7.2 and 8.3 s (duration of the intervention was 1.1 s). Thus, the full operational cycle of the ESC, beginning with initiation of the intervention and ending with the vehicle stabilization, in the considered conditions, takes ca. 1 second.

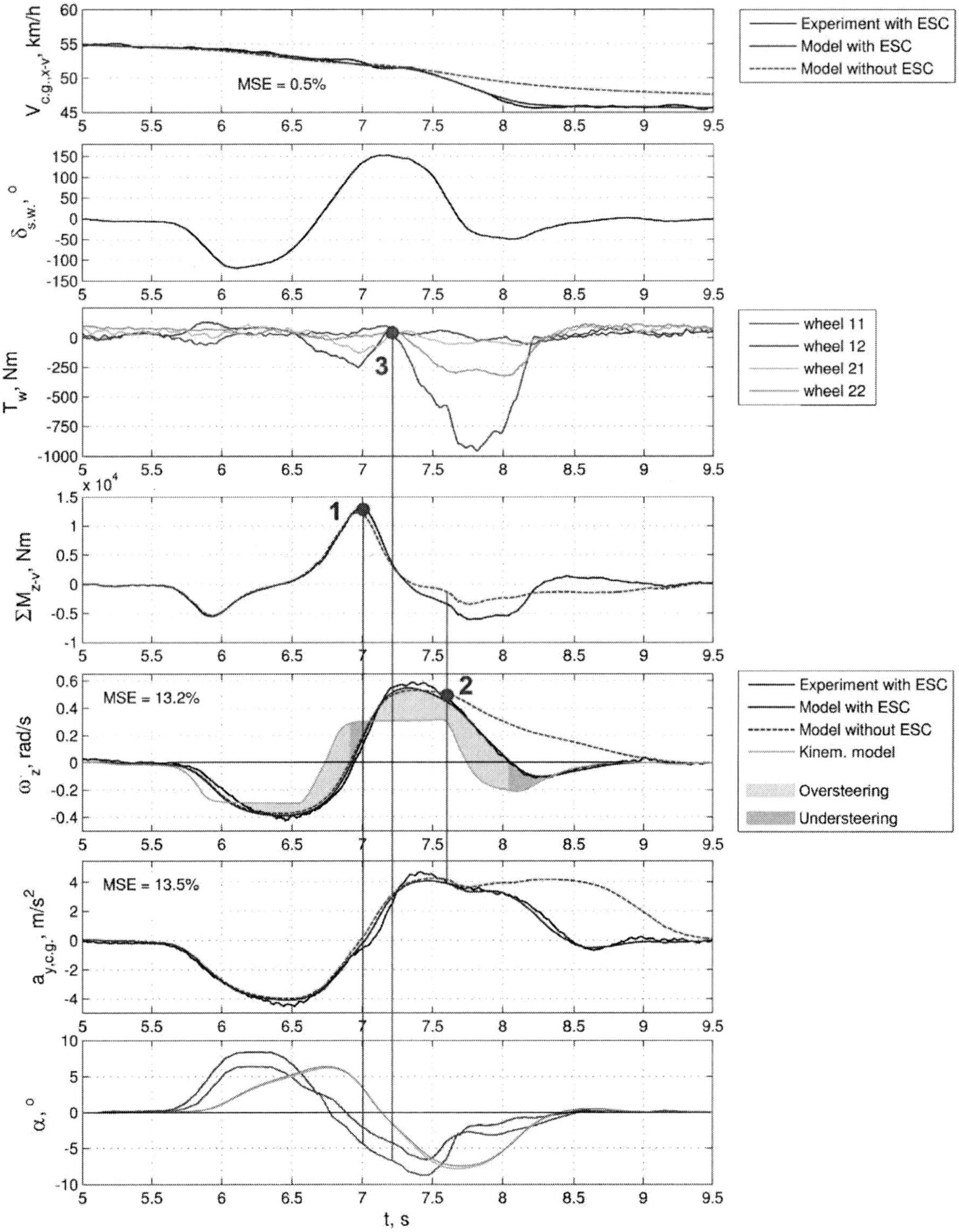

Figure 9. Simulation results: road test with ESC intervention, and results of ESC deactivation ("lane change" maneuver at packed snow).

IOP Conf. Series: Materials Science and Engineering **534** (2019) 012009 doi:10.1088/1757-899X/534/1/012009

In figure 9, additional elements are presented intended for the analysis of the vehicle handling characteristics. As it was mentioned previously, the single-track kinematic model corresponds to a vehicle with the neutral steering behavior. Comparing the yaw rate calculated by this model ($\omega_{z,neutral}$) with the actual yaw rate (ω_z), one can evaluate dynamics of the vehicle steering behavior throughout the maneuver. With these quantities having the same sign, expression $|\omega_z| > |\omega_{z,neutral}|$ suggests that vehicle oversteers, while $|\omega_z| < |\omega_{z,neutral}|$ indicates understeering. However, at some stages of the considered maneuver ω_z and $\omega_{z,neutral}$ have opposite signs. For these cases, the following expression can be proposed for defining the steering behavior: $sgn\{(\omega_z - \omega_{z,neutral}) \cdot sgn(\omega_z)\}$. It makes +1 for oversteering and –1 for understeering. Using this expression allowed dividing the maneuver into the stages with different steering behavior. The difference between ω_z and $\omega_{z,neutral}$ is marked by areas of two colors: green for oversteering, and blue for understeering.

Additionally, there are three specific points located on the graphs. Point 1 corresponds to the maximum of yawing moment. It emerges because the rapid counter-steering changes direction of the front tire lateral forces steeply, while the rear lateral forces change their direction with a time lag. As a result, lateral tire forces at the front wheels and the rear wheels, for a certain time, have opposite directions (it can be seen by opposite signs of front and rear sideslip angles), while yawing moments created by these forces become acting in the same direction. The sum of these moments forms a peak, which can be seen at the point 1 of ΣM_{z-a} graph. In the point 2, yaw rates corresponding to the enabled and disabled ESC start to diverge. Therefore, point 2 was assumed an evident beginning of a skid. In the point 3, ESC commences building braking torques, preventing vehicle from the right skid.

It can be concluded that in emergency maneuver, rapid counter-steering entails an excessive yawing moment. Having certain amplitude and duration, this excessive moment brings vehicle into a skid. To prevent this, ESC should intervene at the point located between the maximum of the yawing moment and the beginning of the skid. It can be assumed that in the said point, the process of stability loss has its origin, although no evident signs of a skid are observed. Appearing of known skid indications (rear sideslip angles substantially exceed the front ones) means that rear tire forces provided their maximum ability for counteracting skid, however, the moment exerted by them was insufficient for skid prevention, because excessive yawing moment, emerged previously, has achieved its critical magnitude and duration.

From the simulation results, one can notice specific indications of the stability loss process and their non-consistence with known criteria of the skid:

– The maximum of yawing moment appears when rear tire forces are acting in opposite to the direction of the turn (and of the skid), while rear sideslip angles are diminishing and approaching to zero.

– Stability loss begins (7.15…7.2 s) when rear tire forces and rear sideslip angles are small (in the case at hand, sideslip angles are ca. 1.35°).

From the aforementioned, it can be concluded that elimination of skid requires ESC intervention to be preventive (i.e. commencing before the skid indications appear), which is actually implemented by the studied production ESC.

Figure 10 shows simulation results for the third step of the procedure (i.e. introducing the experimental yaw rate regulator) along with results of the second step (skid). Graphs are arranged similarly to those in figure 9. Note that in this case, the yaw rate calculated by the kinematic model is also a command signal for the experimental regulator, with $\omega_{z,neutral} - \omega_z$ being the tracking error.

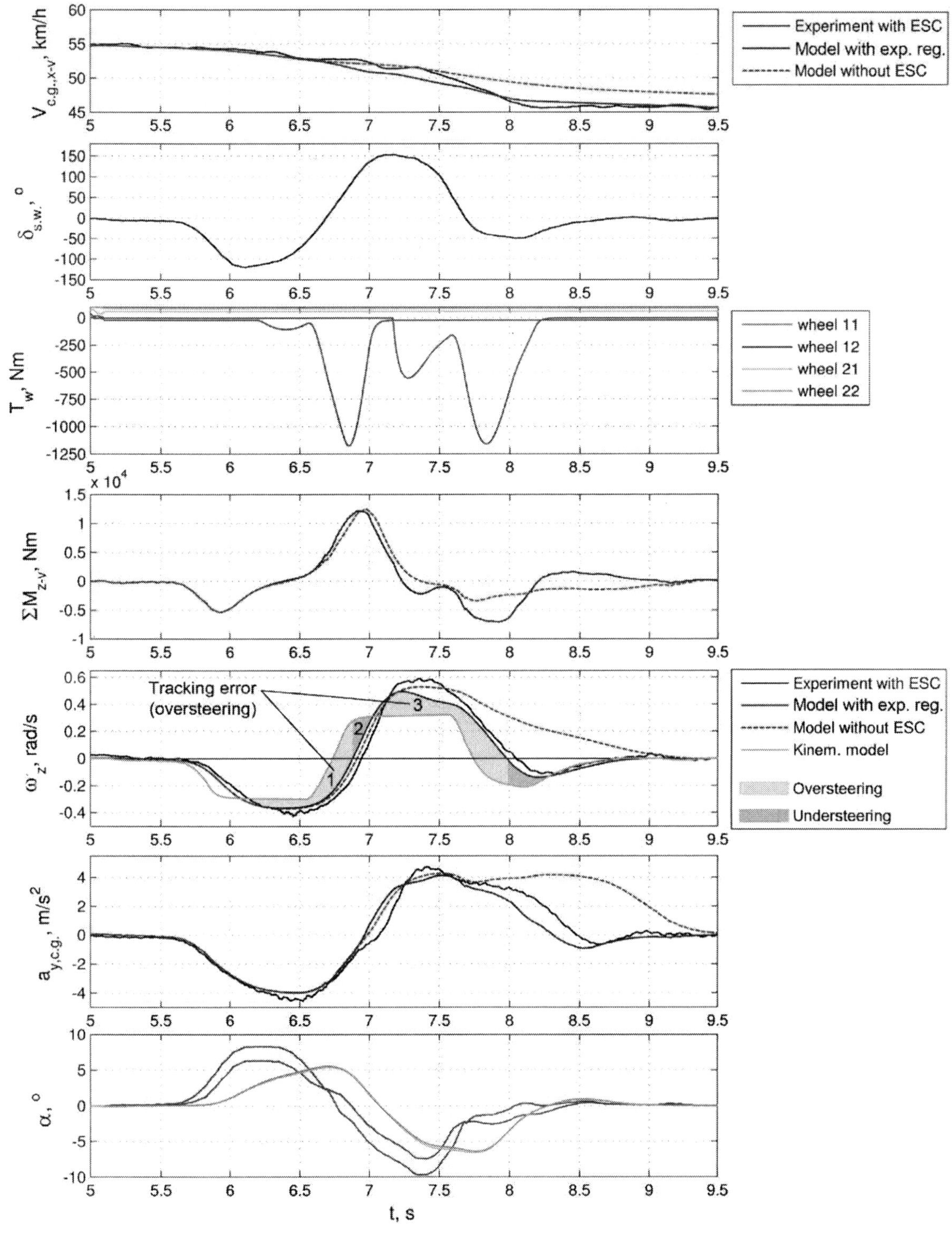

Figure 10. Simulation results: introduction of the experimental regulator, and results of its deactivation ("lane change" maneuver at packed snow).

IOP Conf. Series: Materials Science and Engineering **534** (2019) 012009 doi:10.1088/1757-899X/534/1/012009

Graphs show that both intervention time and operating duration of the regulator are similar to those of the production ESC. However, the regulator exerts higher braking torques than the production system, which is due to simplified operation of the former and its "stiffer" adjustments. One can notice that regulator tracking error (oversteering area marked green) is rather large – usually regulators track command signal with higher precision. However, ESC controller should be relatively "soft" due to physical limits of yawing moment correction. A "stiff" controller with higher precision would lock the wheel whatever intensity of the maneuver is; however, the effect of that would hardly be different from, for example, 50…60% braking, since as it is shown by the adhesion characteristics in figure 7, the most significant decrease of lateral adhesion (i.e. the bulk of tire force vector rotation) takes place below 60% slip.

The experimental regulator is only intended for correction of the oversteering. Due to this, transition from the left turn with the oversteering phase (area 1 in figure 10) to the right turn with the short understeering phase (area 2) makes the regulator decrease braking torque at the right wheel (wheel 11) down to zero. The filter at the regulator output does not allow abrupt dropping of the torque, instead bringing it down to zero through a transient process. In the area 3, the right turn with an oversteering phase takes place; therefore, the regulator becomes operating at the left wheel (wheel 12), but not before the torque at the right wheel is reduced to zero.

To verify the assumption about the point 3 (see figure 9) being an origin of the stability loss process, a number of simulations were performed with various presets of the regulator dead zone. It was found that making this dead zone wider weakens initial regulator response in the area 3 (see figure 10) down to the zero torque. As a result, regulator intervention moves in time towards point 2 where the skid takes place. Figure 11 shows the essential variables featuring this experiment, namely the torque at the front left wheel (T_{w12}); vehicle yaw rate (ω_z); vehicle center of gravity coordinates ($X_{c.g.}$, $Y_{c.g.}$) projected on the road plane and thus defining the vehicle trajectory.

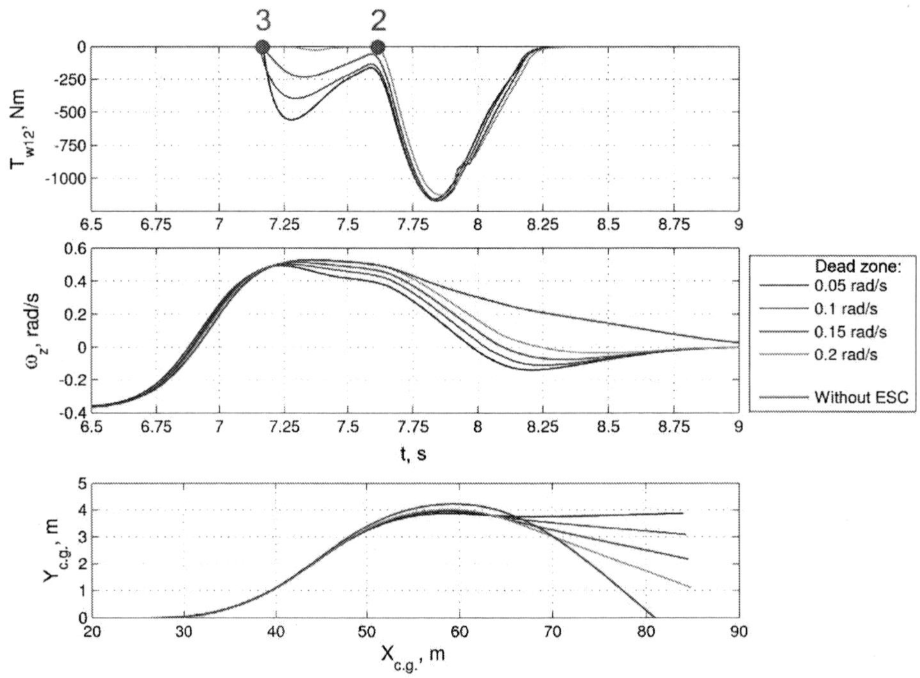

Figure 11. Influence of the yaw rate regulator dead zone on the vehicle directional stability.

Presented graphs show that weakening of the initial regulator response (dead zone > 0.05 rad/s) results in increased oversteering. With dead zone 0.2 rad/s, the regulator intervened after skid indications have appeared (from 7.6 seconds on), and after that, even exerting of an appropriate correcting torque has not prevented the vehicle from the loss of stability. Therefore, conducted simulations verified that ESC operates effectively when it intervenes at the beginning of the stability loss process. The latter can be recognized from the difference between the required and the actual yaw rate, rather than from known indications of the skid (i.e. large rear sideslip angles and the maximum adhesion seen at the rear wheels). It was shown that the production ESC provides such preventive intervention, as well as the experimental regulator having dead zone 0.05 rad/s.

Figure 12 shows vehicle trajectories resulted from each step of the above-described research procedure. Trajectory of the vehicle with the experimental regulator was obtained with the dead zone 0.05 rad/s. Intervention of the experimental regulator resulted in a lower vehicle lateral accelerations than those obtained with the production ESC. This prevented vehicle from excessive drifting towards the margin of the left lane (however, this should not be considered as some advantage of the experimental regulator over the production system, but rather a mere observation made from the simulations).

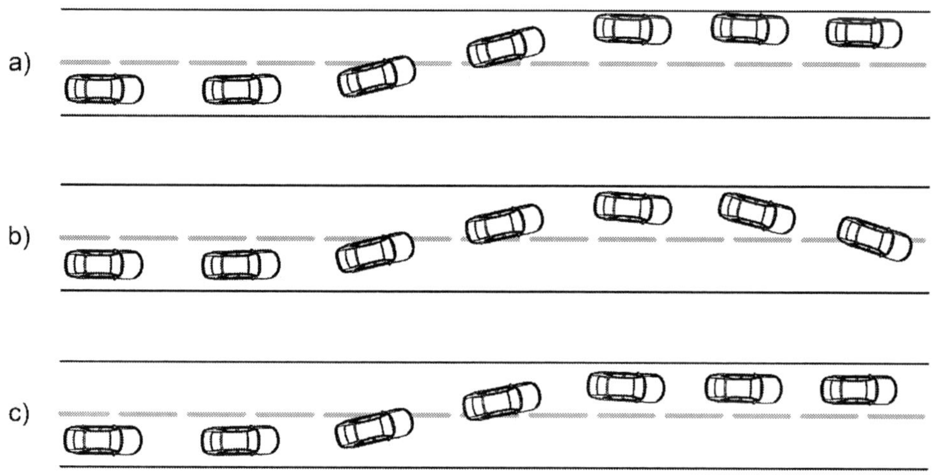

Figure 12. Vehicle trajectories obtained in simulations. a – road test simulation with ESC intervention, b – disabling of the ESC, c – with the experimental regulator.

Studies similar to the described in this article were conducted for a number of maneuvers at different vehicle speeds, both at snow and ice, with the results alike. In most experiments, MSEs of the following orders were obtained: 10...20% for both lateral acceleration and yaw rate, at most 12% for the longitudinal acceleration, and at most 2% for the longitudinal velocity. Based on literature analysis on vehicle dynamics modeling, and also considering non-rigidity (to a certain extent) of the snow surface as well as involvement of studded tires (which adhesion characteristics are known to be rather scattered [15]), obtained MSEs were considered proving a sufficient accuracy of the model.

11. Conclusions

Elaborated model of vehicle dynamics allows simulation of vehicle tests at snow-covered surfaces with sufficient accuracy substantiated by reasonable relative mean square errors. The method employed for identification of unmeasured wheel torques allowed disclosing the ESC interventions occurred during the road test. Identification results were validated indirectly, i.e. by MSEs of the essential vehicle dynamic parameters, and by comparing identified interventions with descriptions given in the literature by the ESC producers (e.g. Bosch).

Proposed 3-step modeling procedure allowed to discover that at snow-covered surfaces both production ESC and the experimental regulator prevent the vehicle from the loss of directional stability. Disabling the ESC or said regulator brings vehicle into a skid seen from its trajectory and specific changes in yaw rate and lateral acceleration.

Analysis of the simulation results revealed stages of the emergency maneuver associated with directional stability (whether it is maintained or lost). Three specific points can be distinguished. The first one corresponds to the maximum yawing moment resulted from opposite directions of front and rear lateral tire forces. The second point indicates the beginning of a skid, which can be recognized either by a substantial growth of rear sideslip angles or by distinctive increase in yaw rate and lateral acceleration. In the third point, temporally located between the first one and the second one, the ESC should intervene to prevent vehicle from a skid. In this point, lateral tire forces (or sideslip angles) of the rear wheels are close to zero. Time locations of said points suggest that ESC should intervene preventively, i.e. when stability loss originates, not when it reveals itself. Analysis of the modeling results showed that both production ESC and the experimental regulator satisfy this requirement.

Skid prevention effectiveness depends on adjustments of the yaw rate controller, in particular its dead zone. Conducted study has shown that the experimental regulator possessed the effectiveness comparable to that of the production ESC, when the dead zone was ca. 0.05 rad/s. Increasing of this threshold brings the point of regulator intervention closer to the beginning of the skid deteriorating the system performance.

Appendix. Nomenclature of variables and parameters

Variable/parameter designation	Variable/parameter name	Value and/or dimension
a_1, a_2	vehicle front and rear tracks	1.6, 1.66 m
$a_{x,c.g.}$, $a_{y,c.g.}$	components of the vehicle center of gravity acceleration vector projected on vehicle axes	m/s^2
b	vehicle wheel base	3.07 m
b_1, b_2	longitudinal distance from the vehicle center of gravity to the centers of front and rear axles respectively	1.5, 1.57 m
c_φ	suspension roll stiffness	N·m/rad
d_φ	suspension roll damping coefficient	N·m/rad/s
e_ω	yaw rate controller tracking error	rad/s
$F_{w,x}$, $F_{w,y}$	components of the air drag force vector projected on the vehicle axes	N
g	gravity constant	9.81 m/s^2
G_{xy}, G_{yx}	correction functions for longitudinal and lateral tire forces taking into account combined slip (MF model)	–
h_{ij}	tire force arm	m
$h_{c.g.}$	vehicle center of gravity height	0.639 m
$\mathcal{J}_w$	wheel inertia	1.62 kg·m^2
$\mathcal{J}_{x,sprung}$	roll inertia of the sprung mass	920 kg·m^2
$\mathcal{J}_z$	vehicle yaw inertia	5500 kg·m^2
k	number of wheels (tires) at i-th axis	2
m	vehicle mass	2558 kg
M_{z-v}	yawing moment exerted by the wheel	N·m
M_{z-w}	tire aligning moment	N·m
n	number of vehicle axes	2
r	wheel radius	m
r_e	wheel rolling radius	m
r_{e0}	effective rolling radius	m
R	turning radius of the rear axle center	m
R_{ij}	tire force	N
R_{y-v}, R_{x-v}	components of the tire force vector projected on the vehicle axes	N

R_{x-w}, R_{y-w}	components of the tire force vector projected on the wheel axes	N
R_z	tire normal force	N
S_x	tire longitudinal slip	–
T_w	wheel torque	N·m
$V_{c.g.,x-v}$, $V_{c.g.,y-v}$	components of the vehicle center of gravity velocity vector projected on the vehicle axes	m/s (km/h for plots)
V_w	linear velocity of the wheel center	m/s
$V_{w,x-v}$, $V_{w,y-v}$	components of the wheel center velocity vector projected on the vehicle axes	m/s
u_{steer}	vehicle steering ratio	–
u	control signal	–
$X_{c.g}$, $Y_{c.g}$	vehicle center of gravity coordinates projected on the road plane	m
y_{w-v}	lateral displacement of the wheel with respect to the vehicle axes	m
α	sideslip angle of the wheel	rad (° for plots)
$\delta_{s.w.}$	steering wheel angle	rad (° for plots)
δ	steer angle of the wheel	rad
ΔM_z	correcting yawing moment	N·m
μ_x, μ_y	longitudinal and lateral tire-surface adhesion coefficients	–
φ	roll angle	rad
ψ_z	yaw angle	rad
ω_w	wheel angular speed	rad/s
ω_z	yaw rate	rad/s
ω_z *	yaw rate command	rad/s
$\omega_{z,neutral}$	yaw rate calculated by the single-track model with rigid wheels	rad/s

References

[1] Ivanov A M Kristalnyi S R Popov N V and Fomichev V A 2018 Results of tests of the vehicle ESP system efficiency equipped with studded tires (in Russian) *Journal of Automotive Engineers* (Russia) **2** (109) pp 40-44

[2] Fomichev V A 2017 *A method for performance assessment of electronic stability control systems for vehicles equipped with studded tires* PhD thesis (Russian)

[3] Wiessalla J Mao Y and Esser F 2015 *From Tyre Measurements on Low μ Surface to Full Vehicle ESC Simulations under Winter Conditions*. Proc. of the 4th Int. Tyre Colloquium pp 324-329

[4] Liebemann E K Meder K Schuh J and Nenninger G. 2005 *Safety and Performance Enhancement: The Bosch Electronic Stability Control (ESP).* 19th Int. Technical Conf. on the Enhanced Safety of Vehicles (ESV)

[5] Isermann R (Hrsg.) 2006 *Fahrdynamik-Regelung. Modellbildung, Fahrerassistenzsysteme, Mechatronik* (Friedr. Vieweg & Sohn Verlag | GWV Fachverlage GmbH, Wiesbaden) pp 169-211

[6] Lata V N 1989 *Selection and investigation of vehicle handling ability criteria by means of frequency analysis* PhD thesis (Russian)

[7] Goggia T Sorniotti A De Novellis L Ferrara A Pennycott A Gruber P and Yunus I 2015 Integral sliding mode for the yaw moment control of four-wheel-drive fully electric vehicles with in-wheel motors *Int. Journal of Powertrains* **4** pp 388-419

[8] Pandit C A *Model-free Approach to Vehicle Stability Control* 2013 PhD Thesis

[9] Zang H, Zhang S Dai Y and Congna D 2017 The Research of Vehicle Handling Performance and Stability Based on the Integrated Control of Electronic Stability Program and Anti-block

Braking System *Boletín Técnico* **55** pp 634-640

[10] Du H Zhang N and Smith W 2009 Robust Yaw Moment Control for Vehicle Handling and Stability *SAE Technical Papers* **2009-01-0578**

[11] Mirzoev G K and Peshkilev A G 1980 Computer Analysis of Suspension Kinematics (in Russian) *Automotive Industry* (Russia) **2** 12-14

[12] Pacejka H B and Besselink I 2012 Tire and vehicle dynamics. Third Edition (Elsevier Ltd.) pp 165-183

[13] Kulikov I A Bakhmutov S V and Barashkov A A 2016 An Investigation of Vehicle Dynamics Concerning Active Safety Systems by Simulations and Driving Tests *Proc. of NAMI* **265** pp 53-65

[14] Kulikov I A 2016 Advanced Tools for Research and Development of Hybrid Vehicles Based on the Component-in-the-Loop Technology PhD thesis (Russian Federation) pp 65-71

[15] Svendenius J 2007 *Tire Modeling and Friction Estimation* Lund University pp 130-133

IOP Conf. Series: Materials Science and Engineering **534** (2019) 012010 doi:10.1088/1757-899X/534/1/012010

Implementation of the interaction of the steering wheel loader control system of the remote-controlled wheeled vehicle operator interface with a real-time simulation model

N V Buzunov[1], **G O Kotiev**[2] **and V A Gorelov**[3]

Bauman Moscow State Technical University, 105005 Baumanskaya 2th st., 5, Moscow, Russian Federation

E-mail: [1] buzunovnv@bmstu.ru, [2] kotievgo@yandex.ru, [3] gvas@mail.ru

Abstract. A wheeled vehicle remote control workstation should include systems and devices that would provide steering feedback for the operator. The electric loader of the steering wheel with digital control can act as a system of this kind. Development and adjustment of the steering wheel loader can be performed with real-time simulation models sufficiently reflecting the behavior of the remote-controlled vehicle. The paper describes interaction of the real-time simulation model of an 8x8 wheeled chassis with the prototype of a steering wheel loader control system. The authors demonstrate compatibility of the software and hardware components in question, show adequacy of the model during interaction with the steering wheel loader control system, and propose using such models during the development of the remote control operator workstations.

When a vehicle is remotely controlled, the interface between the control object and the operator (who is located remotely) should provide the same information about the state of the vehicle and the environment as when the vehicle is locally controlled.

It is known that during a maneuver, the driver turns the steering wheel at an angle that ensures the minimum deviation of the wheeled vehicle (WV) from the desired direction. However, during the movement, the ratio between the current steering angle and the corresponding change in direction of travel is constantly changing. This occurs with the following sequence of events of the turning WV: «Turning the steering wheel (SW) – changing the angle of rotation of the steered wheels – forming lateral forces in the contact patch of the tires on the road – changing the direction of movement - adjusting the control action depending on the position of the WV» [1].

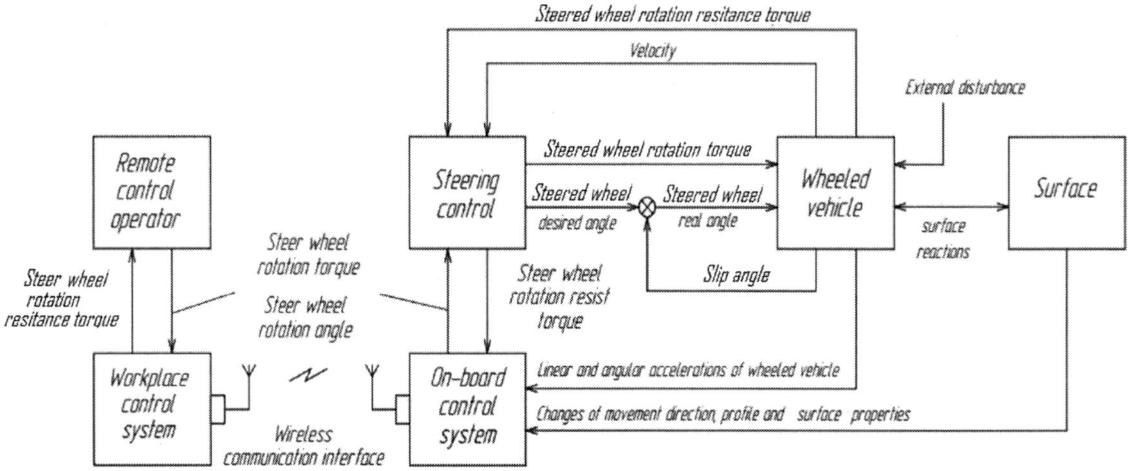

Figure 1. Structural scheme of interaction of the remote-controlled vehicle operator with the control object.

The visual channel for obtaining information has relatively high, but still limited possibilities. The reaction time of the driver to visual information is greater than when the information is perceived by the auditory or tactile channel. In this regard, the operator's workplace for remote control must ensure the transmission of information via both the visual, tactile, and auditory channels.

The transmission of information via a tactile channel when driving a vehicle is realized by forming the moment of resistance to steering wheel rotation in accordance with the current steering angle and the vehicle speed (figure 1) [1].

When implementing the workplace for remote control of the vehicle, the «rigid» connection between the steering and steered wheel, which is typical for local control, is absent. When working under current conditions, the operator does not feel the moment of the resistance to the steered wheels turn.

To provide feedback on the steering wheel, additional elements and systems must be introduced into the design of the remote control workstation providing the required rotation resistance. An example of such a device is the electric loader of the steering wheel (ELSW) with an electronic control system (figure 2). The steering wheel in this case plays the role of a control device without direct force transfer from the operator to the steered wheels.

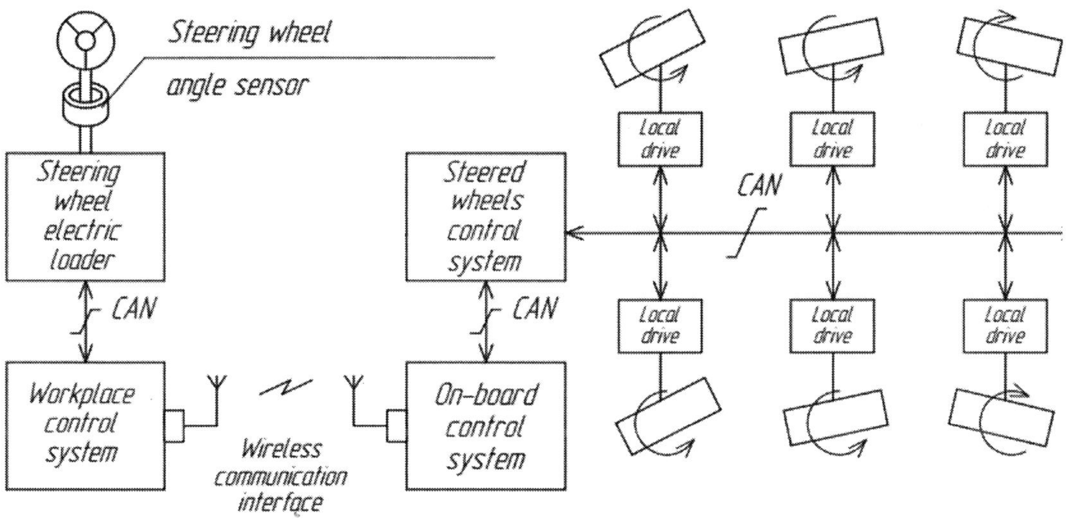

Figure 2. Implementation of the steering system of the remote operator's workplace.

Because of the specific purpose of the ELSW, the development of such systems, the correction of the initial control laws, the adjustment of the parameters of interaction between the workplace and the control object are currently possible at all stages of design only if there is a prototype of the wheeled vehicle object. This fact, in turn, increases the cost and timing of development work.

The specified problem can be solved by performing the adjustment and debugging of the software and hardware parts of the ELSW control system using imitation mathematical models of the movement dynamics of wheeled vehicles operating in the real-time mode. With this approach to the development of a control system, there is no need for a prototype object. It is sufficient that the applied mathematical model describes the behavior of the vehicle with an accuracy sufficient to solve the problems posed.

Based on the submitted data, the following requirements can be formulated for the simulation model used in the development of remote control workstations:

- functioning in the real-time mode due to teamwork with the workplace and operator management system implemented at the physical level;
- identity to the replaced object from the point of view of the physical control system interacting with the model being developed.

The purpose of this work is to determine the possibility of the steering wheel loader control system functioning from the operator's workplace of the remote-controlled vehicle in interaction with the simulation real-time model of the plane curvilinear movement of the wheeled vehicle.

The virtual object in this study is the model of the planar curvilinear motion of the wheeled chassis along a rigid road described in [2, 3, 4]. The adequacy of this model was repeatedly confirmed by comparing its simulation results with the results obtained during test runs of prototypes of different wheeled vehicles [2, 5, 6]. In order to comply with the objects of wheeled vehicles, the remote control of which is planned using the workplace considered, this model was adapted to describe the dynamics of the chassis with the wheel formula 8x8 and the steering formula 1-2-3-4. Other parameters of the virtual chassis are presented in Table 1. In the course of this study, the model of motion dynamics of a multi-axle wheeled chassis is realized in the real-time mode to ensure joint operation with the ELSW control system.

Table 1. Parameters of the virtual chassis.

Parameter name	Value
Mass, kg:	29600
Length, m:	6,15
Track width, m:	2,6
Center of mass height, m:	1,38
Loaded tire radius, m:	0,725
Moment of inertia around the vertical axis, $kg \cdot m^2$	110106
Maximum power of the wheel hub motors, W	60000

System of equations (1) makes it possible to calculate current accelerations, based on the values of the forces and moments acting on the vehicle.

The movement of the wheeled vehicle as a rigid body is considered in a horizontal plane on an even, unreformed bearing surface and is composed of translational motion of the center of mass and rotational motion around the center of mass (figure 3) [7,8].

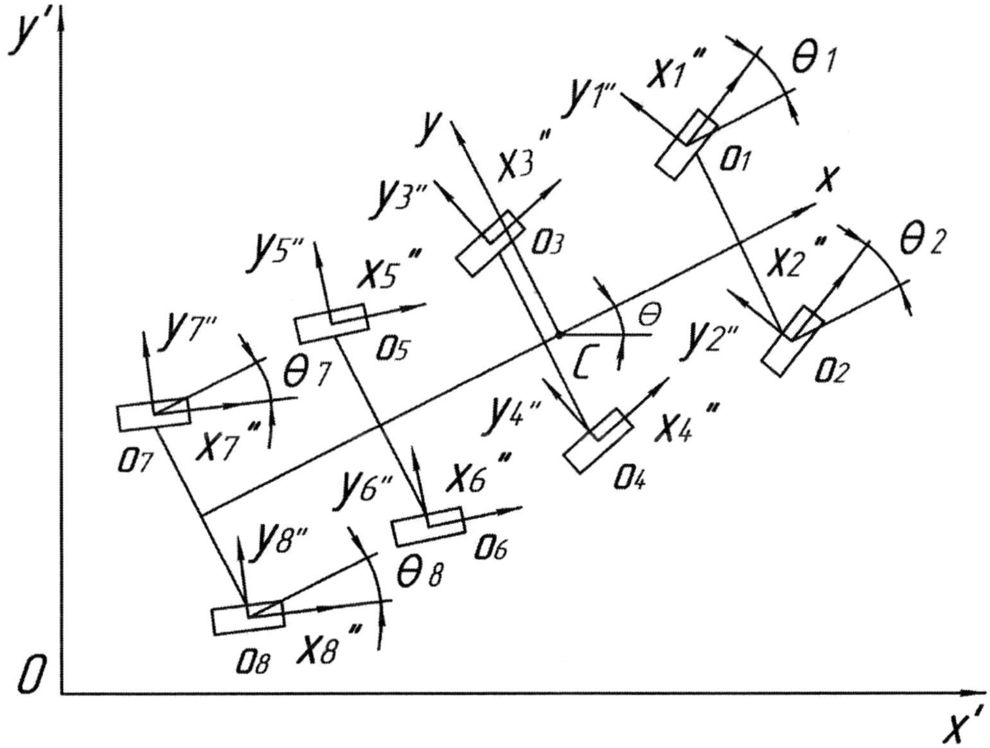

Figure 3. Coordinate systems used in modeling the curvilinear motion of the 8x8 wheeled vehicle:
x' - O - y' is the fixed coordinate system (FCS);
x - C - y is the movable coordinate system (MCS), associated with the object center of mass;
x_i'' - O_i - y_i'' is the moving coordinate system associated with the axis of rotation of the i-th wheel;
Θ is the heading angle of the wheeled vehicle model;
θ_i is the angle of the i-th wheel rotation.

System of equations (1) describing the vehicle motion provides calculation of the current accelerations according to the values of forces and moments acting on the wheeled vehicle from the surface and the environment [7,8,9]:

$$\begin{cases} a_x = \dfrac{dV_x}{dt} - \omega_z \cdot V_y = \dfrac{1}{m} \cdot \left(P_{wx} + \sum\limits_{i=1}^{6} R_{xi} \right) \\ a_y = \dfrac{dV_y}{dt} + \omega_z \cdot V_x = \dfrac{1}{m} \cdot \left(P_{wy} + \sum\limits_{i=1}^{6} R_{yi} \right) \\ J_z \cdot \dfrac{d\omega_z}{dt} = \sum\limits_{i=1}^{6} M_{tri} + \sum\limits_{i=1}^{6} M(\vec{R}_i) \\ V_{x'} = \dfrac{dx'}{dt} = V_x \cdot \cos\theta - V_y \cdot \sin\theta \\ V_{y'} = \dfrac{dy'}{dt} = V_x \cdot \sin\theta + V_y \cdot \cos\theta \\ \omega_z = \dfrac{d\theta}{dt} \end{cases}, \qquad (1)$$

IOP Conf. Series: Materials Science and Engineering **534** (2019) 012010 doi:10.1088/1757-899X/534/1/012010

where m is the mass of the wheeled vehicle; J_z is the moment of inertia of the wheeled vehicle with respect to the z axis (center of mass); $\vec{V}$ is the velocity vector of the center of mass of the wheeled vehicle; $\vec{a}$ is the acceleration vector of the center of mass of the wheeled vehicle (the absolute derivative of the velocity vector of the center of mass of the car); $\frac{d\vec{V}}{dt}$ is the relative derivative of the velocity vector of the center of mass of the wheeled vehicle; ω_Z is the angular speed of rotation of the wheeled vehicle around the center of mass; θ is the angle of rotation of the wheeled vehicle relative to the axis x'; $\vec{R}_i$ is the vector of the interaction force with the support base acting on the i-th wheel; $\vec{P}_w$ is the vector of the air resistance force; M_{tri} is the torque of resistance to the rotation of the i-th wheel.

The diagram used for calculation of the forces and velocities of an individual wheel is shown in figure **4** [6, 8].

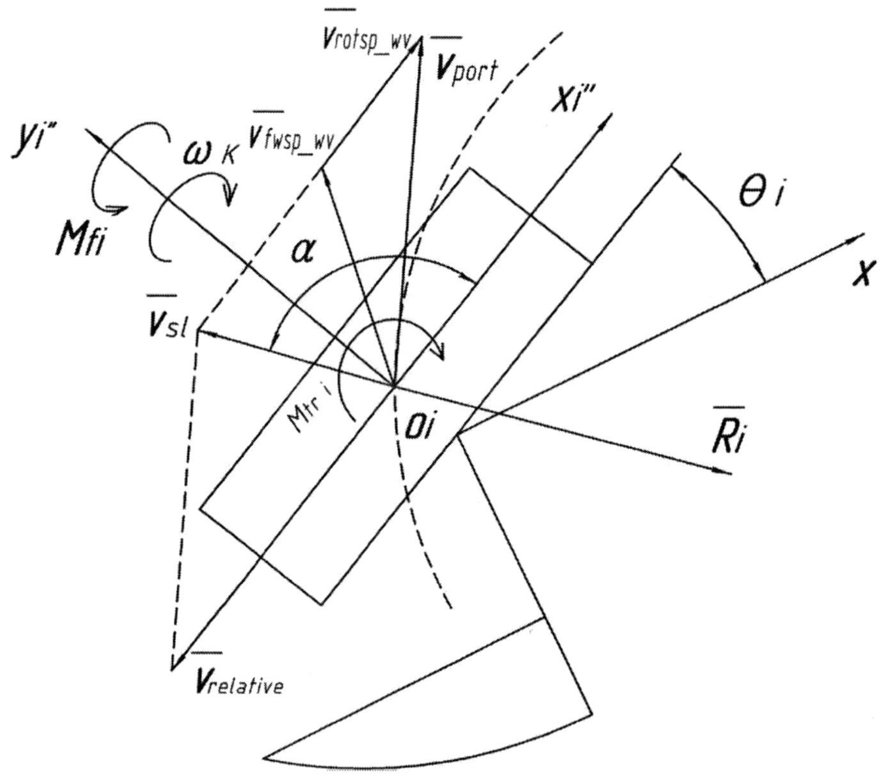

Figure 4. Velocities and forces of an individual wheel.

In figure 4 the following elements are presented:

$\vec{V}sl$ is the velocity vector of the sliding point of the contact of the wheel with respect to the support surface;

$\vec{V}relative$ is the vector of relative wheel speed in the contact patch;

$\vec{V}port$ is the vector of the transport velocity of the rotation center of the wheel O_i (velocity vector of MCS point x - C - y, with which the center O_i of MCS x_i''- O_i - y_i'' is currently connected, about FCS x' - O - y');

$\vec{V}fwsp\_wv$ is the vector of the linear velocity of the translational motion of the center of mass in the FCS x' - O - y';

$\vec{V}rotsp\_wv$ is the vector of the linear velocity of the rotational motion of the i-th wheel center about the center of mass;

α is the angle of the wheel speed vector rotation relative to the axis x_i'' of the system $x_i'' - O_i - y_i''$;

ω_κ is the angular speed of the i-th wheel rotation;

$\vec{R}_i$ is the force of interaction with the support surface of the i-th wheel;

θ_i is the angle of the i-th wheel rotation;

M_{fi} is the moment of the i-th wheel rolling resistance;

M_{tri} is the moment of the i-th wheel resistance to the rotation.

The results of the verification of this real-time model, performed by comparison with the results of the «reference» model, showed a high degree of compliance with the replaced object of wheeled vehicles.

After confirming the adequacy, realization of the interaction of the real-time model with the control system of the steering wheel loader was carried out. The order of information exchange is presented in figure 5.

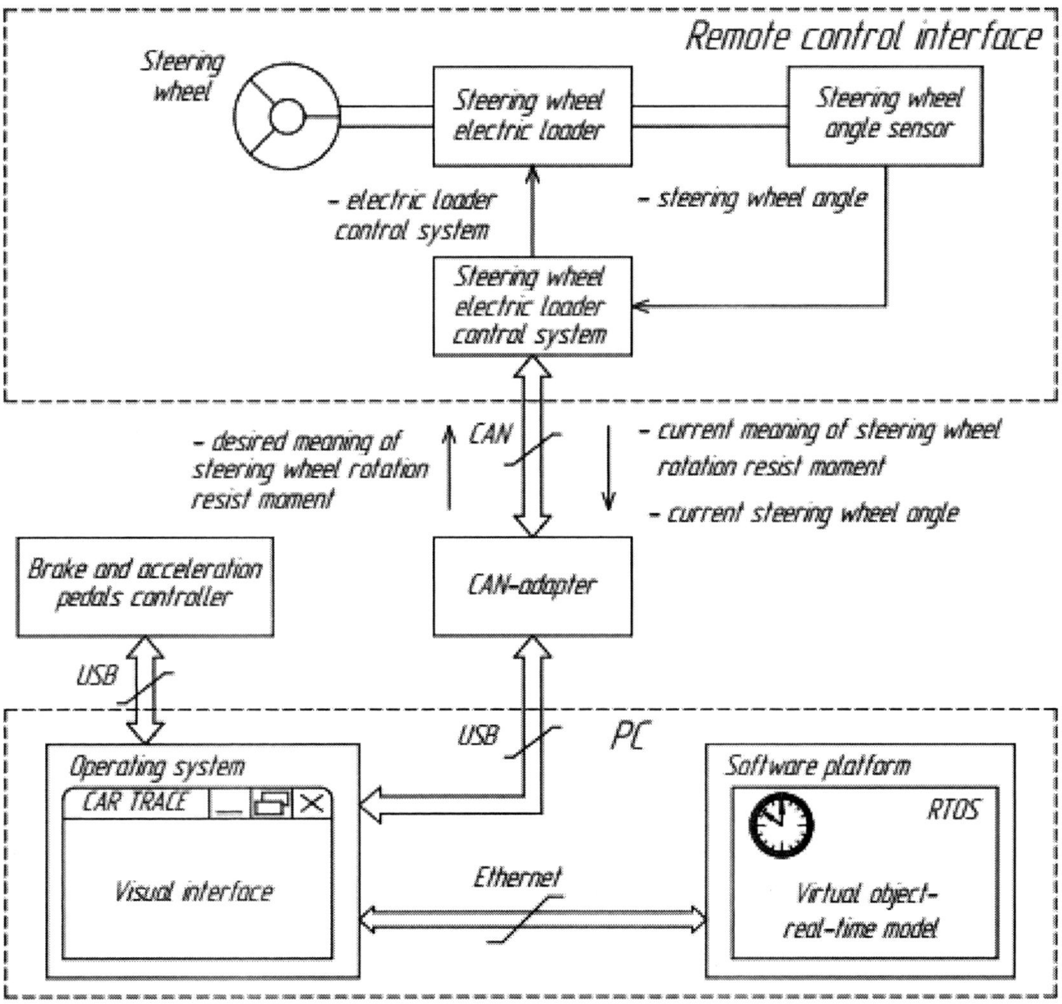

Figure 5. Structural diagram describing the interaction of the control system of the steering wheel electric loader with the real-time simulation model of the wheeled chassis.

IOP Conf. Series: Materials Science and Engineering **534** (2019) 012010 doi:10.1088/1757-899X/534/1/012010

When the real-time model and the steering wheel loader control system of the remote control operating station directly cooperate, the following relationship is realized as the law of forming the moment of resistance on the steering wheel:

- the value of the moment of resistance on the steering wheel depends on the angle and speed of the steering wheel rotation, as well as the discrepancy between the steering wheel angle and one of the steered wheels:

$$M_r = K \cdot \left(-k_1 \cdot \alpha - k_2 \cdot \dot{\alpha} - k_3 \cdot \left(\alpha - \frac{\theta_1}{i_{sg}} \right) - k_4 \cdot \left(\dot{\alpha} - \frac{\dot{\theta}_1}{i_{sg}} \right) \right), \tag{2}$$

where M_r is the moment of resistance on the steering wheel, N·m; α is the value of the angle of rotation of the steering wheel, rad; $\dot{\alpha}$ is the time derivative of the angle of rotation of the steering wheel, rad/s; θ_1 is the value of the angle of rotation of the front left wheel, rad; $\dot{\theta}_1$ is the time derivative of the angle of rotation of the front left wheel, rad/s; i_{sg} is the transmission ratio of the steering gear; K, $k1$ ... $k4$ are the gain factors, $k_1 = 3$ N·m / rad, $k_2 = 1$ N·m /s^{-1}, $k_3 = 0.2$ N·m/rad, $k_4 = 1$ N·m/s^{-1}, $K = 3$.

The value of the current steering angle is fed into the model from the control system of the ELSW (in accordance with figure 5), and the calculation of the steering wheel speed, the angle of rotation of the selected steering wheel and steering angle mismatch are performed while the model is running.

In order to control the rotation of the steered wheels during the implementation of the real-time model, the control law described in paper [9] was chosen. This control law is described by the following relationship:

$$X_p = L/2 \cdot \left[\frac{\theta_a - \theta_{del}}{\theta_{max} - \theta_{del}} \right]^n, \tag{3}$$

where X_P is the coordinate of the steering pole shift along the base; L is the base of the chassis; θ_a is the average angle of the front axle, proportional to the angle of rotation of the steering wheel; θ_{max} is the maximum angle of rotation of the steering wheel; θ_{del} is the lag angle - the limiting angle of rotation of the front axle, within which the steering pole lies on the rear (last) axis.

To check the correctness of the real-time model operation during interaction with the ELSW control system, two simulations of the lane change on a distance Sr = 20 m were performed with the same parameters of motion. The first run was carried out without any moment of resistance on the steering wheel. During the second run, a resistance torque was formed on the SW, the value of which was determined in accordance with dependence (2). Figure 6 and figure **7** show the trajectories obtained at each simulation. Figure 8 shows the dependence of the required and current moment on the SW resistance.

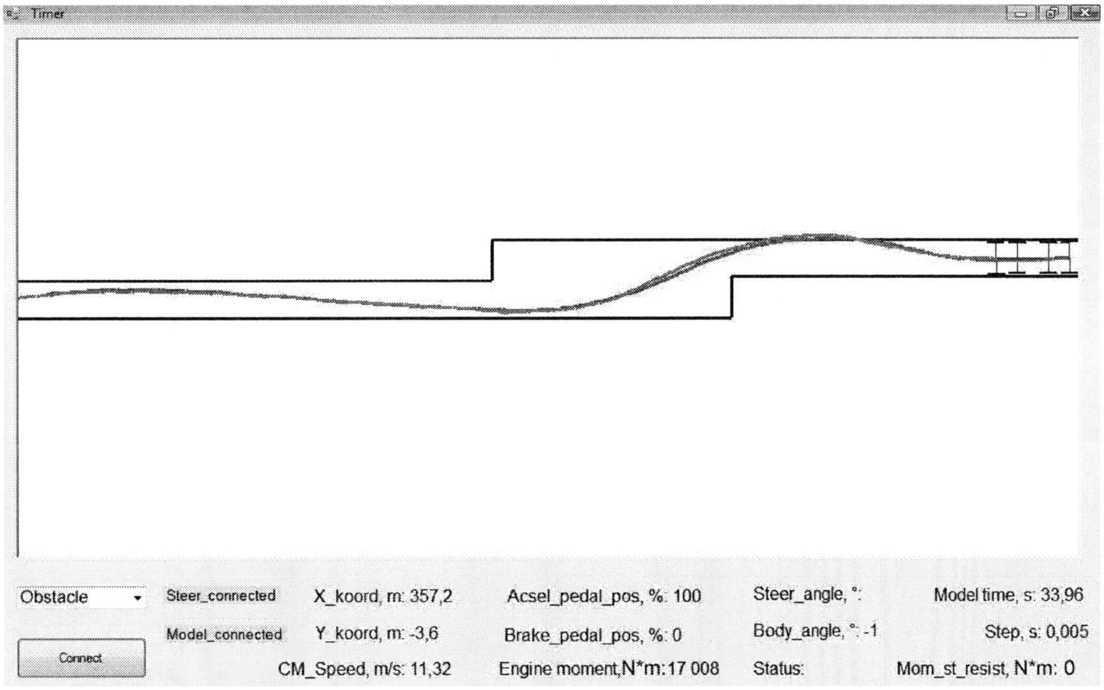

Figure 6. Simulation results for the lane change without resistance on the steering wheel.

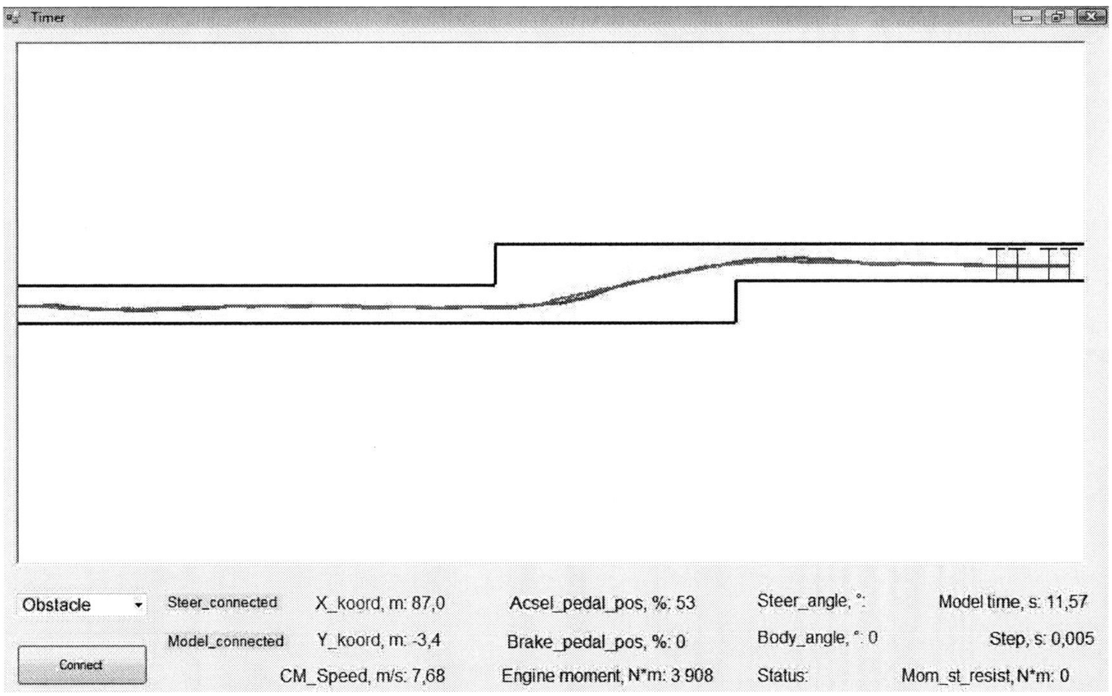

Figure 7. Simulation results for the lane change with the moment of resistance on the steering wheel.

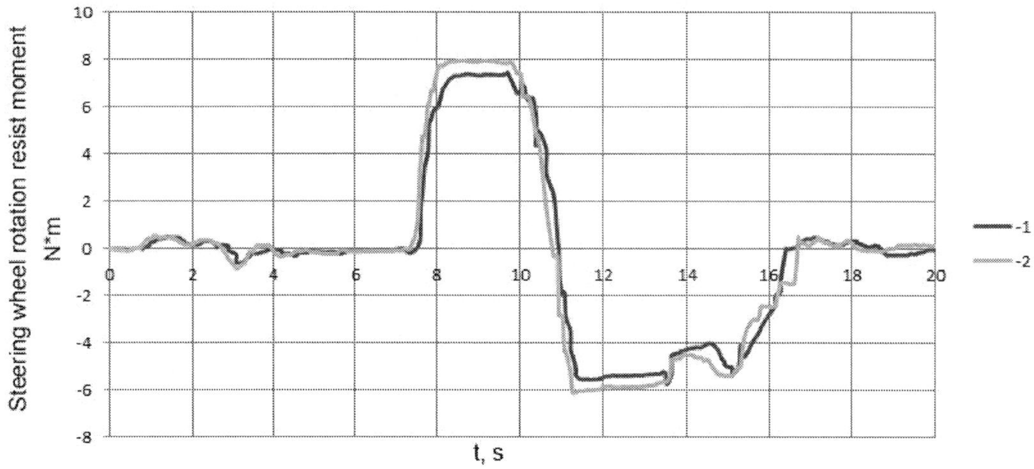

Figure 8. Variation of the moment of resistance on the steering wheel during the lane change:
1 – current moment of resistance on the steering wheel;
2 – required moment of resistance.

According to the results of the lane change simulation it was noted that the ELSW control system correctly receives and executes the specified control action with the required resistance moment value from the real-time model. Generation of the required moment of resistance is carried out in a timely manner, without significant phase shifts, the actual value corresponds sufficiently to the specified value. When maneuvering, there was noticeable resistance on the steering wheel, as when changing the angle of rotation, and when it was held in a position other than neutral. The presented fact points to the possibility of applying real-time models of traffic dynamics in the development of remote control systems for wheeled vehicles.

The next stage of the experiment was to perform series of simulations of typical maneuvers for various parameters of the chassis movement. Lane change simulation was carried out at speeds of 15 km/h and 25 km/h; maneuver «Cornering with a radius of 35 m» was carried out at speeds of 15 km/h and 20 km/h.

Verification of the results of the real-time model when it interacts with the control system is carried out using a «reference» imitation mathematical model implemented in Matlab/Simulink. The correctness of the considered «reference» model was confirmed in the course of full-scale tests of the wheeled vehicle during typical maneuvers [2,3].

The general algorithm for verifying the real-time model consists of the following stages:

- simulation of the wheeled chassis movement in the real-time mode with recording the control actions to the file;
- modeling the movement of the wheeled chassis using a «reference» mathematical model, where the input dependencies are obtained using the dependencies of the control parameters;
- comparison of the results of mathematical modeling in real-time mode with the results of the work of the «reference» model with the purpose of estimating modeling errors in real-time;
- conclusions about the correctness of the real-time model by the results of the error estimation.
- To confirm the adequacy of the real-time model when working in conjunction with the ELSW control system of the remote control workstation, the following criteria are used:
- the maximum value of the relative error of modeling in the real-time mode should not exceed 10%;
- the value of the average relative error should not exceed 3%.

IOP Conf. Series: Materials Science and Engineering **534** (2019) 012010 doi:10.1088/1757-899X/534/1/012010

In accordance with the presented conditions, the following parameters are analyzed based on the results of the lane change an cornering simulations:

- the trajectory of the center of mass (CM);
- heading angle.

Figure 9 and figure 10 show the time histories of the parameters of the virtual chassis motion when performing the lane change with a speed of 15 km/h in real-time mode.

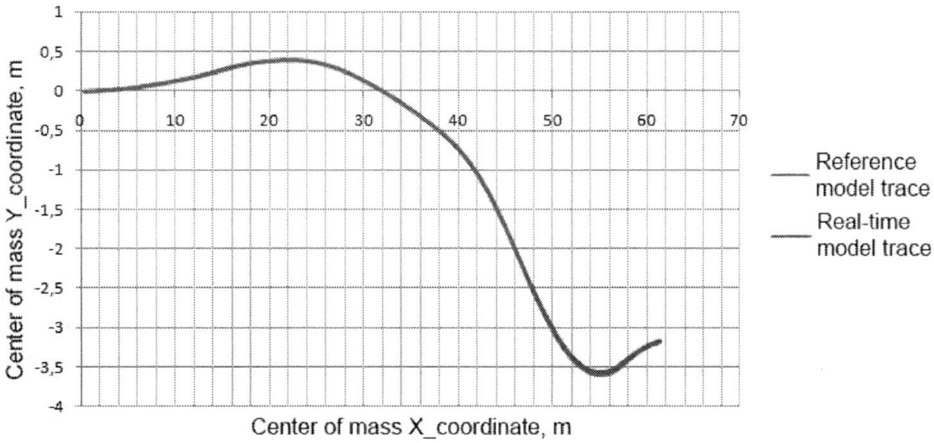

Figure 9. Centers of mass trajectories of the real-time model and the «reference» model during the lane change at speed of 15 km/h.

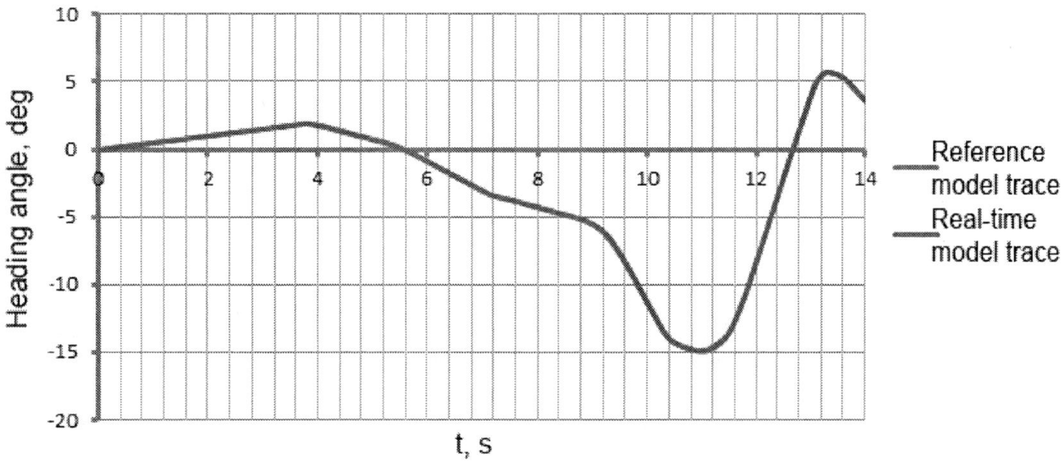

Figure 10. Heading angle of the real-time model and the «reference» model during the lane change at speed of 15 km/h.

Table 2 shows the maximum and average values of the relative errors of the output parameters obtained during simulation of typical maneuvers.

Table 2. Simulation errors.

Type of error	Lane Change $S_R = 20$ m, 15 km/h	Cornering $R_T = 35$ m, 15 km/h	Cornering $R_T = 35$ m, 20 km/h	Lane Change $S_R = 20$ m 25 km/h
Maximum relative error of the X-coordinate of the CM, %	0,5	0,34	0,5	0,57
Average relative error of the X-coordinate of the CM, %	0,0388	0,048	0,192	0,0388
Maximum relative error of the Y-coordinate of the CM, %	0,43	0,7	3,51	6,51
Average relative error of the Y coordinate of the CM, %	0,194	0,293	1,187	2,55
Maximum error of the heading angle, %	1,45	5,81	9,91	5,2
Average error of the heading angle, %	0,406	0,626	1,34	0,815

The presented data has proved adequacy of the real-time model on the basis of accepted criteria. This confirms the correctness of the real-time model work during interaction with the ELSW control system. The model can be used for the development of the workstations for remote control of vehicles.

The article is written based on the results of the work carried out with the financial support of the Ministry of Education and Science of the Russian Federation under agreement No. 14.574.21.0178 (Unique identifier: RFMEF57417X0178).

References

[1] Rajmpel' J 1987 Vehicle chassis: steering control (Moscow: Mashinostroenie) p 232

[2] Gorelov V A, Kotiev G O and Beketov A A 2008 Movement mathematical model of all-wheeled vehicle *J Automobile engineers association* **1**(48) p 50-54

[3] Keller A V, Gorelov V A, Anchukov V V 2015 Modeling truck driveline dynamic loads at differential locking unit engagement *Procedia Engineering* Vol. 129 pp 280–287

[4] Kotiev G O, Cernyshev N V and Gorelov V A 2009 Curvilinear motion mathematical model of 8x8 wheel arrangement vehicle with different way of turning *J Automobile engineers association* **2** pp 34–40

[5] Gorelov V A, Komissarov A I and Miroshnichenko A V 2015 8x8 wheeled vehicle modeling in a multibody dynamics simulation software *International Conf on Industrial Engineering, ICIE 2015* **129** pp 300-307

[6] Kotiev G O, Butarovich D O and Kositsyn B B Energy efficient motion control of the electric bus on route 2018 IOP Conference Series: Materials Science and Engineering **315 issue 1** 012014

[7] Keller A V, Gorelov V A, Vdovin D S et al. 2015 Mathematical model of all-terrain truck *Proceedings of the ECCOMAS Thematic Conference on Multibody Dynamics* pp 1285-1296

[8] Keller A and Alyukov S 2016 Power Distribution in Transmissions of Multi-Wheeled Vehicles *SAE Technical Paper* 2016-01-1103

[9] Keller A and Aliukov S 2016 Efficient Power Distribution in an All-Wheel Ground Vehicles *SAE Technical Paper* 2016-01-1105

[10] Zhileykin M M, Kotiev G O and Nagatsev M V 2018 Comparative analysis of the operation efficiency of the continuous and relay control systems of a multi-axle wheeled vehicle suspension *IOP Conference Series: Materials Science and Engineering* **315 issue 1** 012030

[11] Kupreyanov A A, Morozov M V, Belousov B N et al. 2014 Experimental research of tire elastomer-surface tribological properties *ASME International Design Engineering*

IOP Conf. Series: Materials Science and Engineering **534** (2019) 012010 doi:10.1088/1757-899X/534/1/012010

Technical Conferences and Computers and Information in Engineering Conference, IDETC/CIE 3

[12] Gorelov V A, Kotiev G O and Tropin S L 2012 "Fan" control law for all-wheel steering of multi-axle vehicles *Herald of the Bauman Moscow State Technical University* **2** pp 102-116

IOP Conf. Series: Materials Science and Engineering **534** (2019) 012011 doi:10.1088/1757-899X/534/1/012011

Efficiency of photoelectric converters intellectual system application on ground electric transport

A F Kolbasov, K E Karpuchin, A S Terenchenko and O I Girutskiy

FSUE "NAMI", Moscow, Russian Federation

E-mail: aleksey.kolbasov@nami.ru

Abstract. The development of environmentally friendly road transport is now directly related to the introduction of electric actuators, high-voltage storage systems and the accumulation of electricity in transport. Thus, a significant variety of hybrid traction drive circuits appears, which can be charged from an external power source. However, all of them use hydrocarbon fuel for charging and storage of electricity, which emits CO_2 while burning. Clean electric vehicles are the most efficient in terms of environmental indicators, since they use only electricity produced in power plants located far from the operation aureole of the electric vehicle, which allows to improve the environmental situation in megapolises significantly. As you know, energy complexes in different countries have different environmental efficiency and CO_2 emissions during power generation are also present. Transition to the use of renewable energy sources, such as solar power plants, can give its effect, since at present, their productivity in the world reaches 400 GW, but the transmission of energy over distances reduces the efficiency. Thus, based on the development trends of solar power plants, it is important to consider the use of photovoltaic cells, the so-called "solar cells" directly on the electric vehicle itself in order to improve environmental friendliness.

1. Introduction

In the era of electric drive development in motor vehicles, both in cooperation with internal combustion engines, with electrochemical fuel cells on hydrogen and in its pure form, the question sooner or later arises: what will be the next stage? The next step in the development of environmentally friendly transport can be considered the stage of energy replenishment systems development with the help of various converters that provide power generation, both from mechanical energy and from a renewable source, for example, the Sun [1,2].

Also, due to the worldwide struggle to reduce greenhouse gas emissions, a lot of research has been carried out [3], which showed that the energy produced by solar power plants has CO_2 emissions no more than 48 g/kWh. This quantity includes emissions over the entire life cycle. It means that solar energy is quite clean compared to other options that are available for installation on a vehicle, for example, oil or gas-fueled heat engines (Figure 1).

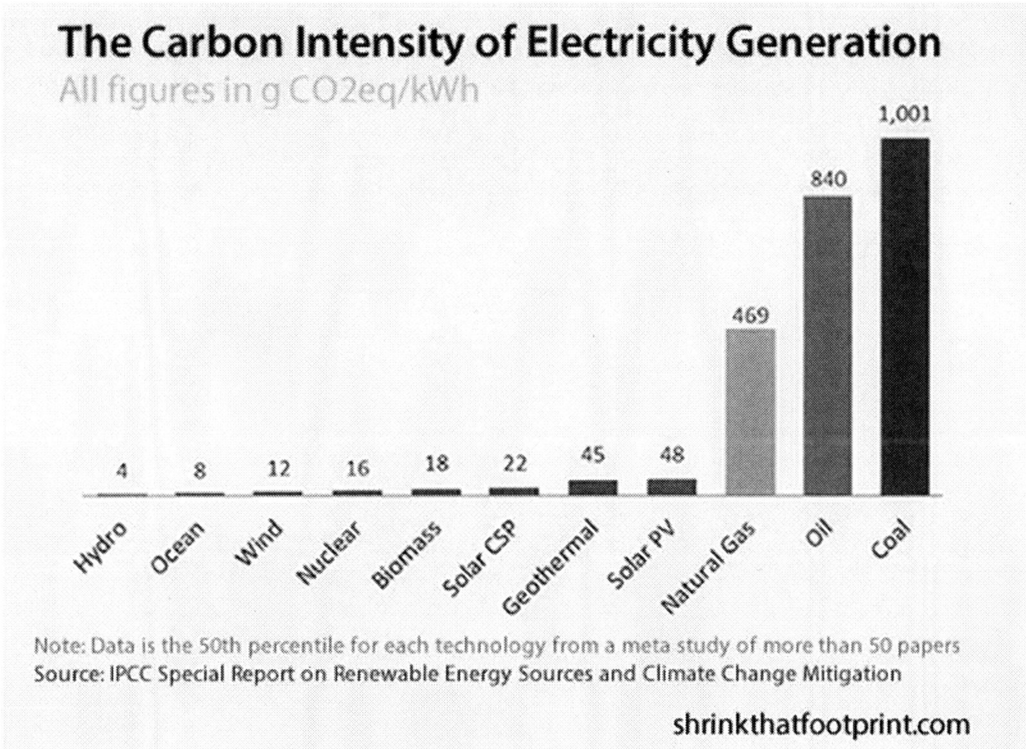

Figure 1. The level of greenhouse gas emissions at various ways of electricity generation.

The share of solar energy in total electricity generation is growing rapidly and, over the past five years, its capacity has increased almost four times. Solar power plants that were commissioned in 2017 increased electricity production by 97 GW, which increased the total power output of this method to 400 GW, which is 32% more compared to the end of 2016 [4].

2. Main part

In the field of photoelectric converters (PV), a lot of discoveries occur that lead to an increase in the efficiency of single-piece solar cells in laboratory conditions, as well as under operating conditions. Back in 2009, Spectrolab research company, part of the Boeing corporation, developed PV with a conversion efficiency of 41.6% [5], then in 2011, Solar Junction (USA) set a new level of PV efficiency – in the laboratory an efficiency of 43.5% was obtained on 5.5 × 5.5 mm sample which is 1.2% higher than the previous achievement.

In 2013, Sharp Company manufactured a three-layer PV with dimensions of 4×4 mm on an indium gallium-arsenide basis with an efficiency of 44.4% [6].

At the end of 2013, a new level was taken in the field of PV development: scientists from Soitec and CEA-Leti (France), together with the Fraunhofer Institute for Solar Energy Systems ISE (Fraunhofer Solar Power Systems, Germany) developed a PV with an efficiency of 46% [7].

A new milestone was set in November 2017 by Russian scientists from the Institute of Semiconductor Physics (ISP) named after Rzhanov. With the support of the ISP Screen company, they developed a new type of vacuum photodiodes, which will improve the efficiency of PV batteries, which, according to the developers of the device, can reach 50% or more, which, in the future, will allow vacuum photo cells to compete with the best multi-stage ones [8].

Thus, the use of PV in electric transport is becoming increasingly relevant and aspects of their usage in road transport need to be studied and researched.

IOP Conf. Series: Materials Science and Engineering **534** (2019) 012011 doi:10.1088/1757-899X/534/1/012011

In the FSUE "NAMI", as a result of the work on the "Solar Battery" project, an electric vehicle was manufactured with a system of additional energy sources based on photovoltaic cells, which includes an electricity transmission channel to a high-voltage battery as well as a thermal control system for the main components of an electric vehicle (Figure 2).

Figure 2. General view of an experimental electric vehicle with PV, developed and manufactured in the FSUE "NAMI".

Calculations preceding the start of the work showed that at the latitude of St. Petersburg a similar PV system can generate up to 260 kWh per year, at the latitude of Moscow up to 280 kWh per year, at the latitude of Sochi and Astrakhan up to 380 kWh per year at the latitude of Vladivostok up to 400 kWh. Therefore, installation of an additional power source based on photovoltaic cells with an area of 1.8 m^2 on an electric vehicle is capable to reduce CO_2 emissions by 153,000 – 240,000 grams per year [9,10] compared to a similar electric vehicle with the same mileage, and this is not accounting for power transmission losses at the standard method of electric vehicle charging [11].

During the work, a mathematical model of an electric vehicle with a photoelectric converters battery was created, which was implemented on a computer in the Simulink environment (Figure 3).

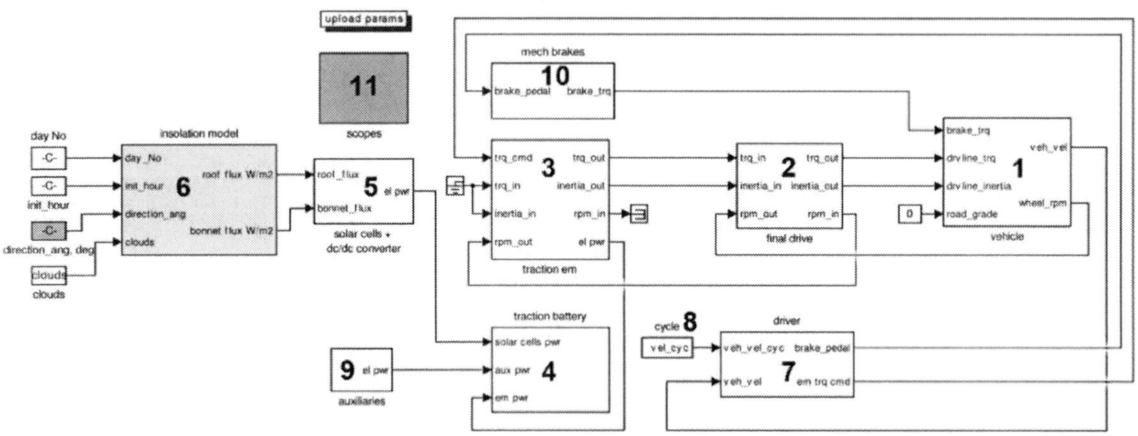

Figure 3. Implementation of an electric vehicle model with a PVs battery in the Simulink software environment (the upper level of the model).

Subsystem 1 contains the equation for the linear motion of the car and the equations for resistance calculation to the motion forces. Subsystem 2 is a model of the main transmission, which reproduces its

IOP Conf. Series: Materials Science and Engineering **534** (2019) 012011 doi:10.1088/1757-899X/534/1/012011

power and kinematic transfer functions. Subsystem 3 contains the model of the traction motor. Subsystem 4 represents the traction battery. Block 5 is a model of energy transformations. Sub-model 6 generates two signals: roof and hood surfaces density that enters the photocell battery block (5), 7 and 8 are sub-models of the vehicle speed control system. Block 9 simulates energy consumption by additional onboard devices; block 11 contains means for variables monitoring.

The model of the electric car included the following models:

- solar radiation model;
- driver model;
- power plant components model;
- vehicle movement model.

To determine the power generated by the photo cells, it is necessary to calculate the density of solar radiation when a car moves. It depends on a number of factors: time of day, season, angle of inclination and orientation of the light-receiving surface. These factors are taken into account in the model for an adequate assessment of the energy effect obtained from the use of photo cells. To calculate the density of solar radiation Bird's method was used [12], since it provides a good adequacy of the results using relatively simple dependencies for calculating the transmittance of solar radiation and requires a small amount of input data, which is important for design calculations.

The dependence underlying the calculations of all particular cases of surfaces and a solar radiation (SR) flux interaction is the expression of the SR density falling on a horizontal platform without taking into account the atmosphere (space SR):

$$\Psi^{0}_{hor} = \Psi_{C\perp} \cdot \cos\theta_z \text{ [13]} \qquad (1)$$

where $\Psi_{C\perp} = 1367$ W/m^2 – solar constant; θ_z – sunlight incidence angle.

The cosine of the sunlight incidence angle is calculated according to the equation:

$$\cos\theta_z = \cos\delta_n \cdot \cos\varphi_l \cdot \cos\omega_s + \sin\varphi_l \cdot \sin\delta_n \qquad (2)$$

where δ_n – sun declination angle; φ_l – location latitude; ω_s – sun hour angle.

The geometric meaning of the parameters used in this expression is shown in Figure 4.

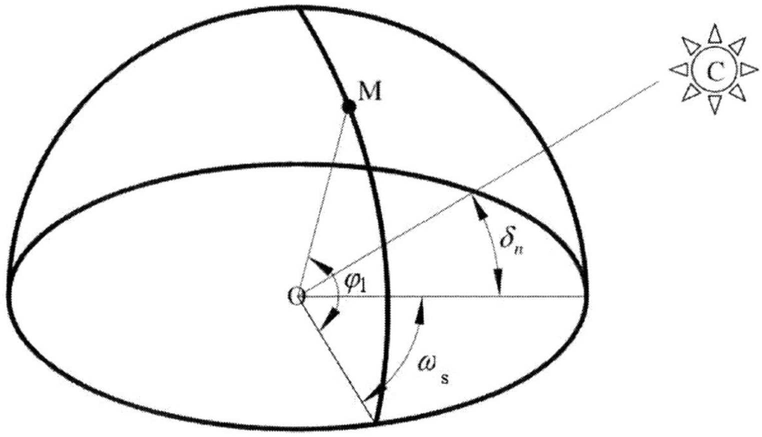

Figure 4. Scheme for determining the location latitude, the sun hour angle and the sun declination angle.

The sun declination angle is calculated using the equation

$$\delta_n = 23.5^o \sin\left(360^o \frac{n-81}{365}\right) \quad (3)$$

where n is the ordinal number of a day, counted from the 1st January.

To calculate the sun hour angle, the following expression is used:

$$\omega_s = \frac{15^o}{hour}\left(t_l - \Delta t_{decr} - 12 + t_t\right) + \left(\lambda_l - \lambda_m\right) \quad (4)$$

where t_l – local time; Δt_{decr} – amendment to the decree time used in the territory of the Russian Federation (1 h); t_t – time amendment; λ_l – longitude of the place; λ_m – longitude of the middle meridian of the time zone.

The time amendment is the difference between the true solar time and the average solar time on the considered meridian, which is associated with uneven movement of the Earth around the Sun due to the elliptical orbit of the planet:

$$t_t = 229.2 \cdot (0.000075 + 0.001868 \cdot \cos B - 0.032077 \cdot \sin B - 0.014615 \cdot \cos 2B - 0.04089 \cdot \sin 2B) \quad (5)$$

where $B = 360^o \frac{n-1}{365}$.

Cosmic radiation is absorbed and scattered by molecules of atmospheric gases (nitrogen, oxygen, carbon dioxide, water vapor, ozone). The degree of absorption and scattering is determined by the fundamental properties of these gases: the natural frequencies of atomic vibrations in molecules, the rotational-vibrational spectra in molecules and the frequencies ratio of these vibrations and the radiation frequencies in the solar spectrum. When calculating the radiation density with regard to the atmosphere, these properties are taken into account by empirical coefficients [14].

A common case is the determination of the solar radiation density perceived by the oriented surface on which it acts, passing through the atmosphere in cloudy conditions. Figure 5 shows the radiation components that are taken into account. The scattered part of direct SR is called diffuse solar radiation. If the light-receiving surface is at an angle to the horizon line, it is also affected by the SR, reflected from the Earth's surface. Modeling close to real climatic conditions requires the calculation of SR both in clear weather and in the presence of clouds, which reflects part of the SR, reducing its flux falling on the light-receiving surface.

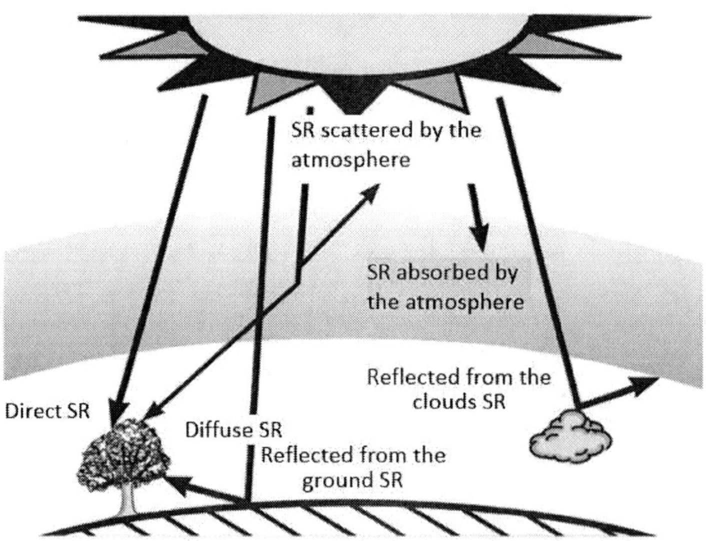

Figure 5. Components of solar radiation passing through the atmosphere of the Earth and falling on an oriented surface.

The initial version of the SR calculation taking into account the atmosphere is its definition for clear weather. The density of the direct SR, falling on a horizontal platform, taking into account the atmosphere, is calculated according to the equation:

$$\Psi_{hor}^{dir} = \Psi_{C\perp} \cdot \cos\theta_z \cdot \tau_\Sigma \cdot K_{hor}^{dir} \qquad (6)$$

In this formula, the correction factor is used, the values of which for the territory of the Russian Federation were calculated by the authors of the article [12]. For direct SR, it is equal to 1.14 in winter and 0.91 in summer. In this paper, linear interpolation is used to calculate values between these seasons.

The designation τ_Σ is called the transmittance of direct solar radiation, the value of which depends on the absorption and light scattering in the atmosphere and defined as the product of several partial transmittances:

$$\tau_\Sigma = \tau_{O3} \cdot \tau_{gas} \cdot \tau_{H2O} \cdot \tau_R \cdot \tau_A \qquad (7)$$

where, τ_{O3}, τ_{gas}, τ_{H2O} take into account the absorption of solar radiation by ozone, gas mixture (O_2, N_2, CO_2) and water vapor respectively; and take into account Rayleigh and aerosol dispersion. These coefficients are calculated using the formulas presented below.

$$\tau_R = \exp(-0.0903 \cdot (M^*)^{0.84} \cdot (1 + M^* - (M^*)^{1.01}) \qquad (8)$$

where M^* is the atmospheric effective mass.

Atmospheric mass is a measure of the sunlight path length in the atmosphere. Unit of atmospheric mass is the path of the sun's rays in the atmosphere when the Sun is at its zenith:

$$M = \frac{1}{\cos\theta_z + 0.15 \cdot (93.885 - \theta_z)^{-1.25}} \qquad (9)$$

In calculations, the atmospheric mass is used, taking into account the pressure correction:

$$M^* = \frac{p \cdot M}{p_0} \qquad (10)$$

p_0 is the normal pressure of 1 atm.

$$\tau_A = \exp(-\tau_{A'}^{0.873} \cdot (1 + \tau_{A'} - \tau_{A'}^{0.7088}) \cdot M^*)^{0.9108} \qquad (11)$$

where $\tau_{A'} = 1.832 \cdot \alpha$, $\alpha = 0.0314$ – spectral turbidity coefficient of Angstrem.

$$\begin{aligned}\tau_{O3} = 1 - 0.1611 \cdot d_{O3} \cdot M^* \cdot (1 + 139.48 \cdot d_{O3} \cdot M^*)^{-0.3035} - \\ 0.002715 \cdot d_{O3} \cdot M^* \cdot (1 + 0.044 \cdot d_{O3} \cdot M^* + 0.0003 \cdot (d_{O3} \cdot M^*)^2)^{-1}\end{aligned} \qquad (12)$$

where d_{O3} = 0.0032 m is the thickness of the ozone layer at normal temperature and pressure for middle latitudes.

$$\tau_{gas} = \exp(-0.0127 \cdot (M^*)^{0.26}) \qquad (13)$$

$$\tau_{H2O} = 1 - 2.4959 \cdot d_{H2O} \cdot M^* \cdot (1 + 79.034 \cdot d_{H2O} \cdot M^*)^{0.6828} + 6.385 \cdot d_{H2O} \cdot M^*)^{-1} \qquad (14)$$

where d_{H2O} is the indicator of water vapor content in the vertical column of the atmosphere, equal to 1.3 g/cm^2 for winter and 3 g/cm^2 for summer. In this paper, linear interpolation is used to calculate values between these seasons.

The calculation of the diffuse solar radiation density incident to a horizontal area is performed using the following formula:

$$\Psi_{hor}^{dif} = \Psi_{C\perp} \cdot \cos\theta_z \cdot \tau_{O3} \cdot \tau_{gas} \cdot \tau_{H2O} \cdot \tau_{AA} \cdot (0.5 \cdot (1 - \tau_R) + B_a \cdot (1 - \tau_{AS})) / (1 - M^* + (M^*)^{1.02}) \cdot K_{hor}^{dif} \qquad (15)$$

In addition to the coefficients presented above, it contains the transmittance coefficient τ_{AA}, which takes into account aerosol dispersion, the absorption coefficient τ_{AS}, which takes into account the scattering of dry air particles and B_a – the ratio of the scattered direct radiation to the total scattered radiation. The correction factor for diffuse SR is K_{hor}^{dif}= 1.05. Additional factors are calculated as follows:

$$\tau_{AA} = 1 - 0.1 \cdot (1 - M^* + (M^*)^{1.06}) \cdot (1 - \tau_A) \quad (16)$$

$$\tau_{AS} = \tau_A / \tau_{AA} \quad (17)$$

$$B_a = 0.5 \cdot (1 + \cos\theta_z) \quad (18)$$

The density of the total radiation incident to a horizontal area, taking into account the atmosphere, is determined by the sum of direct and diffuse SR:

$$\Psi_{hor}^{tot} = (\Psi_{hor}^{dir} + \Psi_{hor}^{dif}) / (1 - r_E \cdot r_A) \quad (19)$$

where r_E and r_A are the albedo (characteristic of the diffuse reflectivity of the surface) of the earth's surface (the surface along which the car moves) and the atmosphere respectively.

To assess the effect of the car hood using to accommodate photo cells, it should be noted that its surface is at an angle to the horizon, and when the car is moving, it is also oriented in azimuth direction. Taking into account these features, the cosine of the sunlight incidence angle is calculated as follows:

$$\begin{aligned}\cos\theta_{z,inc} = {} & \sin\delta_n \cdot \sin\varphi_l \cdot \cos\beta - \sin\delta_n \cdot \cos\varphi_l \cdot \sin\beta \cdot \cos\gamma + \cos\delta_n \cdot \cos\varphi_l \cdot \cos\beta \cdot \cos\omega_s + \\ & + \cos\delta_n \cdot \sin\varphi_l \cdot \sin\beta \cdot \cos\gamma \cdot \cos\omega_s + \cos\delta_n \cdot \sin\beta \cdot \sin\gamma \cdot \sin\omega_s\end{aligned} \quad (20)$$

where β is the angle of the light-receiving surface to the horizon inclination; γ – the astronomical azimuth of the light-receiving surface (measured from the direction to the south). The value $\cos\theta_{z,inc}$ is used to calculate the direct SR for the inclined platform (Ψ_{inc}^{dir}) according to the similar formula.

To calculate the density of diffuse SR in the case of an inclined surface, the value of diffuse SR for the horizontal surface, previously calculated from the above formulas, is used:

$$\Psi_{or}^{dif} = \Psi_{hor}^{dif} \left(\frac{1 + \cos\beta}{2} \right) \quad (21)$$

For an inclined surface is also determined by the density of the reflected SR:

$$\Psi_{or}^{ref} = \Psi_{hor}^{tot} \cdot r_E \cdot \left(\frac{1 - \cos\beta}{2} \right) \quad (22)$$

The total density of the SR for an inclined surface is determined by the sum of the direct, diffuse and reflected components:

$$\Psi_{or}^{tot} = \Psi_{or}^{dir} + \Psi_{or}^{dif} + \Psi_{or}^{ref} \quad (23)$$

To take into account the conditions of cloudiness, [11] proposed to use solar radiation correction coefficients calculated for clear weather. These coefficients are determined by the ratio between the observed (experimental) SR in cloudy ($\Psi_{ob,cloud}^{tot}$) and clear ($\Psi_{ob,clear}^{tot}$) weather:

IOP Conf. Series: Materials Science and Engineering **534** (2019) 012011 doi:10.1088/1757-899X/534/1/012011

$$K_{\text{cor}} = \frac{\Psi^{tot}_{ob,cloud}}{\Psi^{tot}_{ob,clear}} \tag{24}$$

Experimental SR values for a known geographic location and time of year can be obtained from climate guides or databases. Using the correction factor, the full SR for cloudy weather is calculated as follows:

$$\Psi^{tot}_{theory,cloud} = \Psi^{tot}_{theory,clear} \cdot K_{cor} \tag{25}$$

Based on the calculation of the electric vehicle mileage increase per NewEuropean Driving Cycle (NEDC), currently used in the certification procedures of UNO Regulation №101 and WLTC (WorldwideHarmonizedLightVehicleTestCycle) – a mixed driving cycle, which was developed as a replacement for the outdated NEDC through the use of a system based on PVs were obtained indicators presented in table 1.

Table 1. The results of computational experiments.

Configuration	NEDC. S, km	DS, %	NEDC-city. S, km	DS, %
Day 15 (January), to the South, 9 o'clock				
Basic electric car	152.2	0.00	225.4	0.00
Electric car + PV	153.5	0.82	229.8	1.95
Day 15 (January), to the South, 12 o'clock				
Electric car + PV	155.0	1.84	237.0	5.17
Day 15 (January), to the South, 15 o'clock				
Electric car + PV	152.7	0.29	227.4	0.89
Day 105 (April), to the South, 9 o'clock				
Electric car + PV	159.9	5.04	257.5	14.24
Day 105 (April), to the South, 12 o'clock				
Electric car + PV	161.5	6.13	266.8	18.37
Day 105 (April), to the South, 15 o'clock				
Electric car + PV	157.7	3.62	249.8	10.85
Day 105 (April), to the South, 18 o'clock				
Electric car + PV	152.6	0.24	227.0	0.74
Day 196 (June), to the South, 9 o'clock				
Electric car + PV	160.8	5.62	261.8	16.14
Day 196 (July), to the South, 12 o'clock				
Electric car + PV	162.3	6.64	270.5	20.03
Day 196 (July), to the South, 15 o'clock				
Electric car + PV	159.2	4.58	256.3	13.73
Day 196 (July), to the South, 18 o'clock				
Electric car + PV	153.9	1.13	233.1	3.44
Day 288 (October), to the South, 9 o'clock				
Electric car + PV	156.2	2.64	241.3	7.04
Day 288 (October), to the South, 12 o'clock				
Electric car + PV	157.3	3.37	247.3	9.70
Day 288 (October), to the South, 15 o'clock				
Electric car + PV	153.7	0.95	232.0	2.95
Day 288 (October), to the South, 18 o'clock				
Electric car + PV	152.2	0.00	225.4	0.00

The photovoltaic indicators system, made for an electric vehicle, consists of two solar panels with 84 elements on the roof surface and 36 elements on the hood surface, which together make up 1.85 m^2.

IOP Conf. Series: Materials Science and Engineering **534** (2019) 012011 doi:10.1088/1757-899X/534/1/012011

These batteries are protected from environmental influences by glass, which repeats the shape of the body elements of an electric vehicle. Solar cell battery was installed on a protective glazing to minimize the sun's loss occur when the beam overcome the air gap between the protective glass and the LET system. The back of the LET battery was sealed to prevent the contact with metal body elements, for this goal a special heat-resistant (up to + 250 °C) two-component compound - "Silagerm-2104" encapsulant was used. Thus, the air gap between the surface of PVs and protective glazing is completely excluded. The used compound has absolute resistance to UV light and allows the solar cell system to maintain full operation in an open air environment and conditions of high humidity (Figure 6).

Figure 6. Elements of a PV system mounted on a protective glazing.

A schematic diagram of the photovoltaic cells system components relationship on the vehicle is shown in Figure 7.

The scheme has two main modes of operation:

- stationary;
- dynamic.

The first option assumes that the charge of the high-voltage battery occurs when the vehicle is idle in automatic mode. A battery of photoelectric converters, using a controller that monitors the point of battery maximum efficiency, supplies a buffer lithium-ion drive with a nominal voltage of 12V. When the buffer drive is fully charged, the charging of a high-voltage battery occur.

In the dynamic version, when the vehicle is in motion, the circuit operates according to the following algorithm: the battery of PV feeds the buffer battery through the charge controller, which is the main battery of the entire low-voltage circuit of the experimental vehicle. With the help of a current sensor, the consumption of on-board low-voltage electronics is monitored and, with a lack of power produced by the PV battery, electricity is supplied from the high-voltage battery to the buffer drive through the DC converter, but in a smaller amount compared to the disconnected PV cell battery.

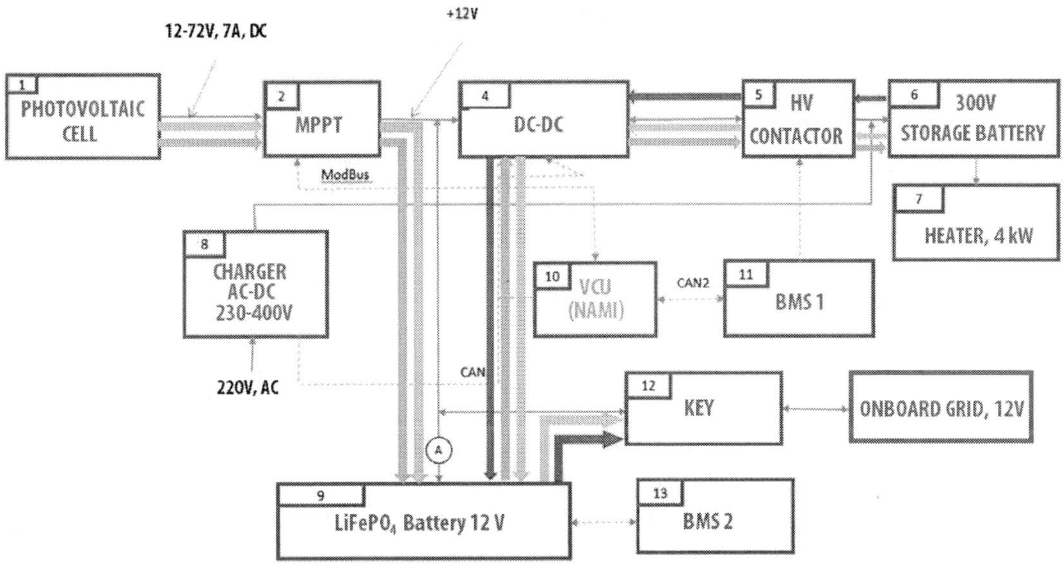

Figure 7. Scheme of an additional energy sources system based on solar cells.

Figure 8. Main control unit.

Figure 8 shows the main control unit designed and manufactured as part of the Solar Panel project. The input and output interfaces of this unit meet the requirements for compiling into C ++ code. The unit controls the following devices: an MPPT controller (control converter of the PV battery) and a DC/DC converter that exchanges energy between the buffer and traction batteries and also controls the elements of the temperature control system. To work as a part of a hardware controller, the control system was compiled into C++ program code. Compilation is performed by the built-in tools of the MATLAB/Simulink software environment.

To monitor and record the input and output parameters of an electric vehicle with a system of additional power sources based on photovoltaic cells in LabVIEW, an interface was developed, the working menu of which is shown in Figure 9. This interface provides for the manual mode of controlling individual devices in order to receive experimental data separately.

IOP Conf. Series: Materials Science and Engineering **534** (2019) 012011 doi:10.1088/1757-899X/534/1/012011

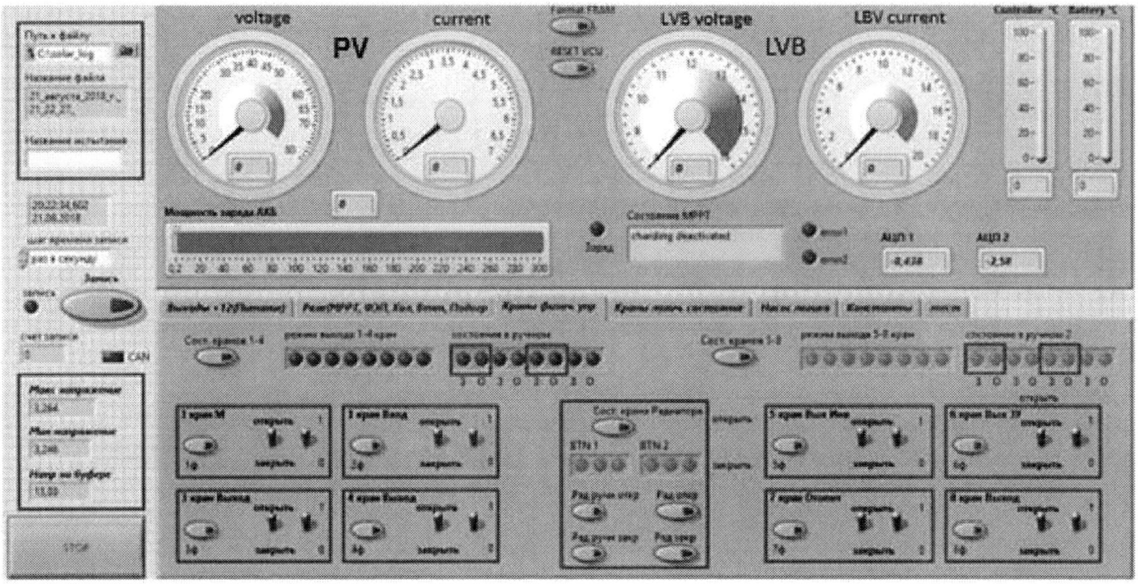

Figure 9. The system for indicators of electric vehicle with a PV system monitoring.

The obtained results at the FSUE "NAMI" testing ground (Figure 10) for the NewEuropeanDrivingCycle (NEDC) were partially shown in table 2. During field tests the solar energy density was about 600-750 W/m^2 (Figure 11), which is far from the maximum possible value for the climatic zone of Moscow city.

Figure 10. Road tests conduction at the FSUE "NAMI" testing ground.

Figure 11. Measuring of the solar energy intensity.

Table 2. Results of the road tests.

Configuration	NEDC S, km	DS, %	NEDC-city S, km	DS, %
Day 227 (August), to the South, 12 o'clock				
Basic electric car	142.4	0.00	211	0.00
Electric car + LET	149,6	5.08%	230,3	9,15%

3. Conclusions

The effect of increased mileage primarily depends on the intensity of traffic. In case of its increase, what characterizes WLTP cycles, which reproduce modern road traffic conditions, the advantages of using photocells are reduced by 5-10% compared with the outdate NEDC cycle, which has a low dynamics of the European movement of the 1980s. In mixed cycles in summer conditions, the increase in mileage is up to 6%.

Obviously, even now, with PV efficiency of 20-22%, the increase in the mileage of an electric vehicle can reach relatively high rates, and, at the large-scale development of electric transport with an additional power supply system based on PV, will provide an additional opportunity to reduce greenhouse gas emissions.

Undoubtedly, with the development of technologies that increase the efficiency of PV that can be used in the automotive industry, we will see significant advantages of this method for replenishment of energy over the currently used mileage multipliers, both in terms of environmental and energy efficiency. In addition using the vehicle according to the Vehicle-to-Grid protocol (V2G, car-electric grid) will allow returning excess electricity to the network, since, as a rule, most of the personal transport is not used throughout the daylight hours.

Acknowledgments

This scientific article was prepared based on the results of an applied scientific research, which was carried out with the financial support provided by the Ministry of Education and Science of Russia, agreement № 14.624.21.0047. The unique identifier of the project is RFMEFI62417X0047.

References

[1] Solomatin N S, Shavrin P A, Lata V N 2017 Experimental model of autonomous, fully-managed electric vehicle *Journal of automotive engineers* **3** 8–11
[2] Panasonic's photovoltaic module HIT adopted for Toyota Motor's new Prius PHV Information on

http://news.panasonic.com/global/press/data/2017/02/en170228-3/en170228-3
[3] Is photovoltaic energy better than BC's grid? Vancouver renewable energy cooperative (VREC) solar Information on http://www.vrec.ca/blog/2014/11/05/photovoltaic-energy-better-bcs-grid/
[4] Solarenergy Information on https://www.bp.com/en/global/corporate/energy-economics/statistical -review-of-world-energy/renewable-energy/solar-energy.html
[5] Jones R K 2010 Status of 40% production efficiency concentrator cells at Spectrolab *IEEE PVSC* 35
[6] Datas A NGCPV a new generation of concentrator photovoltaic cells, modules and systems *Conference 2013 28th EU PVSEC* (Paris)
[7] Soitec team pushes solar record to 46% Information on http://optics.org/news/5/12/1
[8] Tereshchenko O E, Golyashov V A, Rodionov A A, Chistokhin I B, Kislykh N V, Mironov A V and Aksenov V V 2017 Solar energy converters based on multi-junction photoemission solar cells *Scientific reports*
[9] Information on http://www.interrao.ru/upload/docs/Inter_RAO_Annual_Report_2016.pdf
[10] Tuan N K, Karpukhin K E, Terenchenko A S, Kolbasov A F 2018 World trends in the development of vehicles with alternative energy sources *ARPN Journal of engineering and applied sciences* **13** 2535–42
[11] Kolbasov A, Karpukhin K, Terenchenko A, Kavalchuk I 2018 Concept of intellectual charging system for electrical and plug-in hybrid vehicles in Russian Federation *IOP Conference Series: Materials Science and Engineering*
[12] Elistratov V V, Aronova E S 2012 Solar power plants. Estimation of solar radiation *SPb Publishing house of the Polytechnic University* 165
[13] Tiwari G, Dubey S 2010 *Fundamentals of Photovoltaic Modules and Their Applications* RSC Publishing pp 29-71
[14] Karpukhin K E, Terenchenko A S, Kolbasov A F, Kondrashov V N 2017 Using photoelectric converters in road transport in order to improve energy efficiency in Russia *International Journal of Mechanical Engineering and Technology* **8** pp 529–34

IOP Conf. Series: Materials Science and Engineering **534** (2019) 012012 doi:10.1088/1757-899X/534/1/012012

Development of the upper-layer electronic control unit based on a production-type unit for a commercial vehicle

D A Astapenko[1], D O Butarovich[2] and A A Smirnov[3]

Bauman Moscow State Technical University, 105005 Baumanskaya 2th st., 5, Moscow, Russian Federation

E-mail: [1] astap-astap@yandex.ru, [2] buta73@mail.ru, [3] smr_a@mail.ru

Abstract. The development of a modern vehicle is associated with the need to co-operate with a variety of components from different manufacturers, due to the fact that companies mainly specialize in the manufacture of specific equipment (powertrain, steering, braking system, dashboard, etc.). Majority of manufacturers of automotive electronic units do not allow changes to their finished products, therefore, when integrating components into a single vehicle, there is a problem with the coordination of control signals and communication protocols. In response to this problem, it is required to use an additional electronic unit - the electronic control unit of the upper layer (ECU). This device generates signals for each component in accordance with the requirements of manufacturers. This unit may have different components and different functionality depending on the requirements. The unit is based on a microcontroller with a control program. To control the power discrete output, power transistor MOSFET keys are used. Optocouplers or specialized microchips are used for processing discrete digital signals. Moreover, the unit is equipped with specialized microcircuits, which allow the exchange of information with other electronic units via digital protocols.

The modern development of motor vehicles is characterized by the extensive use of electronic devices to control various units and systems: powertrain, steering, braking system, cushion system, etc. These motor vehicles electronic control systems are most often integrated into a single communication network to enable data exchange and effective work with the ability to implement intersystem functions.

The most common way to integrate electronic control systems in commercial motor vehicles is the CAN communication bus, for which the link-layer and physical layer protocols are standardized by the ISO 11898 standard group.

American standard SAE J1939-71 [1] is most prevalent in commercial motor vehicles for the protocols of the upper (applied) layer, in which standard parameters are determined and are transmitted as bus messages. However, despite the presence of the SAE J1939-71 standard, when integrating ECU from different manufacturers in one motor vehicle, there is a problem with the coordination of control signals and communication protocols.

There is a practice in the Russian automotive industry in which, when integrating ECU of foreign manufacturers into a newly developed motor vehicle, foreign partners, as a rule, do not provide an opportunity to correct (parameterize) the ECU internal software. Often, only the so-called CAN matrix of messages received and sent by the ECU is provided. As a result, a modern commercial motor vehicle equipped with a large number of control systems contains several ECUs with a list of messages that are defined in accordance with SAE J1939-71, but often there are situations when CAN messages from

ECUs of different manufacturers have Identical identifiers (IDs), different definitions of the beginning of the bytes and bits counting, etc. As a result, the integration of these units in the motor vehicle becomes difficult to implement.

It is necessary to use an additional electronic unit in the motor vehicle - the electronic control unit of the upper layer (ECU) to solve this problem. This device generates signals for each ECU component in accordance with the requirements of manufacturers.

There are several solutions to the problem of integrating components: use a ready-made commercial product with extensive parametrization capabilities or the possibility of adjustments; order on a specialized enterprise the development of a new electronic control unit for a vehicle; develop a prototype electronic control unit for debugging algorithms of work on a prototype vehicle [2, 3, 4, 5].

All of these methods require time and financial resources, while it should be noted that most of the integrated foreign ECU have abundant characteristics for processing input and output signals [6, 7].

Currently, the specialists of Bauman Moscow State Technical University have been tasked with integrating several ECUs aboard a city bus with a combined power plant application consistent with the use of modern ECUs of type safety ADAS level 1 in the design of this vehicle. The solution of these tasks is carried out in accordance with the technical requirements of the work "Development of scientific and technical solutions for the creation of a Russian combined power plant for small-class urban and suburban buses", carried out under the Agreement No. 14.574.21.0178 (a unique work identifier: RFMEFI57417X0178) and with the financial support from the Ministry of Higher Education and Science of Russia.

Earlier, the self-engineered ECU based on the STM32 microcontroller was used on the first prototypes during the electric bus LiAZ-6274 development performed by LLC "LiAZ" in cooperation with Bauman Moscow State Technical University.

The use of the self-engineered ECU allowed integration of many ECUs aboard the electric bus and performing factory tests in the shortest period.

According to the results of the electric bus factory tests, the hardware and software of the self-engineered ECU were completed due to the need to increase the number of processed signals [6].

Based on the test results and experience gained during the electric bus test operation, technical requirements were formed for the ECU for a modern commercial motor vehicle: bus, electric bus, combined power plant bus, trolley bus, etc.

Electronic control unit may have different components and different functionality depending on the requirements. The unit is based on a microcontroller with a control program. Power transistor MOSFET keys are used to control the power discrete outputs. Optocouplers or specialized microchips are in operation for processing discrete digital signals. In addition, the ECU can be equipped with specialized microcircuits, which allow the exchange of information with other electronic units using digital protocols.

Due to limited demand for such ECUs when performing the above tasks in cooperation with Industrial Partner LLC "United Engineering Center" we made a decision to select and use commercial control system electronic units produced in Russia in our future work.

As a result of the review and analysis, the serial engine control unit MD22 produced by NPP "ITELMA" (Figure 1) was selected as the basic unit.

IOP Conf. Series: Materials Science and Engineering **534** (2019) 012012 doi:10.1088/1757-899X/534/1/012012

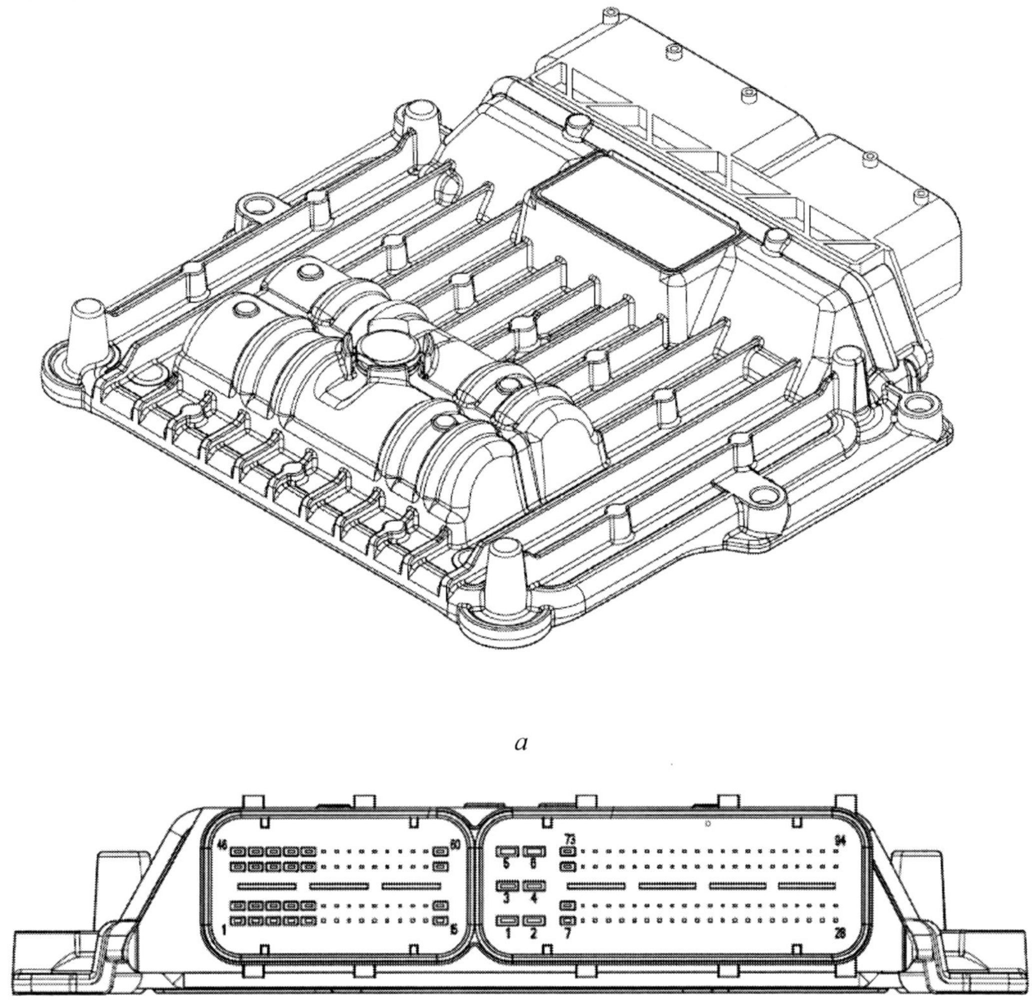

a

b

Figure 1. ECU MD22: a – case MD22, b – Unit connector MD 22.

Control system electronic unit MD22 is used in the internal combustion engine control system for automotive truck with a supply voltage of 24V. Brief characteristics control system electronic unit MD22 are shown in Table 1.

Table 1. SCEU MD22 brief characteristics.

№	Interfaces	Quantity	Note
1	Digital Input	15	
2	Analog Input	20	Voltage 0…5 V
3	Frequency Input	3	
4	Electric output («back-ups»)	20	0,2A…2A
5	Electric output («upper key»)	8	up to 5 A
5	Electric output PWM	6	
6	CAN 2.0b	2	Two-direction

IOP Conf. Series: Materials Science and Engineering **534** (2019) 012012 doi:10.1088/1757-899X/534/1/012012

Previously arising questions regarding the requirements of tightness, vibration resistance, temperature range, the use of electronic components of the AEC (Automotive Electronics Council) standard and safety, dependent on the control system electronic unit architecture, were settled automatically when using the MD22 unit.

The unit electronic circuit and the software of the lower layer remained unchanged when developing a control system electronic unit based on MD22, and the upper layer software was developed newly in the MSE Pro environment provided by NPP "ITELMA". The MSE Pro environment is based on MATLAB / Simulink software and uses the same user interface as Simulink.

Figure 2 shows the block diagram of the on-board control system electronic unit of the city bus, in which the control system electronic unit MD22 is presented.

As an example of working with the unit software, the description of the top-layer algorithms implementation is given below: receiving and sending CAN messages between different control system electronic unit with data frame correction and control of the high-voltage contactors for the traction battery charging.

Receiving and sending CAN messages is carried out using higher layer functions. In order to work with a function, it is required to set its input and output parameters in the correct format (see Figure 3). A function is called at different layers of software by the name of this function (Action name), and it does not matter how the function is implemented at the lower layer.

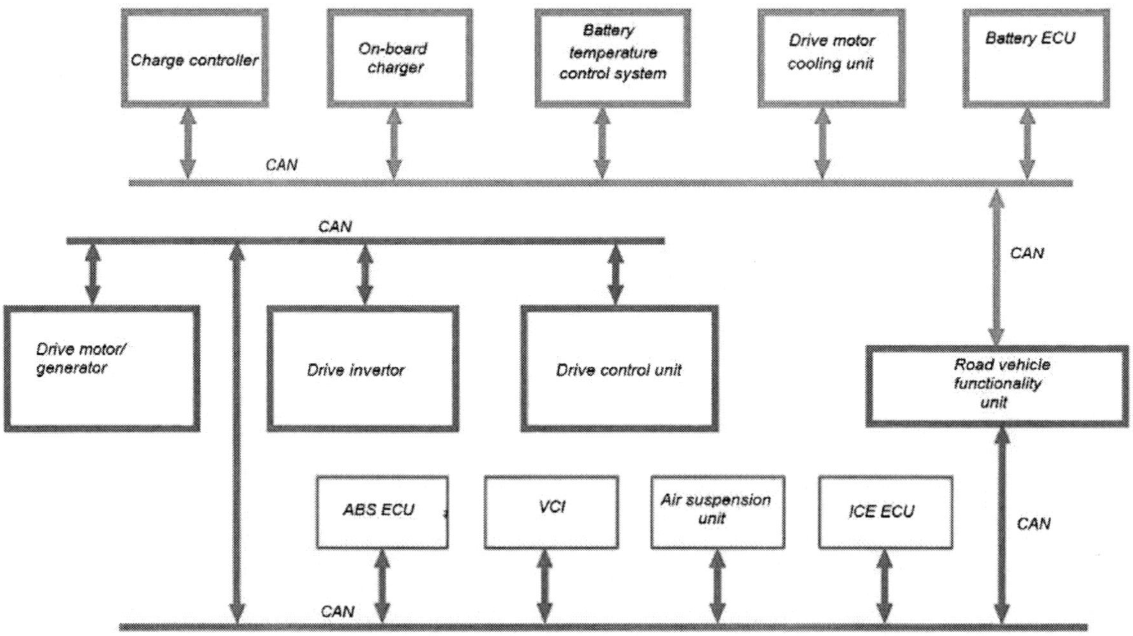

Figure 2. City bus electronic control system flow diagram.

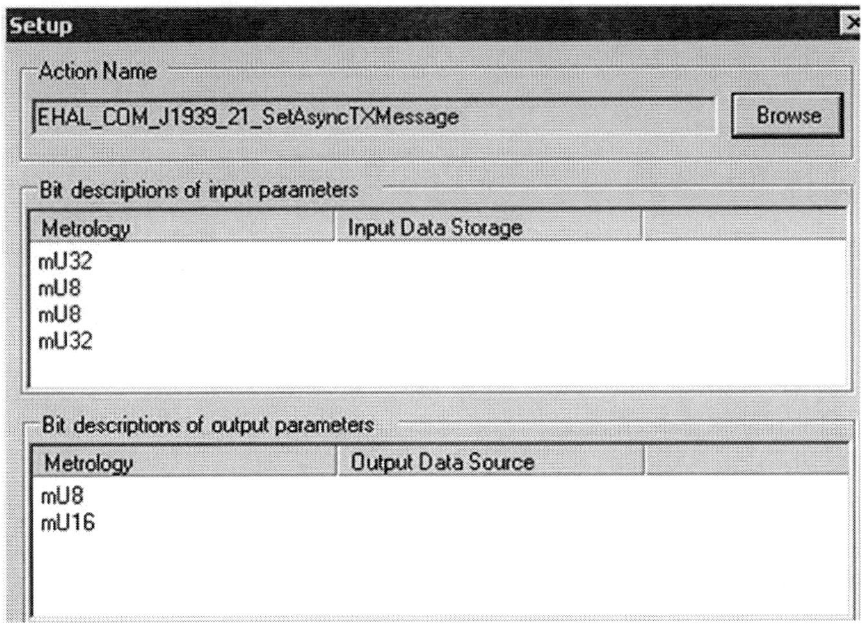

Figure 3. Called function description.

It is necessary to initialize receiving and sending CAN messages to get started (Figure 4).

It is required to submit the number of the parameter group (PGN) of the message (c_canRxPgnBMUtoVCU_1) and the bus number (q_canNetGAZ_CAN) from which it will be received when receiving the initialization function. The function of sending input parameters requires more: PGN messages, bus number, message priority (c_canPriority_0x18), an indication to a callback function (c_NULL_POINTER). The initialization function returns its result (0 - completed successfully, 1 - an error occurred), and an indication to the CAN message (RxHandler). Further work with the CAN message occurs at the returned pointer.

In the real-time system, the function of receiving and sending CAN messages is triggered with a frequency of 10 ms. Working with a CAN message consists of three subtasks: a counter (to receive and send messages with the desired frequency), work with a buffer (reading or writing), and clearing the buffer. Each subtask has its own priority (Figure 5).

IOP Conf. Series: Materials Science and Engineering **534** (2019) 012012 doi:10.1088/1757-899X/534/1/012012

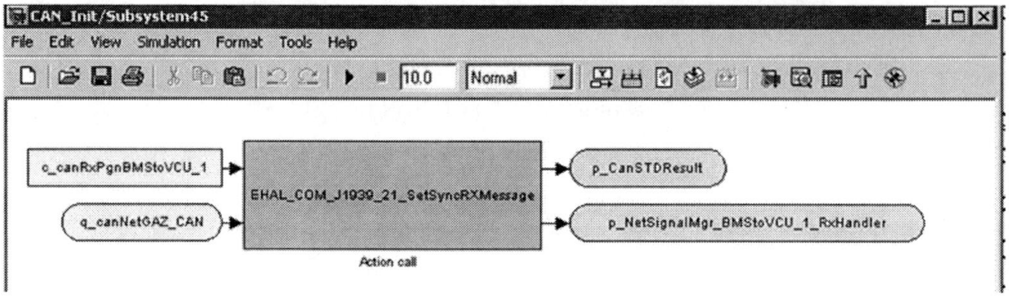

a

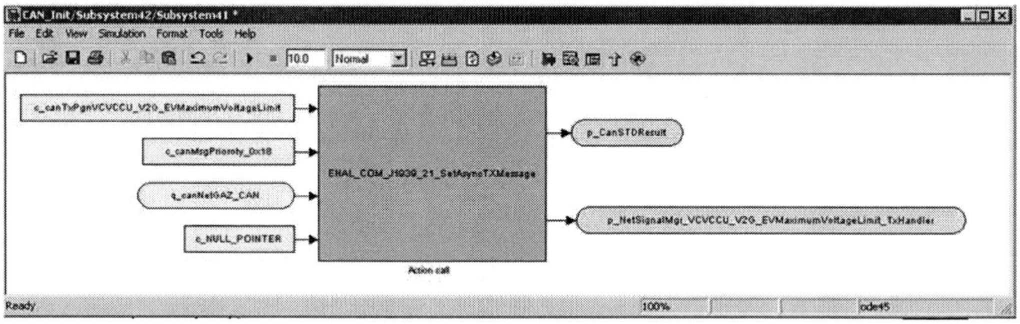

b

Figure 4. CAN messages initialization: *a* – receiving, *b* – sending.

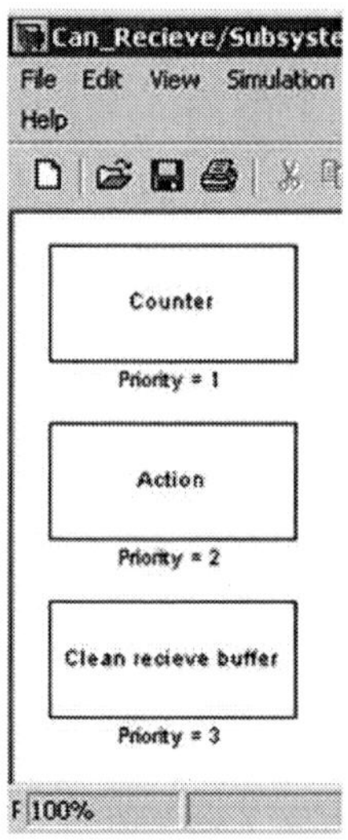

Figure 5. Subtasks call during work with CAN messages.

The counter implementation is shown in Figure 6. The p_counter variable stores the current value of the counter, it is incremented by the value of p_Time_t_Period_10msLv for each function call, while the top is limited by the limit (c_delay) ¬. The limit is the period of sending the desired CAN message.

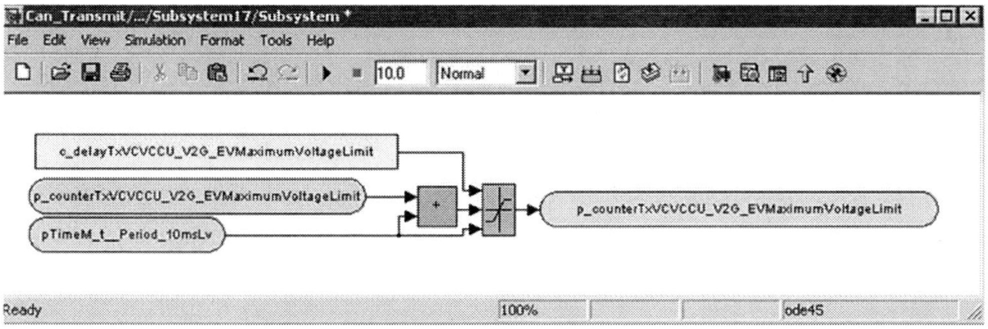

Figure 6. Counter realization.

It checks whether the counter reaches the desired value, and if it is reached, work with the buffer begins at each task invocation (see Figure 7a). When a message is received, the upper layer function is called EHAL_COM_21_J1939_SyncReceive, and if the function returns 0, then the received message is available for further work (see Figure 7b).

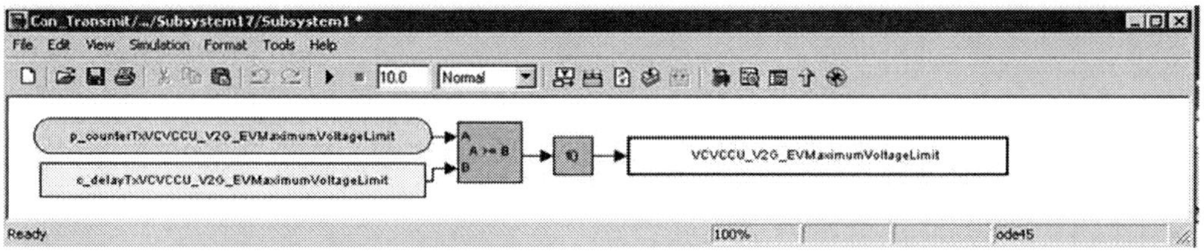

a

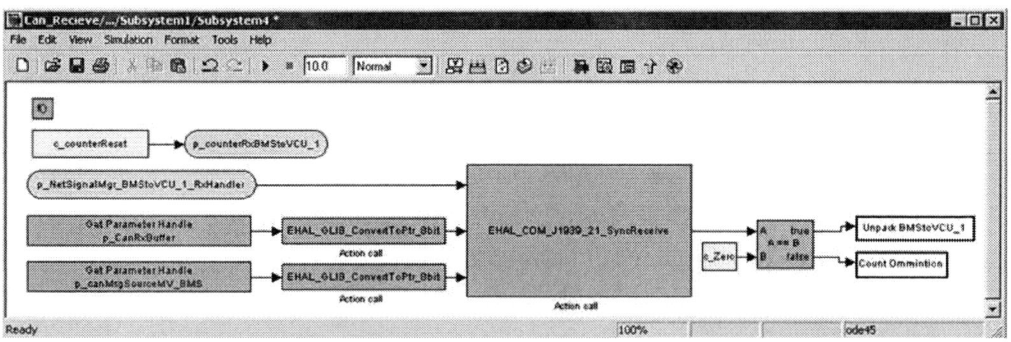

b

Figure 7. Receiving CAN message function: *a* – counter check, *b* – receipt function call.

A variable with the required metrology is required for the purpose of reading the required parameter from the CAN message correctly (Figure 8a). The parameter "Maximum battery voltage" considered as an example. The parameter "Maximum battery voltage" is contained in the 0th byte, has a length of 8 bits, and the resolution is 3.2 V per bit. A variable p_canRxMv_MaxBatSystemVoltage with the metrology BatteryVoltage_0_800_U8 is created to store the parameter (Figure 8b, 8c). Further, the value of the variable p_canRxMv_MaxBatSystemVoltage is written to the variable p_volBatteryMax, which has a different metrology, and the conversion of values takes place in the program automatically.

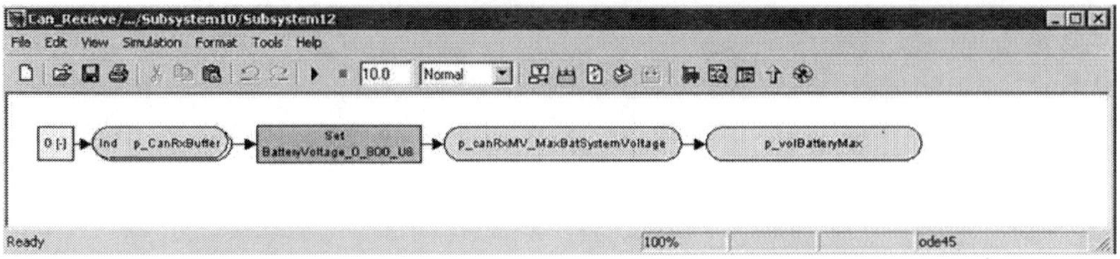

a

3. Data properties	
Metrology	BatteryVoltage_0_800_U8 (3,2 [V])
Minimum value	0
Maximum value	816

b

1. Description	
Name	BatteryVoltage_0_800_U8
Title	Напряжение батарей от 0В до 800В с шагом 3.2В
Comments	
Author	bmstu (04.12.2017)
Modifier	bmstu (11.01.2018)
2. Code generation	
Built-in type	8 Bit Unsigned Integer
3. Data properties	
Coefficients	32 / 10
Minimum value	0
Maximum value	816
Value step	3.2
Transfer coefficient	1 / 1
Units	bVoltage
4. Appearance	
Derived unit name	V
Format string	%3.2

c

Figure 8. Signal receipt: *a* – parameter reading from CAN buffer, *b* – metrology choices, *c* – metrology realization.

A reverse operation is applied to send a CAN message. The value of the variable p_volBatteryMax is written to the variable p_canTxVCVCUU_V2G_EVMaximumVoltageLimit_Value (Figure 9a). with Bat-teryVoltage_S16 metrology (Figure 9b). The resulting variable p_canTxVCVCUU_V2G_EVMaximumVoltageLimit_Value is written to the buffer for sending via the CAN interface.

IOP Conf. Series: Materials Science and Engineering **534** (2019) 012012 doi:10.1088/1757-899X/534/1/012012

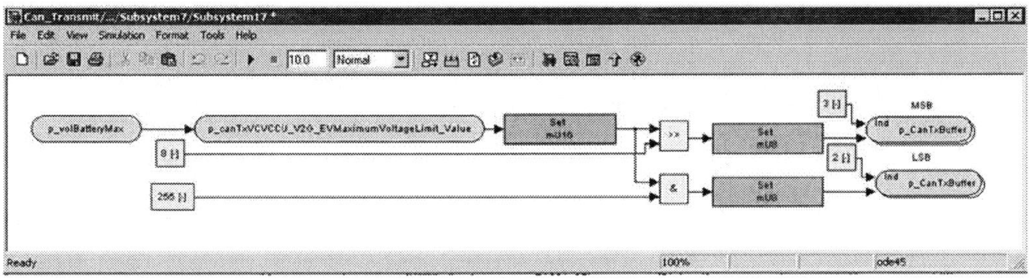

a

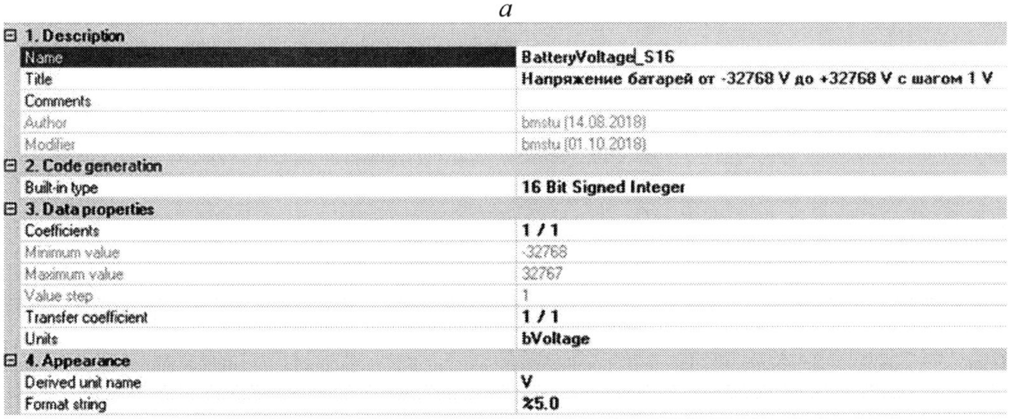

1. Description	
Name	BatteryVoltage_S16
Title	Напряжение батарей от -32768 V до +32768 V с шагом 1 V
Comments	
Author	bmstu (14.08.2018)
Modifier	bmstu (01.10.2018)
2. Code generation	
Built-in type	16 Bit Signed Integer
3. Data properties	
Coefficients	1 / 1
Minimum value	-32768
Maximum value	32767
Value step	1
Transfer coefficient	1 / 1
Units	bVoltage
4. Appearance	
Derived unit name	V
Format string	%5.0

b

Figure 9. Signal sending: *a* – parameter recording in CAN buffer, b – metrology realization.

As the result, the interface of the upper layer software allows one to get rid of the errors of converting quantities from one format to another by creating different metrologies.

The work with digital input and digital output of control system electronic unit is considered on the example of the operation of a high-voltage circuit contactor. The contactor has one control signal and one feedback signal. Feedback allows to control the current state of the contact, which is necessary because a powerful electric arc may strike and the contactor will stick when the contactor is opened under load. In normal condition, the contactor is open. In order to close the contactor, it is required to apply a voltage of 24 V to the control signal, and if the contactor successfully closes, a voltage of 24 V is also set at the feedback signal.

It is essential to configure a digital output to supply a control signal and a digital input to read the feedback signal to implement the algorithm in the program. The function EHAL_IO_MGR_D_SetChannelState is used to control the digital output (Figure 10a) and it takes two parameters as input: the output contact number is q_nOutputUltrafastChargeContactor1 and the output state is p_pinUltrafastChargeContactor1 (0 is off, 1 is on). The function EHAL_IO_MGR_D_GetChannelState is used for reading digital inputs (Figure 10b) and it takes one parameter as input: the number of the input contact is q_nInputUltrafastChargeContactor1, and has two output parameters: the state of the input is p_inUltrafastChargeContactor1Raw (0 - off, 1 - on) and the result of the function (0 - successful execution, 1 - error). The input signal has an unstable state due to bounce at the first moment of time. To get rid of it, the additional AntiBounce function is used (Figure 10c), at the output of which the filtered value of the digital input p_stUltrafastChargeContactor1 is set.

IOP Conf. Series: Materials Science and Engineering **534** (2019) 012012 doi:10.1088/1757-899X/534/1/012012

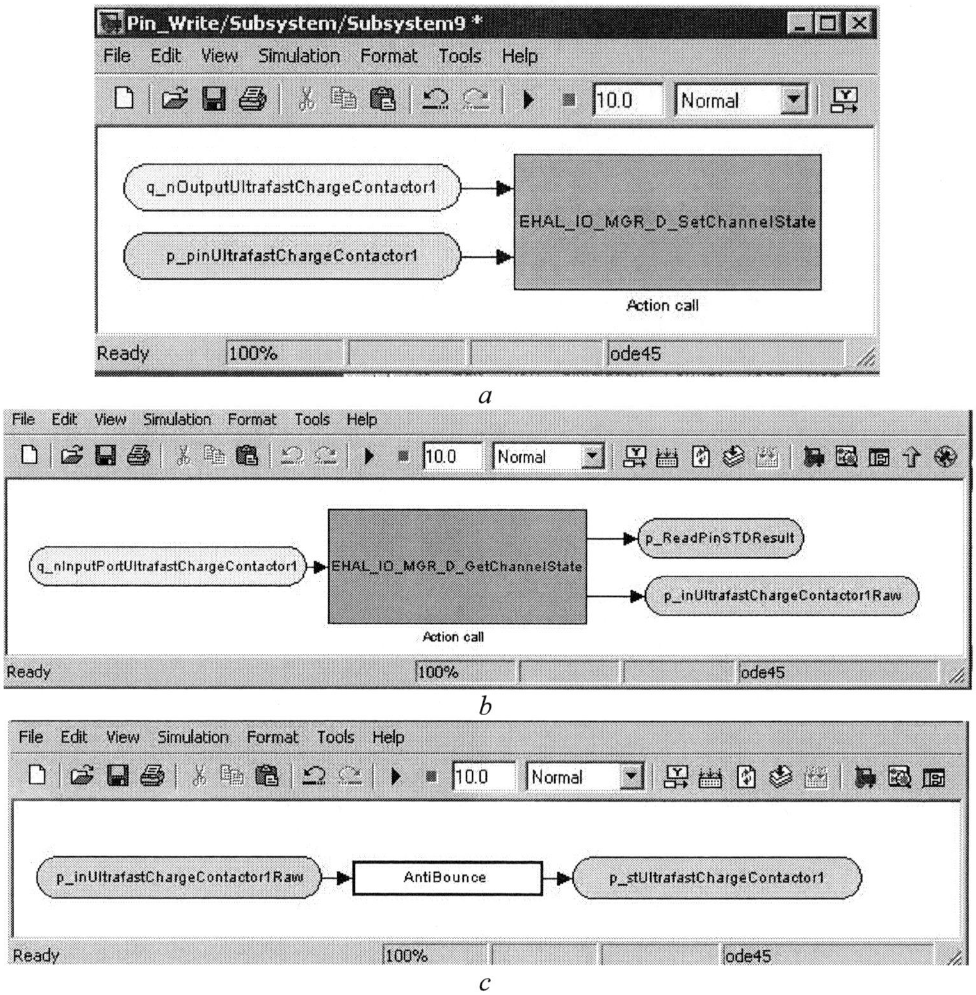

Figure 10. Functions for digital input/output operation:
a – digital output control, b – digital input read operation,
c – contact bounce screening.

The digital input read and digital outputs record are functions that are performed in a separate subtask. The contactor functionality is usually implemented in another subtask, which also implements the main control flow of the system. Figure 11 shows the part of the algorithm that is responsible for working with the contactor.

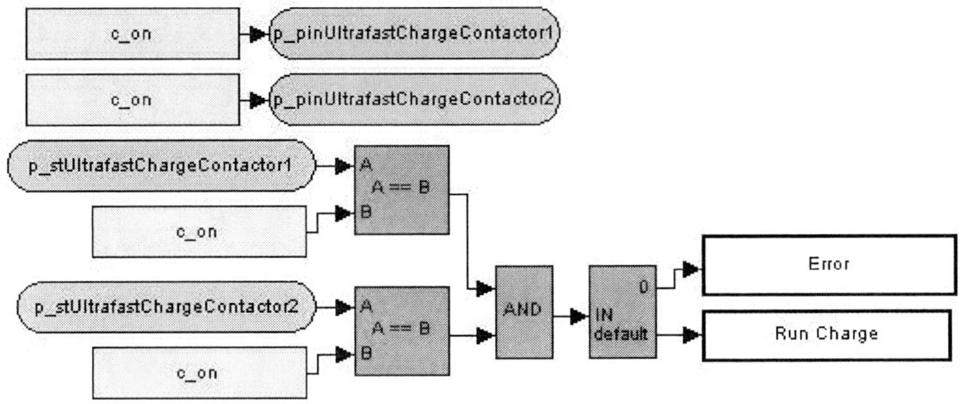

Figure 11. Power contactor operation algorithm.

A real-time operating system (RTOS) is implemented in the MSE Pro environment. RTOS allows implementing preemptive multitasking, i.e. the system operation in which several tasks are performed almost simultaneously due to the rapid switching between them. Distribution of the developed individual algorithms (receiving / sending CAN messages, operation of a power contactor, etc.) into tasks with a specific call period (10 ms, 100 ms, etc.) is as appropriate condition for the program correct operation. The implementation of the RTOS and an example of the inclusion of individual algorithms in the tasks are presented in Figure 12.

Consequently, upper layer software is developed that implements the new required functionality: receiving and sending CAN messages, digital outputs control, reading the status of digital and analog inputs, etc by means of visual programming tools in MSE Pro. Automatic C code generation is performed in the MSE Pro environment after working on upper layer algorithms has been finished. The compiler is used to extract the executable code (firmware) generated from the C code as a hex file. Having downloaded the new firmware to the control system electronic unit, originally intended for controlling the operation of the internal combustion engine, it can be used in another motor vehicle for other purposes, for example, as a controller of the electric bus upper layer.

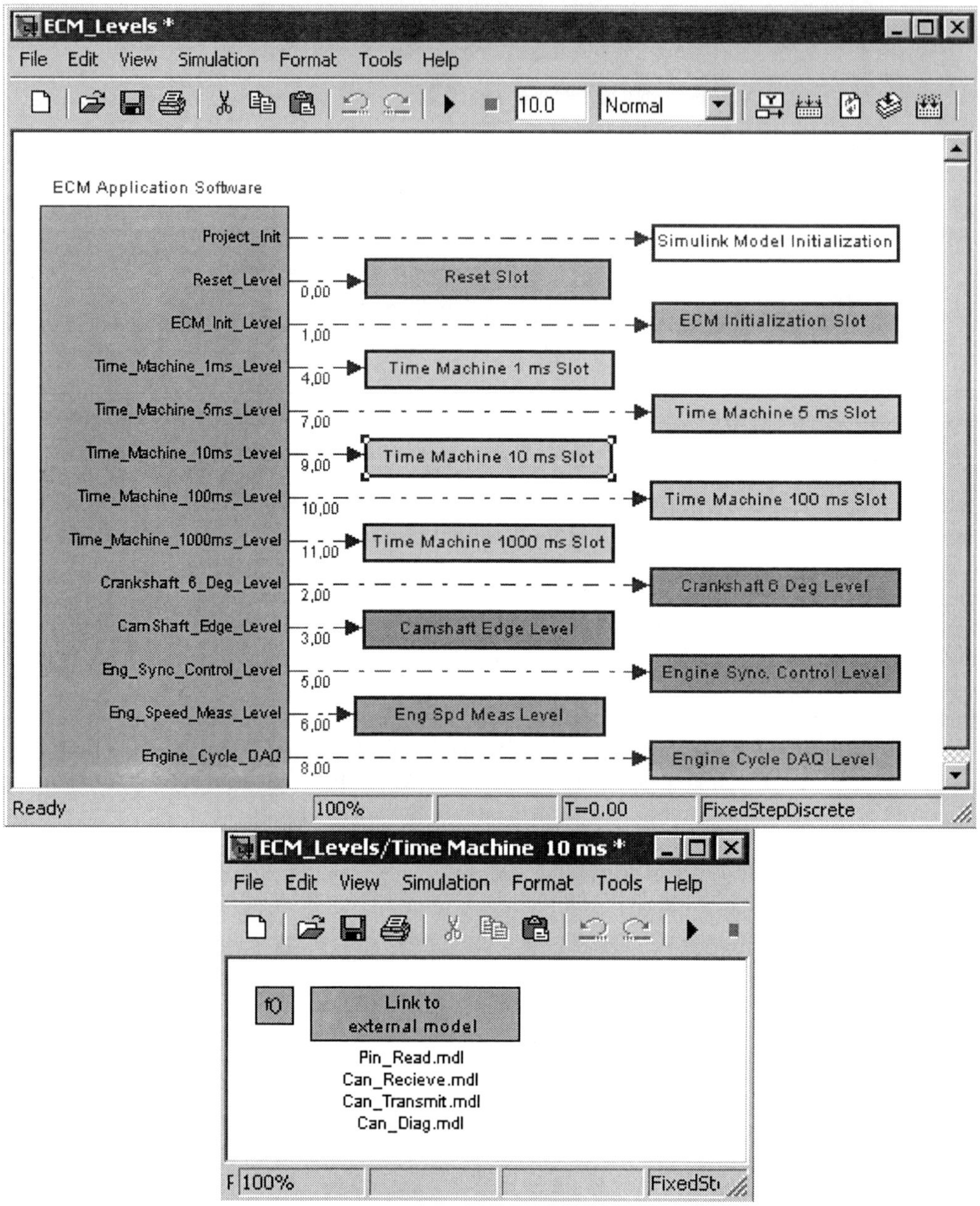

Figure 12. RTOS in MSE Pro environment.

The results of the preliminary tests of the vehicle with the installed control system electronic unit MD22 confirmed the possibility of using serial EBSUs, originally developed for the control system of the internal combustion engine as the ECU of the entire vehicle. This approach is much cheaper than ordering the development of a new electronic unit from Western manufacturers.

It should be noted that if a commercial control system electronic unit has more than 3 CAN interfaces it can be used for many different intelligent onboard security systems of the ADAS type for wide range data exchange with many onboard control system electronic units.

IOP Conf. Series: Materials Science and Engineering **534** (2019) 012012 doi:10.1088/1757-899X/534/1/012012

References

[1] SAE J1939-71 Vehicle Application Layer

[2] Buzunov N V, Kotiev G O and Miroshnichenko A V 2015 Improving the development of onboard information management systems of wheeled and tracked vehicles *Automotive Engineer Journal* **4**(93) 11-15

[3] Keller A V, Gorelov V A, Vdovin D S, et al. 2015 Mathematical model of all-terrain truck *Proceedings of the ECCOMAS Thematic Conference on Multibody Dynamics* pp 1285–1296

[4] Keller A V, Gorelov V A and Anchukov V V 2015 Modeling truck driveline dynamic loads at differential locking unit engagement *Procedia Engineering* Vol. **129** pp 280–287

[5] Buzunov N V 2016 Analysis of «real time» plane nonlinear automobile motion model with wheel configuration 6x6 *Moscow State Technical University MAMI News* **3**(29) 2-8

[6] Philippskiy A V, Shedrin I V and Artyushkov A Y 2016 Development issues of software for working with an electronic vehicle control unit *Modern trends in the development of science and technology* **3-3** 120-123

[7] Galin D A and Nazarkin A S 2016 Electronic control units programming *XLIV Ogarev reading materials of the scientific conference: 3 parts* 168-174 (responsible for the release P.V. Senin)

[8] Butarovich D O, Kositsyn B B and Kotiev G O 2017 A method for developing an energy-efficient law to control an electric bus when driving on a city route *NAMI Proceedings* **2**(269) 16–27

Simulation-based identification of the parameters of a minibus hybrid powertrain

V A Gorelov[1], D O Butarovich[2], A A Staduhin[1] and G I Skotnikov[2]

[1] Multi-purpose Tracked Vehicles and Mobile Robots Department, Bauman Moscow State Technical University, 105005 Baumanskaya 2th st., 5, Moscow, Russian Federation
[2] Wheeled Vehicles Department, Bauman Moscow State Technical University, 105005 Baumanskaya 2th st., 5, Moscow, Russian Federation

Abstract. The bus is one of the most popular types of the road public transport and is much less dependent on the external infrastructure than the tram or the trolleybus. However, buses with conventional powertrains are the active sources of the environment pollution. Buses with hybrid powertrain are known to combine the independence of the motorbuses with a conventional internal combustion engine, a considerable reduction in the environmental pollution and the capability of the braking energy recuperation. This explains the increasing interest to the problems of their construction and development of effective control laws for their units and subsystems. Nowadays, these problems are solved by means of the mathematical simulation of the vehicle operation.

1. Introduction

Currently, the replacement of the conventional buses by the electric buses is one of the most efficient ways to prevent cities from air and noise pollution produced by urban transportation. However, the extensive use of the electric buses is restricted by their disadvantages such as short drive range and a low number of charge stations. The hybrid powertrain is the middle-ground that makes possible the development of a bus with sufficient range, low noise, and high fuel efficiency.

The hybrid powertrain consists of an internal combustion engine (ICE), one or several electric machines (depending on the type of the hybrid powertrain) and energy storage device. As it is known, there are several types of hybrid powertrains: serial, parallel, and series-parallel. Their main difference is the presence or absence of a mechanical connection between the internal combustion engine and the driving wheels. In a serial hybrid powertrain there is no such a connection and the engine gives all the generated energy to the generator which charges the drive and powers the electric motor.

As a collaborative project of BMSTU and GAZ international LCC, a minibus with a series-type hybrid powertrain is being currently developed. The primary task in the development of a hybrid bus is to determine the parameters of the units and the hybrid powertrain control laws which define the traction, dynamic, braking characteristics, as well as fuel efficiency and environmental friendliness. Now, such problems are solved by methods of computer simulation of the bus operation.

2. Mathematical model description

To determine the parameters of the serial hybrid powertrain aggregates and control system a mathematical model of the bus motion has been developed and implemented in MATLAB/Simulink. The structure of a series hybrid powertrain is shown in Fig. 1. Simulation of the power units operation and the chassis dynamics is carried out using the elements of the Simscape/Driveline and Simscape/Electronics libraries (Figure 2).

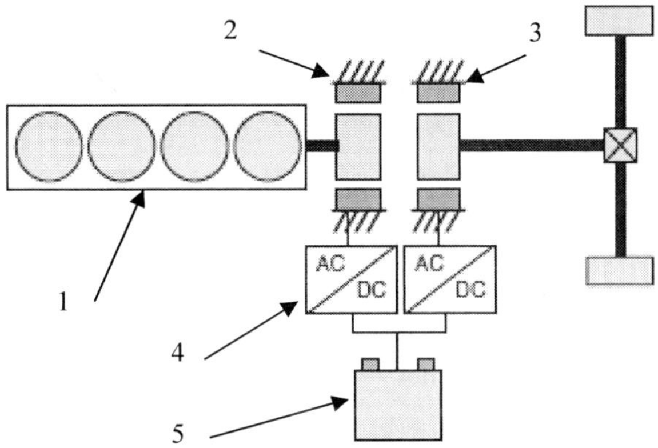

Figure 1. Structure of a series hybrid powertrain:
1 - ICE; 2 - generator, 3 - traction electric motor (TEM); 4 - power converters; 5 - energy storage device.

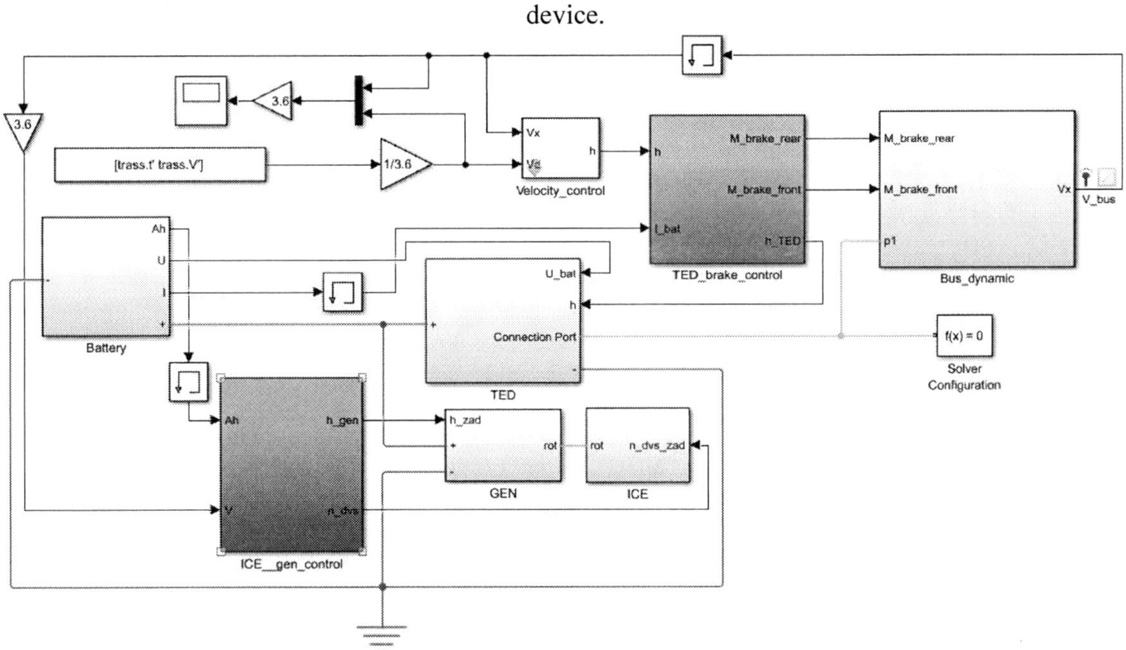

Figure 2. Structure of the computer model of the hybrid bus.

The bus dynamics was simulated only for the case of the linear movement along a rigid road without normal reactions redistribution between the sides of the bus [1] because of the little effect of the curvilinear motion on the values of interest: capacity of energy storage devices, TEM, generator and ICE characteristics. For the same reasons the work of the suspension wasn't taken into account during the creation of the model. Since the bus is operated on hard-surfaced roads, the interaction of the wheel with the road is described by the Magic Formula model with constant coefficients [2], which increases the speed of computational experiments in comparison with the more accurate contact models [3,4,5,6,7].

The work of the internal combustion engine is described by the following differential equation:

$$J_{ICE} \cdot \dot{\omega} = h \cdot M_{ICE} - M_{gen} \tag{1}$$

where J_{ICE} – moment of inertia of the internal combustion engine elements reduced to the crankshaft, $\dot{\omega}$ – angular acceleration of the ICE output shaft, M_{ICE} – torque developed on the ICE output shaft, which is determined by the given engine external speed characteristic (ESC), h – degree of ICE power use, M_{gen} – the generator torque.

The fuel consumption is determined by the interpolation table as a function of the ICE speed and power.

The generator and TEM are modeled by the *Servomotor* element from the Simscape/Electronics library. The Servomotor element initial data are the dependencies of the electric machine maximum moment on the speed, the dependencies of the power of the losses on the torque and speed and the rotor inertial characteristics. The electric machines are controlled by the mechanical torque. The electric motor model is described by the following equations:

$$\begin{cases} N_{el} = I_{cur} \cdot U_{dc} \\ N_{el} = M_{mech} \cdot \omega + N_{los} \end{cases} \tag{2}$$

where I_{cur} – current consumed by the motor, U_{dc} – DC link voltage, N_{los} – power loss, ω – angular velocity of the electric machine shaft, M_{mech} – TEM shaft torque. Power loss N_{los} on each operating mode of the electric machine is determined in accordance with its passport characteristics.

The battery was modeled by the *Battery* element of the Simscape/Electronics library. The model takes into account the following parameters: rated voltage, capacity, internal resistance. The power loss N_{los} is estimated by the following relationship:

$$N_{los} = I_{cur}^2 R_{bat} \tag{3}$$

where I_{cur} – current on the battery, R_{bat} – internal battery resistance.

The ability to simulate both the standard driving cycles and those which are obtained by recording the parameters of the bus on its route is provided by the driver's model. The bus control algorithm aim is to keep the set speed by adjusting the torque of the TEM with the help of a P-controller. Control action varies in the interval [-1; 1]. Negative values correspond to the brake pedal pressing, positive values – to the accelerator pedal pressing.

The hybrid powertrain control system model is based on the input signals: the degree of the accelerator and brake pedals pressing, bus speed, battery charge level, electric current in the battery, etc. The model determines the modes of operation of the internal combustion engine and generator, as well as the distribution of the braking torque between the TEM and the mechanical brakes.

The developed mathematical model makes it possible to estimate the values of currents and voltages in the electrical part of the hybrid powertrain, to determine the fuel consumption and energy storage device SOC (state of charge), to analyze the losses in the hybrid powertrain units, to assess the amount of the theoretically possible recovered energy during traveling along the route given as the function of time for the velocity. The mathematical model makes it possible to evaluate the operational efficiency and to configure various control algorithms for the hybrid powertrain.

The data obtained during the operation of the LiAZ - 6274 electric bus [8] in Moscow on the M2 route between Fili and Kitay-gorod metro station (Fig. 3) were used when studying the parameters of the bus motion and choosing the characteristics of the hybrid powertrain units using the developed mathematical model [11].

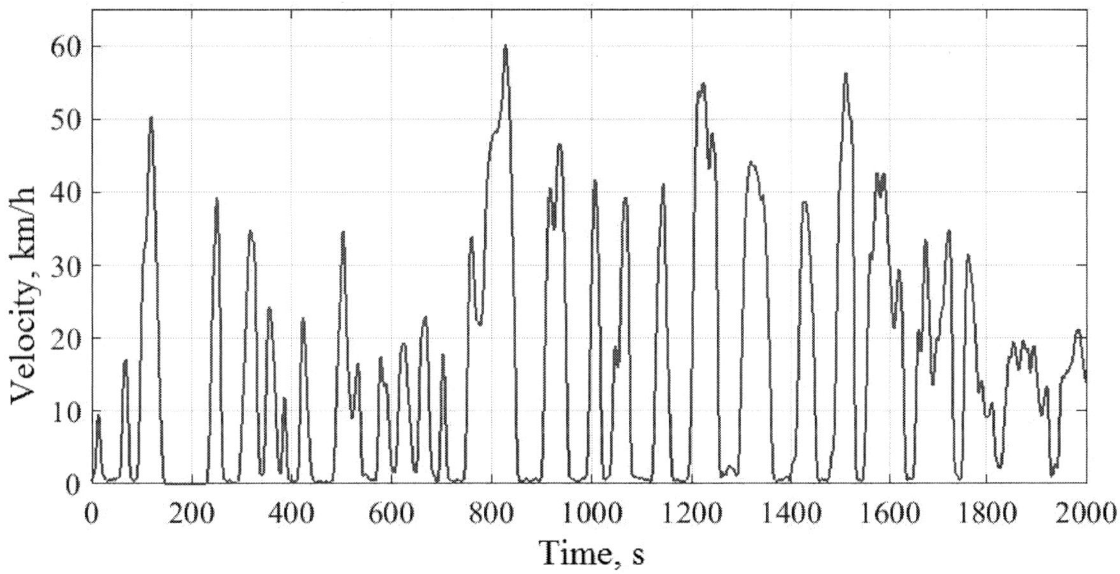

Figure 3. Fragment of the electric bus speed recording on the M2 route (Fili/Kitay-gorod metro station).

3. Parameter identification of the hybrid powertrain units

The primary evaluation of the traction motor parameters for a vehicle with a hybrid powertrain can be carried out using the calculation of the power balance of the bus. The calculation results are used as the initial data for the mathematical model. The lower estimate of the TEM required maximum power is determined by the following equation:

$$N_{TEM}^{nom} = \max(N_{TEM}^{maxV}, N_{TEM}^{max\alpha}) \quad (4)$$

where the powers N_{TEM}^{maxV} и $N_{TEM}^{max\alpha}$ are determined as

$$N_{TEM}^{maxV} = (m \cdot g \cdot f + P_w) \cdot V_{max}/(\eta_{tr} \cdot \eta_{TEM}) \quad (5)$$

$$N_{TEM}^{max\alpha} = (m \cdot g \cdot (f \cdot \cos(\alpha) + \sin(\alpha)) + P_w) \cdot V_\alpha/(\eta_{tr} \cdot \eta_{TEM}^{\alpha}) \quad (6)$$

where m – the bus gross weight, f – the road resistance coefficient, P_w – the air resistance, V_{max} – maximum speed, V_α – the steady-state speed of the up-slope motion given in the technical specification, with the road slope angle α, η_{TEM} – the TEM efficiency at maximum speed, η_{TEM}^{α} – the TEM efficiency at the speed V_α, $\eta_{\text{тр}}$ – the efficiency of the mechanical part of the transmission [9,10].

Based on the simulation results of the motion along the route, the parameters of the TEM are corrected when needed. For example, the power of the TEM can be increased in order to provide better acceleration dynamics or the use of the mechanical braking system can be reduced in order to increase the braking energy recuperation.

The power of the generator is chosen based on the requirements of ensuring its energy balance: positive or zero for a typical hybrid, negative for a plug-in hybrid [12]. The maximum power of the generator is defined as the product of the TEM power by the generator efficiency, while the bus can do without an energy storage device in all driving modes. The lower limit of the generator power for a bus without recharging is defined as the ratio of the energy expended per cycle of traffic to the cycle time.

Figure 4 shows the results of the generator power calculation for the following conditions: long-term movement at speeds of 70 and 90 km/h, maintaining zero energy balance when driving along the city route (Figure 3). The average speed at the city drive cycle is 14.8 km/h. The results were obtained with the use of the developed mathematical model.

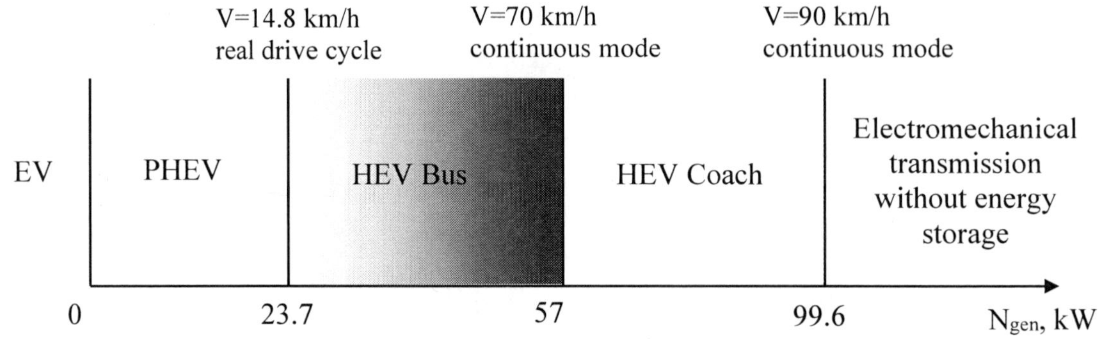

Figure 4. Results of the generator power calculation (EV – electric vehicle, HEV – hybrid electric vehicle).

The internal combustion engine is chosen according to the conditions that its nominal power corresponds to the calculated generator power. Its nominal power will be considered the ICE power at its maximum torque. For the joint operation of the ICE and the generator in optimal conditions it is necessary to ensure the consistency of their torque and rotation speed characteristics.

Table 1 contains the results of simulating the battery charging process by a generator set consisting of an internal combustion engine developing the maximum power of 81 kW (at 4000 rpm) and a generator that continuously develops a power of 50 kW. It can be seen that the internal combustion engine consumes less fuel while operating in the nominal mode than in the higher power mode to provide the same battery charge. However, the less-efficient modes allow charging the battery in less time so they can be used in the development of the hybrid powertrain control system.

Table 1. Parameters of the battery charging on a stopped bus.

ICE rotation speed, rpm	2000	2500	3000
Change of charge, Ah	10	10	10
Charging time, sec	572,9	457,6	382,3
Generator efficiency	0,9193	0,926	0,9287
Battery loss, kWh	0,1259	0,1578	0,1889
Fuel consumption, kg	1,755	1,787	2,07

The task of determining the type of the energy storage device and its capacity for a hybrid bus is multi-parametric. The modes of operation are the initial data for this task, in addition to the mass-dimensional characteristics of the vehicle being designed. As a part of the bus hybrid powertrain, the energy storage device performs the following functions:

- stores the energy produced during braking;
- provides the bus movement with zero emissions;
- provides the operation of the internal combustion engine and generator in nominal modes with high efficiency.

In accordance with the described functions, the energy storage device requirements are the following:

- ability to accommodate the energy produced during braking;
- providing the power required for motion;
- capacity should provide the required autonomous motion.

The ability to receive and release energy during the intense braking and acceleration is limited by permissible charge and discharge currents, which depend on the capacity C (Table 2 [9]) and type of the storage device. Currently, energy storage devices of two types: ultracaps and batteries with cells of various types are the most common.

Table 2. Characteristics of energy storage devices.

	NaNiCl2	LFP	LMO	LTO	NMC	UltraCap
Max Charge Rate	+1C	+1C	+1C	+6C	+2C	+20C
Max Discharge Rate	-1C	-2C	-1C	-6C	-2C	-20C
Max operating SOC	100%	90%	100%	85%	80%	100%
Min operating SOC	20%	10%	20%	15%	20%	10%
Cycle Life (number of 100%DoD* cycles)	4500	2000	3000	20000	1000	50000

*DoD - depth of discharge

To determine the capacity of the energy storage device, the bus motion is modeled along the standard route, for which the time history of the speed is known. Based on the simulation results, it is possible to determine the specific power consumption, the peak values of the charge and discharge currents, which makes it possible to choose the type of energy storage device and its parameters.

The technique of determining the energy storage device parameters, depending on the used generator power, using time history of speed without using the mathematical model of motion, is described in [10, 11]. The authors note the absence of an unambiguous approach to the choice of the vehicle drive cycle, on the basis of which the optimal characteristics of the hybrid powertrain units could be found.

It is known that the selection of the transmission gear ratio is carried out taking into account the requirements for achieving the vehicle maximum speed and the requirements for overcoming the maximum slope. Buses with a series hybrid powertrain and electric buses do not have a gear box, so the selected gear ratio is realized in the final drive or wheel-hub drive, therefore it is necessary for it to ensure TEM operation with high efficiency in the most frequent driving modes.

Let's consider as an example the final drive gear ratio identification of a hybrid bus for which the TEM has been previously selected.

The calculation of the final drive gear ratio was carried out taking into account the data obtained by the results of the measuring of the M2 route (Fili/Kitay-gorod metro station) in Moscow. The speed of the TEM shaft at each moment of time at different gear ratios was determined from the speed of the bus (the wheel radius was assumed to be constant). For each value of the gear ratio, the probability of the TEM speed being within the range of the TEM highest operational efficiency was found (Fig. 5).

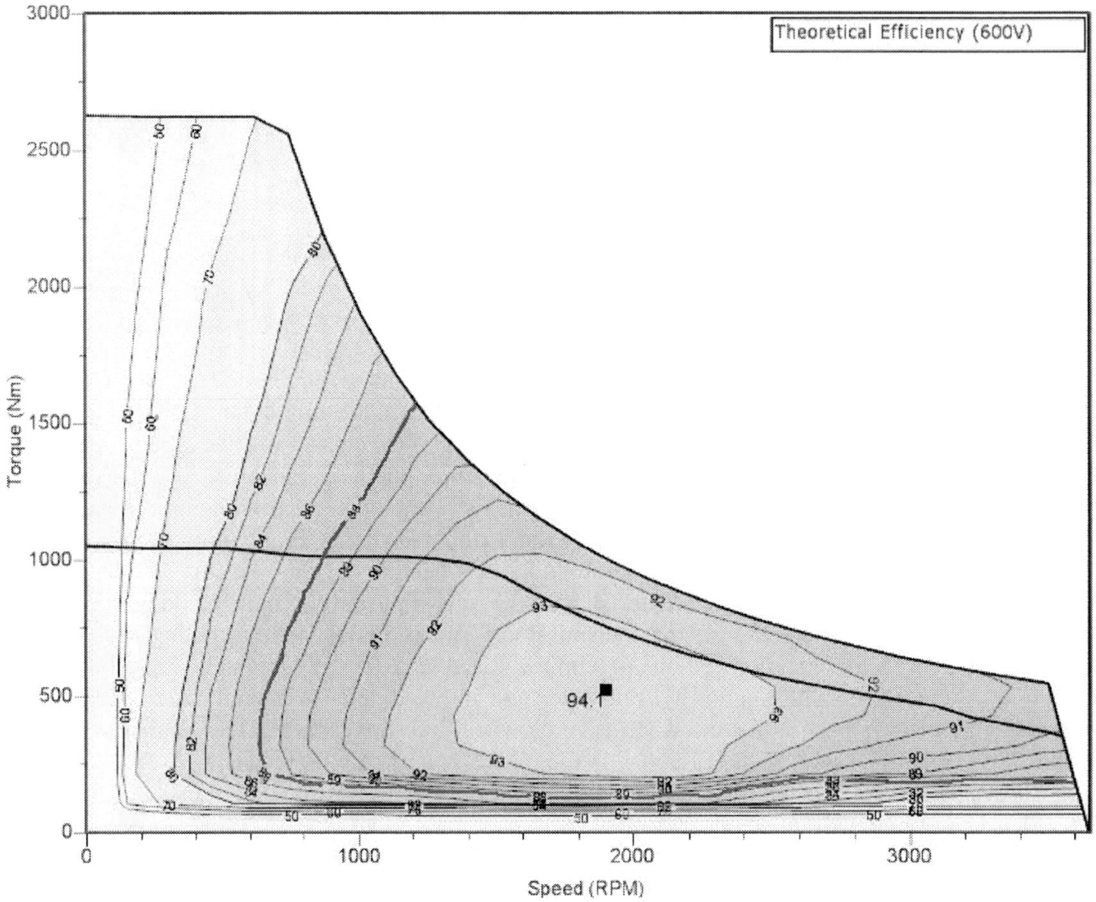

Figure 5. TEM efficiency map.

According to the obtained correlation (Figure 6, a), the final drive gear ratio should be in the interval from 11 to 13 to ensure the TEM operation with a high efficiency on the greater part of the route. However, the condition of reaching a maximum speed of 90 km/h at the maximum speed of the TEM shaft rotation of 3700 rpm does not allow making the gear ratio more than 7.5 (Figure 6, b).

IOP Conf. Series: Materials Science and Engineering **534** (2019) 012013 doi:10.1088/1757-899X/534/1/012013

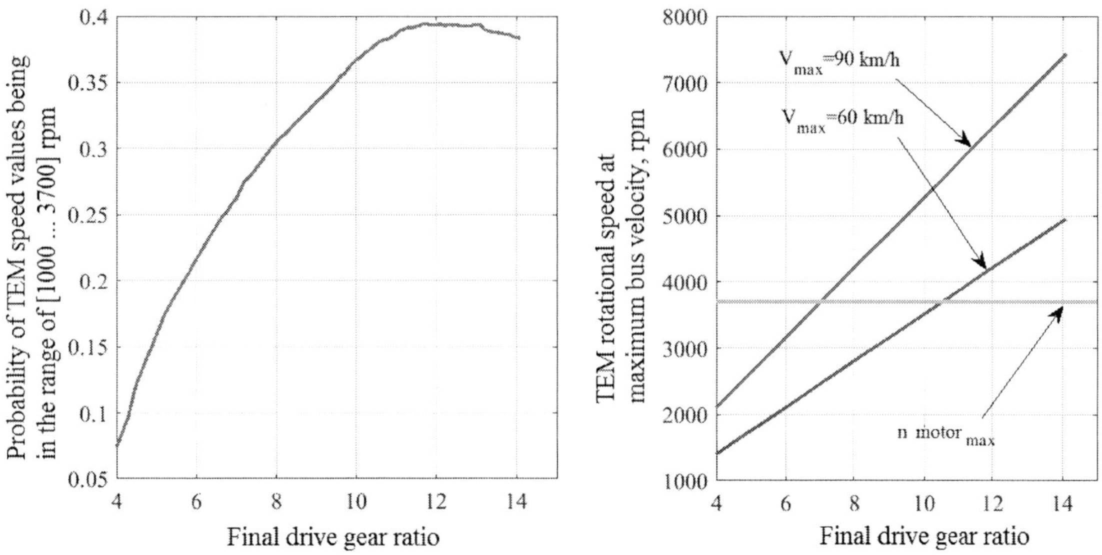

a) b)

Figure 6. (a) probability of TEM speed values being in the range of [1000 ... 3700] rpm at different values of final drive gear ratio; (b) dependence of the TEM rotational speed on the final drive gear ratio at a given speed.

4. Simulation of motion along the route

Figures 7 – 9 show the motion simulation results for a hybrid bus (with 11,500 kg gross weight) operating in the cycle shown in Figure 3. The length of the route is 23.1 km. The average speed on the route was 14.8 km/h. The coefficient of rolling resistance during the simulation was 0.018. The pie chart of the distribution of the energy spent on the route is shown in Figure 7. The average energy consumption was 1.03 kWh/km. The difference between the theoretical and actual recuperated energy equals to the energy of the TEM losses in the generator mode and the energy dissipated in the brake mechanisms and is 8 kWh, i.e. 33.9% of the energy spent on motion. The change in amperage on the battery is shown in Figure 8. The maximum charge and discharge currents are limited by the control system by 200 A, which equals to 3C. The share of the recuperation energy will decrease while using a battery with less permissible charge and discharge currents or with less capacity. That will lead to a decrease in the bus cost-effectiveness. Figure 9 shows the battery charge change.

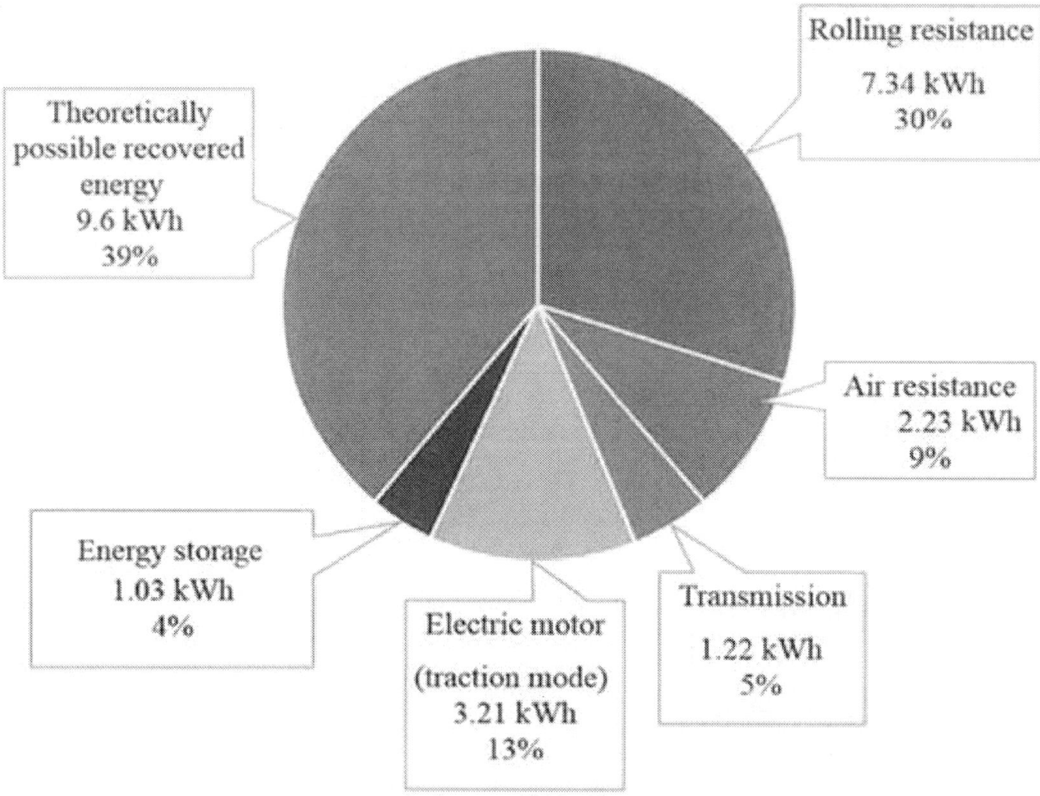

Figure 7. Distribution of the energy losses during the bus motion.

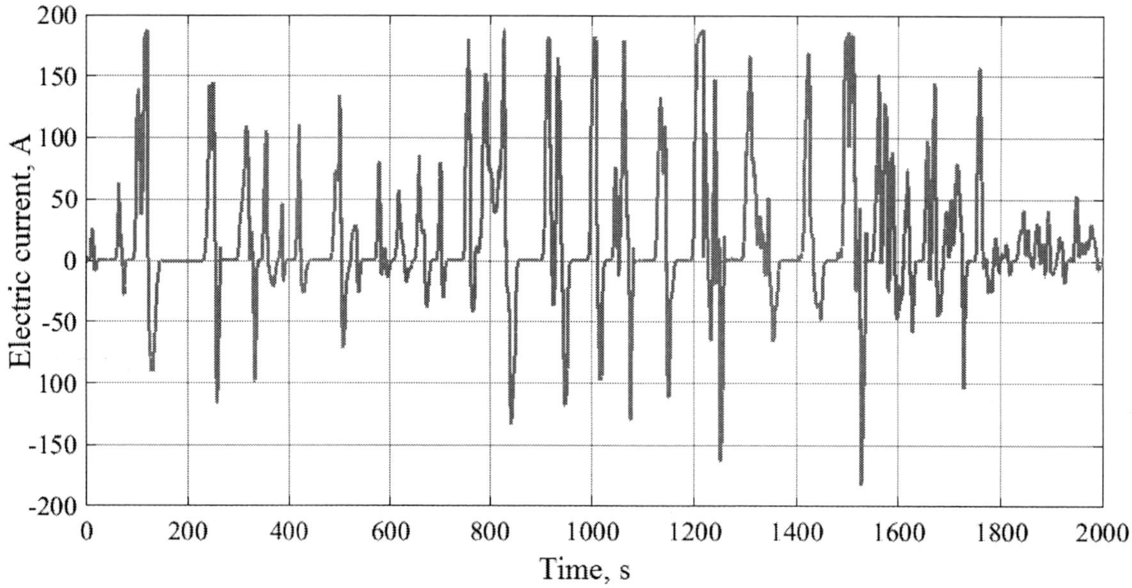

Figure 8. Electric current in the battery.

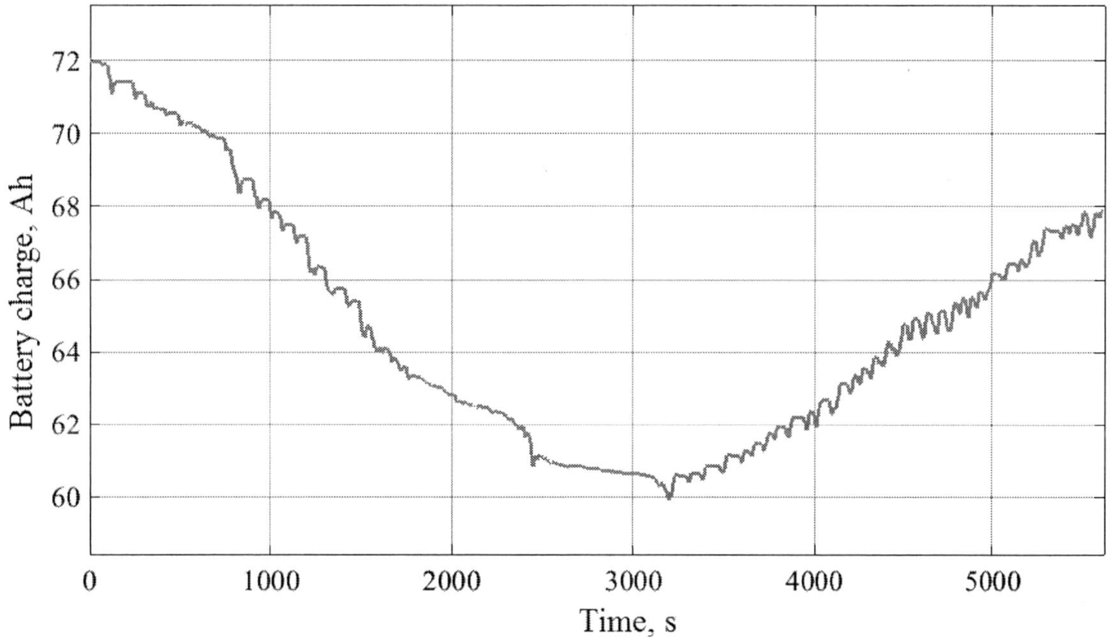

Figure 9. Change of the battery charge.

5. Conclusion

Mathematical simulation provides a comprehensive solution to the problem of the hybrid powertrain parameter identification and its control system development at the design stage. The mathematical model allows carrying out an estimation of the new bus operational properties prior to the creation of the first prototypes, which reduces development costs.

For a hybrid bus with 11,500 kg gross weight and 30 km drive range, a traction motor with 140 kW of maximum continuous power, 50 kW generator, internal combustion engine with 60 kW of rated power (90 kW of maximum power) and storage battery with the capacity of 22.6 kWh have been selected. The parameters of the units of the hybrid powertrain were obtained from the simulation of the bus operation on the existing city route and will provide acceptable traction and dynamic characteristics of the bus, high fuel efficiency and environmental performance.

Acknowledgments

The work was carried out at BMSTU with the financial support of the Ministry of Education and Science of the Russian Federation under Agreement 14.574.21.0178 (Unique identifier: RFMEFI57417X0178).

References

[1] Smirnov A A and Pikalov N A 2016 Simulation methods determination of the required capacity of electrical buses batteries *Engineering bulletin* (Moscow: BMSTU Press) **12** 43
[2] Pacejka H B 2012 Tire and Vehicle Dynamics, 3rd Edition (Butterworth-Heinemann Elsevier)
[3] Volskaia V N, Zhileykin M M and Zakharov A Y 2018 Mathematical model of rolling an elastic wheel over deformable support base *IOP Conf. Series: Materials Science and Engineering* **315**
[4] Keller A V, Gorelov V A, Vdovin D S, et al. 2015 Mathematical model of all-terrain truck *Proceedings of the ECCOMAS Thematic Conference on Multibody Dynamics* pp 1285–1296
[5] Keller A V, Gorelov V A and Anchukov V V 2015 Modeling truck driveline dynamic loads at

differential locking unit engagement *Procedia Engineering* Vol. **129** pp 280–287
[6] Gorelov V A and Komissarov A I 2016 Mathematical Model of the Straight-line Rolling Tire – Rigid Terrain Irregularities Interaction *Procedia Engineering* **150**
[7] Sarkisov P, Prokop G, Gladov G and Popov S 2015 Physical tire model based on understanding of separate components of tire deformation *The Dynamics of Vehicles on Roads and Tracks – Proc. of the 24th Symp. of the Int. Association for Vehicle System Dynamics* (Graz)
[8] Kotiev G O, Butarovich D.O and Kositsyn B B 2018 Energy efficient motion control of the electric bus on route *IOP Conf. Series: Materials Science and Engineering* **315**
[9] Keller A, Aliukov S 2015 Effectiveness of Methods of Power Distribution in Transmissions of All-Wheel-Drive Trucks *SAE Technical Paper* 2015-01-2732
[10] Keller A and Aliukov S 2015 Efficient power distribution in an all-wheel drive truck *Lecture Notes in Engineering and Computer Science* **2218** pp 1201-06
[11] Safonov B A, Eryomin G P and Butarovich D O 2017 The role of Russian Universities in the creation of innovative developments in the field of urban road electric transport *AAE Journal* (Moscow: AAE Press) **6** 51Xiaogang Wu, Han Chen and Demenkov N P 2016 Comparative study on energy consumption for plug-in hybrid electric vehicles based on trip characteristics *Int. J. Electric and Hybrid Vehicles* **8**
[12] MICROVAST 2012 LpTO Battery Technokogy Explained, http://youintel.com/index.php/en/forum/attachment/80
[13] Shtang A A and Yaroslavcev M V 2016 Determination of the main characteristics of the combined power plant for urban non-rail transport *Reports Of The Academy Of Sciences Of The Higher School Of The Russian Federation* **4**
[14] Yaroslavcev M V 2014 Determination of parameters of a power unit of a hybrid vehicle by simulation of energy consumption *Russian Electrical Engineering* **12** pp 724-728

IOP Conf. Series: Materials Science and Engineering **534** (2019) 012014 doi:10.1088/1757-899X/534/1/012014

Basic principles of vehicle suspension control

D Yurlin, S Bakhmutov and O Girutskiy

FSUE "NAMI" State Research Center of the Russian Federation

Abstract. The paper is devoted to the description of the basic operation principles of control systems over suspension (cushioning) system elements of a vehicle. The basic operation algorithms, as well as the basic principles of change of properties of suspension or cushioning components under different vehicle driving conditions, are presented. The list of sensors and signals, which shall be present in the system to ensure its effective operation, is proposed. The results of experimental analysis of the pneumatic elastic element (air bellows) stiffness and the results of computer simulation of the vehicle behavior when turning are additionally presented.

1. Current state of adjustable vehicle suspension systems

Advanced computer-aided design systems are used in designing of modern vehicles [1]. This allows ensuring comfortable driving under normal conditions, however, this does not ensure efficient operation of the suspension under the conditions difficult or unusual for the vehicle. Moreover, change of component properties during operation and their wear may cause inefficient operation of the chassis during normal driving as well. To solve these issues, the adjustable elements of the suspension (cushioning) system are used. The vast majority of modern comfortable vehicles are currently equipped with such systems, however, they all have different component base and control algorithm operation principles. Nevertheless, a number of specific elements and their control methods can be distinguished for all these systems, which will be discussed below.

The standard adjustable suspension has one or two systems:

- adjustable elastic (spring) elements (usually air suspension) [2];
- adjustable shock absorbers.

Besides, a roll (pitch) stabilization system is sometimes applied.

Regardless of which of these three systems is installed, a suspension control system consisting of the following elements is necessary for their operation:

- suspension control unit;
- acceleration and suspension position sensors;
- control interface.

Special signals are required for the suspension control system (the set of which changes depending on the system complexity):

- vehicle speed;
- steering wheel angle / turning speed;
- brake pedal pressing speed;
- accelerator pedal pressing speed;
- transmission gear engagement;
- signal from the interface (driving mode);
- position of each wheel;

IOP Conf. Series: Materials Science and Engineering **534** (2019) 012014 doi:10.1088/1757-899X/534/1/012014

- body acceleration (at the centre of gravity and above each outer corner of the vehicle);
- current rate in the control elements of the shock absorbers;
- road surface profile (from cameras or lidars)
- data on the road friction coefficient.

The information on the wheels position and body accelerations is converted to the information necessary for the system operation, such as body roll or vibration values in individual points.

2. Suspension control principles

When developing a suspension control system, it is necessary to take into account three important aspects:

1) The control system operates depending on the driving conditions.

It is the information on the vehicle driving parameters that shall determine the suspension behavior. There is no general logic able to select the component properties correctly without the information on driving speed, vibrations or environment condition.

2) The control system is designed on the basis of the suspension components.

The control logic is based on the suspension technology and the possibilities of its control which depend on the subsystems, components and quality of their interaction.

3) The control system shall be determined by parameters and target requirements.

Whatever the algorithm, it is the actual control system adjustment and parameterization that will determine the final system response (reaction) and vehicle driving behavior. Generally, the algorithms do not ensure the desired level of the vehicle parameters without adjustment. The system shall be regulated by a certain set of the coefficients, the impact of which is connected to the physical parameters of the vehicle driving.

Vehicles with a suspension control system often feature several driving modes, such as: Auto (Main), Sport or Comfort. Each of the modes has its own set of coefficients, which changes the driver's experience under certain conditions. The additional modes use the signals of pedals pressing and steering wheel turning preceding the longitudinal and lateral roll.

In the modern civil vehicles equipped with the control system, the basic setting is preferred to be based on the condition of the maximum comfort at low and medium speeds. For this purpose, the shock absorbers resistance for the low frequencies of oscillations is decreased to the minimum. As the driving speed or body oscillations increase, the shock absorbers resistance increases, which leads to a decrease in oscillations and increase in stability of the vehicle driving. The control system interference leads to an increase of the shock absorbers resistance and the average level of vibrations in the passenger compartment, therefore, the chassis basic setting is important, as it will allow decreasing the level of involvement of the control system and increasing the driving comfort. During operation, the system changes the suspension components characteristics, therefore, their properties shall be balanced not only within the basic setting, but also within the changed settings, so that the increase of suspension damping or stiffness (rigidity) does not lead to unbalance or re-distribution of operation frequency of the front and rear axles. During suspension operation, it is important to ensure smooth body motion in relation to the centre of gravity without the unbalanced sagging (subsidence) of the right and left sides. For the comfortable vehicles, a big vertical body movement is not unpleasant for the passengers, however, frequent and significant body roll causes discomfort.

3. Basic modes of suspension control

Modern suspension control algorithms are developed on the basis of the equations describing the vehicle body movement. However, the equation parameters ensuring a high level of comfort, as a rule, do not ensure the required steerability/controllability, therefore, additional modules increasing individual consumer properties of the vehicle are applied on top of the basic algorithm. Meanwhile, even without the key algorithm, a set of modules for various driving conditions can ensure a high level of comfort and steerability/controllability of the vehicle. A set of basic modules defining the behaviour of the suspension control system in various driving conditions is presented according to Table 1. A condition is imposed for these modules that adjustable shock absorbers, antiroll bars (adjustable ARB) and pneumatic elastic elements with four variants of stiffness each are installed in the vehicle (Variant 1 – the "softest" parameter, Variants 2, 3, 4 – increase in stiffness by 30, 70 and 100% respectively). The proposed recommendations on algorithms are based on the FSUE "NAMI" expertise in the development of passenger vehicles and on the results of the computer simulation, part of which is presented below.

A change in stiffness occurs due to a change in the internal volume of the pneumatic element [3]. The condition is imposed that there are 3 suspension modes in the vehicle: Main, Comfort, Sport.

Table 1. List of modules under examination.

Section	Condition	Event
4.1	Suspension mode switch	Main mode Additional mode - Comfort Additional mode - Sport
4.2	Roughness (bump/pothole) under one of the wheels	Unexpected change of the road profile under one of the wheels
4.3	Wavy (undulant) road	Great roughnesses (undulation) with low frequency of impact/influence (wave effect)
4.4	Braking	Pressing the brake pedal Keeping the brake pedal pressed
4.5	Turning movement	Beginning and ending of the turning movement Long cornering Braking/acceleration during the turning movement

The principle of changing the resistance of shock absorbers in the modern automotive industry is based mainly on changing the valves section by a forward electric motor. The change in the stiffness of antiroll bars is achieved by way of opening the central clutch and by way of installing an electric motor, which turns the stabilizer creating additional torque of resistance to the lateral roll. The change of the stiffness of a pneumatic element is achieved by way of step change of the internal volume. However, the peculiarity of pneumatic systems operation consists in the inertness of the operation environment being used; therefore, a quick change of the internal volume seems hard to attain. Therefore, a system of opening several air volumes is applied [4].

In order to check the feasibility of considerable change in the stiffness of a pneumatic element by way of reducing its internal volume, an experiment was carried out where its stiffness was measured with various internal volumes. The layout of the experimental unit is shown in Figure 1.

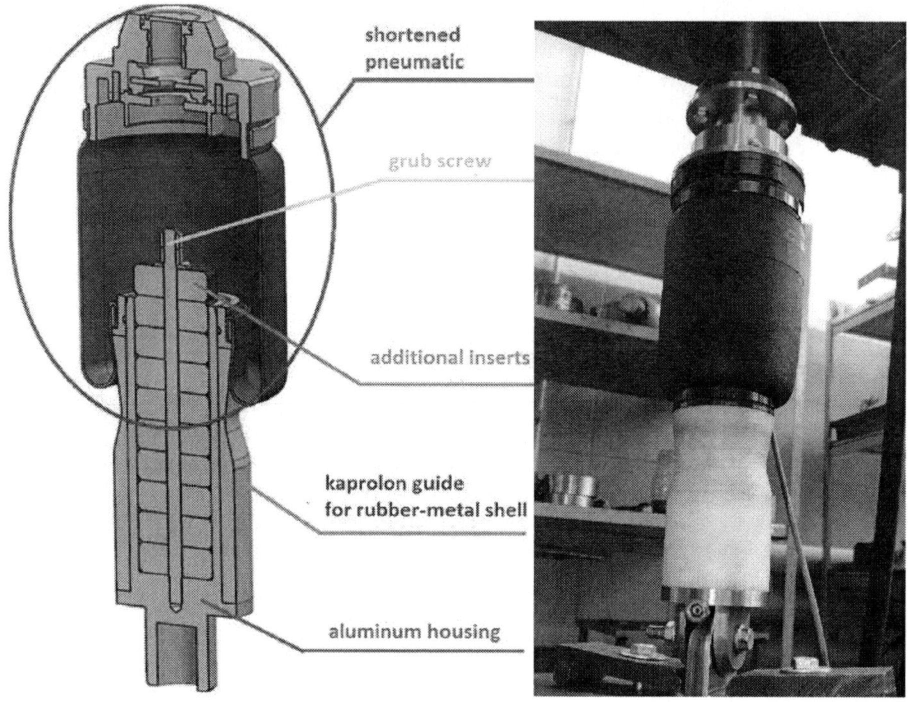

Figure 1. Experimental unit layout.

The test design included special washers the removal of which leads to the increase of the internal volume and reduction of the stiffness in further testing.

The tests were performed with various volumes but the same frequency – 1 Hz.

The resulting force curves for 2.9 l (dash line) and 3.7 l (solid line) are shown in Figure 2.

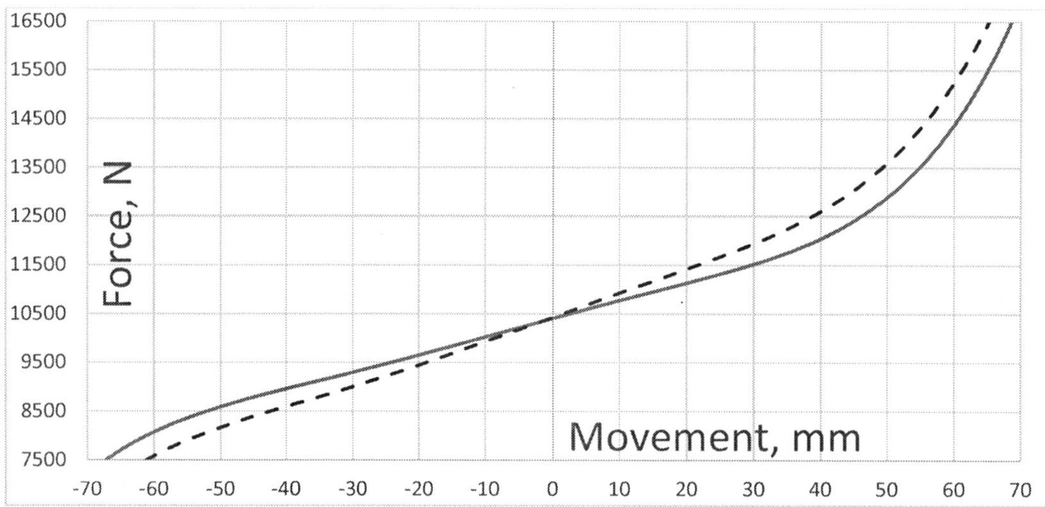

Figure 2. Characteristics of force developed by pneumatic element.

In Figure 3, the resulting dependence of the stiffness of the pneumatic element on the internal volume is shown.

IOP Conf. Series: Materials Science and Engineering **534** (2019) 012014 doi:10.1088/1757-899X/534/1/012014

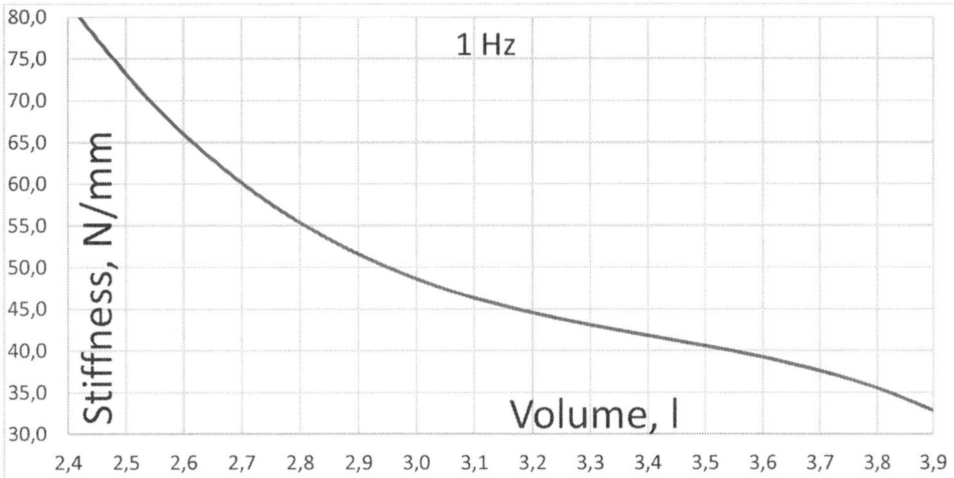

Figure 3. Characteristics of stiffness of pneumatic element in case of internal volume change.

One can see from the shown graph that if a pneumatic element is of the standard 3 l volume, then its volume change by +/- 0.5 l will result in the stiffness change by -17/+52% respectively. Thus, the internal volume change by 0.5 l, which is technically entirely feasible, will lead to the stiffness change by considerable 35%.

Thus, a conclusion may be drawn that the dynamic regulation of stiffness of an elastic element by way of changing the internal volume of the pneumatic element is feasible.

3.1. Suspension operation mode switch

Table 3.1 presents the description for these modules based on the design of the vehicle intended for comfortable city driving. Upon that, comfort is "sacrificed" to stability at high speeds. During suspension development, the most comfortable suspension characteristic is often set for the main mode, then the Sport mode is created deliberately transferring more vibrations to the passengers to create the feeling of dynamic driving. The Comfort mode is created in a similar way allowing greater body pitches than in the main mode, which creates the feeling of smoother driving, which is considered as more comfortable one by nonprofessional drivers.

Table 2. Suspension mode switch module.

System	Wheel under examination				Additional suspension mode	
	Front axle		Rear axle			
	Left (FL)	Right (FR)	Left (RL)	Right (RR)	Sport	Comfort
Suspension height	The suspension height decreases at high speeds				Height decreased	as at the main mode
Elastic elements stiffness	FL Variant 1 or 2 is equally prespecified	FR	RL	RR	Variant 2 or 3 is prespecified	Variant 1 is prespecified, variant 4 is prohibited
Adjustable ARB	No specific operation is required					
Dampers resistance	FL <30% Shock absorbers resistance is kept at the low level except for the cases of significant body pitch or high driving speed	FR <30%	RL <30%	RR <30%	>60% Limitation of the minimum damping is applied	Max < 80% Shock absorbers resistance never reaches the maximum values

Further, the suspension behavior in different driving modes is nearly identical for all modules: there are basic settings for the Auto mode, the decreased prespecified stiffness of the elastic elements and shock absorbers resistance are set for the Comfort mode, the one step "stiffer" characteristics are used for the Sport mode. A softer characteristic of the components stiffness rise is used for the Comfort mode and, on the contrary, more spur increase in stiffness and suspension elements resistance along with the increase of external impact parameters for the Sport mode.

3.2. Roughness (irregularity) under one wheel

Single roughnesses are supposed to cause high-frequency oscillations with a small Z-axis travel. No direct response from the elastic elements of the air suspension, shock absorbers or adjustable ARB is required. Generally, the passive suspension part is adjusted so that these vibrations are transferred to the body minimally [5]. The exceptions can show in the Comfort mode, when the decreased shock absorbers resistance is applied. To solve this problem, the vehicle should be equipped with the wheel position sensors, which would register these oscillations and allow a timely and short-term increase in damping.

3.3. Wavy (undulant) road

The wavy road causes the low-frequency body oscillations: vertical motion, lateral and longitudinal roll.

Table 3. Wavy road module.

Suspension height: Change of height is not required.
Elastic elements stiffness: All wheels are variants 1-4. The stiffness of each element is controlled separately allowing using each of 4 variants. The system stabilizes the body motion (vertical, longitudinal / lateral roll) depending on the body accelerations and wheel movements. The stiffness is added to all 4 wheels equally against the vertical motion. By one or more steps, depending on the oscillations value. Against the longitudinal roll, stabilization is determined by the motion direction and phase. In case of the front axle running into a positive wave, the rear axle stiffness increases at the body rising stage, but both axles reduce the stiffness at the wave decrease. When the body starts to go down, the front axle stiffness increases. Against the lateral roll, the stiffness decreases for the rising side and increases for the sagging one. The phases of the front and rear axles are usually considered separately.
Adjustable ARB: No specific operation is required.
Shock absorbers resistance: all shock absorbers – 0-100%. In case of big lateral waves (equal under the right and left wheel), the system shall adjust the front and rear axle separately, but the approach shall be individual basing on the current vibrations level and the body pitch value. For comfortable vehicles, high amplitudes are allowed, but the generation of the bump (compression) or rebound stroke causes impacts all over the body, which is unacceptable. Therefore, when registering body acceleration and roll angle, which may lead to the suspension breakage, the shock absorbers resistance shall increase.

3.4. Braking module

There is a difference between the braking system activation time and the passive holding of the pressed pedal at slow deceleration.

Table 4. Braking pedal activation module.

Suspension height: Change of height is not required.
Elastic elements stiffness: All wheels – variants 1-4; During braking, a high stiffness of the front suspension is preferable depending on the longitudinal acceleration. The rear axle stiffness shall also be increased (for the same number of steps or one step less), otherwise there may be an issue with the braking forces distribution between the axles and with the rear axle stability. The system responds to the pedal pressing better than to the registration of the longitudinal dive, since the delay is reduced.
Adjustable ARB: No specific operation is required.
Shock absorbers resistance: all shock absorbers – 0-100%. When activating the pedal, damping of all 4 shock absorbers increases sharply before the longitudinal acceleration or body motion occurs. Upon that, the damping value may depend on the speed of pedal pressing. After the pedal pressing, the shock absorbers also become stiffer to decrease the body reverse motion. After the start of deceleration, the force on the shock absorbers decreases to avoid the body vibrations. Change of forces on the shock absorbers when braking shall be coordinated with the ABS and ESP specialists, otherwise it may affect safety.

Table 5. Braking pedal holding module.

Suspension height: Change of height is not required.
Elastic elements stiffness: All wheels – variants 1-3. After the pedal activation, the stiffness shall decrease slowly for all 4 wheels until the pedal is pressed / pressing continues again or until the longitudinal acceleration decreases below the set value.
Adjustable ARB: No specific operation is required.
Shock absorbers resistance: all shock absorbers – 0-50%. The shock absorbers resistance shall be increased until the braking pedal is released. The connection with the vehicle speed is possible.

The system behavior is similar to acceleration, but adjusted for the roll direction.

3.5. Turning movement module

The turning movement module has distinctions for the Beginning of Turning Movement and Long Cornering modes and nuances for the Braking/Acceleration during Turning Movement mode.

Table 6. Beginning of Turning Movement module.

Suspension height: Change of height is prohibited and not required.
Elastic elements stiffness: All wheels – variants 1-2. The stiffness level for the elastic elements depends on the steering wheel turning speed for lateral roll neutralization. At the steering wheel turning speed equal to zero, the stiffness of the elastic elements decreases. Upon that, it is not recommended to increase the stiffness by more than two steps for the comfortable mode.
Adjustable ARB: Changes from 0 to 100%. The antiroll bars stiffness may increase up to 100% to decrease the roll. Upon that, the front and rear axles are controlled separately to keep the balance of roll stiffness of the suspensions and slip angles.
Shock absorbers resistance: all shock absorbers – 0-100%. The stiffness of the elastic elements changes similarly, depending on the steering wheel turning speed as well as on the lateral acceleration. Upon that, in case the steering wheel speed returns to zero, they are kept in the increased state to maintain the possibility of prompt response to the lateral accelerations.

IOP Conf. Series: Materials Science and Engineering **534** (2019) 012014 doi:10.1088/1757-899X/534/1/012014

Table 7. Long cornering module.

Suspension height: Change of height is prohibited and not required.
Elastic elements stiffness: All wheels – variants 1-2. At the steering wheel turning speed equal to zero, yet with the angular deviation from the neutral point (position), the stiffness of the elastic elements remains slightly increased, but lower than at the beginning / ending of the turning movement.
Adjustable ARB: Changes from 0 to 50%. The antiroll bars stiffness increases, yet it is decreased in relation to the beginning of the turning movement.
Shock absorbers resistance: all shock absorbers – 0-100%. The shock absorbers resistance changes in accordance with the change of stiffness of the elastic elements and antiroll bars. Upon that, the shock absorbers characteristic does not play an important role during the long cornering, as the force in the shock absorber depends on the piston speed, which is rather low at the rolls characteristic of the long cornering.

When describing the algorithms of the suspension behavior at acceleration and braking during the turning movement, the main requirement is the safety of the vehicle behavior and, therefore, the correct operation of ABS and ESP. Therefore, keeping the suspension state established for the long cornering in the simple algorithms is preferable.

The graphs from Tables 3.5.3 and 3.5.4 can be used for analysis of the influence of springs, shock absorbers and antiroll bars characteristics on the vehicle behavior at the beginning of the turning movement and during the long cornering. The vehicle response during the lane change, which is similar to the beginning of the turning movement, is shown in Table 3.5.3. The vehicle behavior when moving at a constant radius, which is similar to the long cornering, is shown in Table 3.5.4. Each table shows the execution of these maneuvers with individual changes of the springs, shock absorbers and antiroll bars characteristics (with the stiffness increase by 30% and reduction by 30% relative to the basic value).

Table 8. Vehicle behavior during lane change.

Roll angle change during lane change		
Elastic element stiffness change	Shock absorbers resistance change	Antiroll bars stiffness change
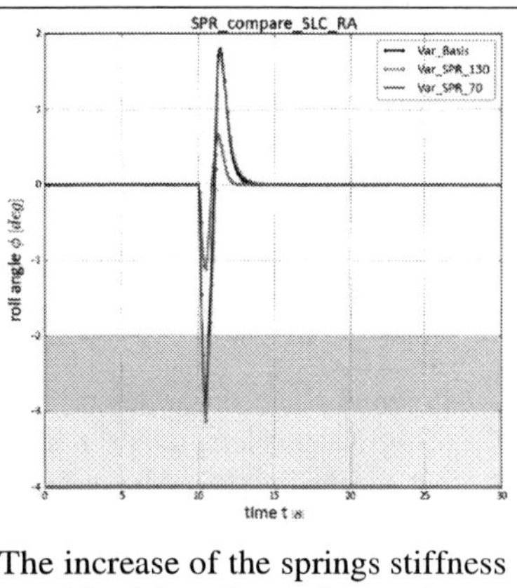	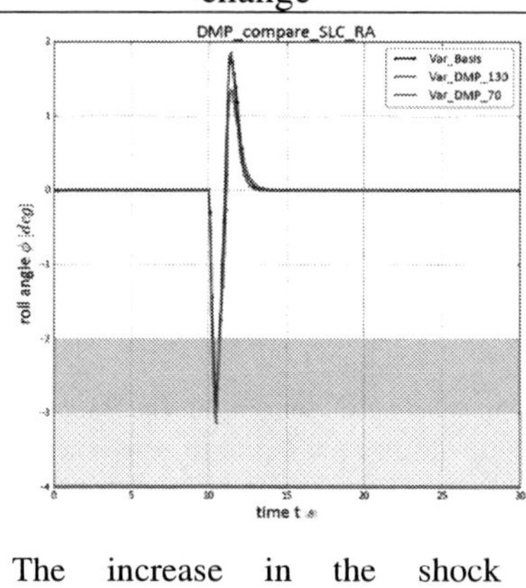	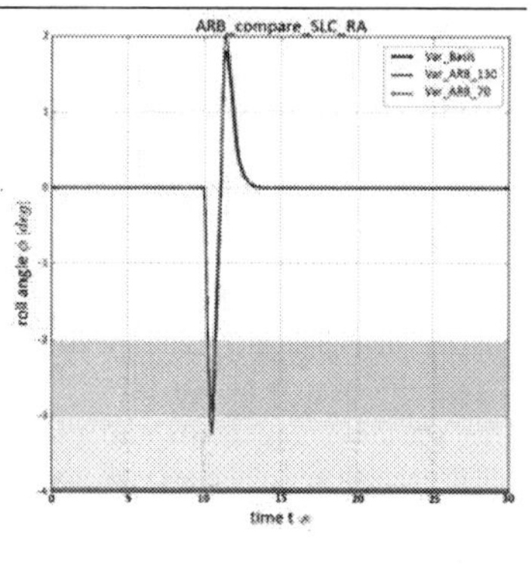
The increase of the springs stiffness considerably reduces the maximum body roll. The stiffness reduction does not lead to a considerable effect due to the antiroll bars functioning.	The increase in the shock absorbers' resistance considerably reduces the maximum body roll (increase by 30% reduced the roll by 20%). The resistance reduction does not lead to a considerable effect, probably due to the antiroll bars functioning.	Change in the antiroll bars' stiffness by 30% does not lead to a significant effect. There is degradation by approximately 9% when the stiffness is reduced by 30%.

+30%	Stability is higher (roll is less)	+30%	Stability is higher (roll is less)	+30%	No significant effect
Basic	meets the requirements	Basic	meets the requirements	Basic	meets the requirements
-30%	No effect (due to antiroll bars)	-30%	No effect (due to antiroll bars)	-30%	Minor effect

These graphs show that in order to increase the stability during sharp turning movements (lane changes), it is necessary to increase the stiffness of elastic elements and shock absorbers.

Table 9. Vehicle behavior when moving at constant radius.

Dependence of the lateral roll angle on transverse acceleration, with change of components' properties

Elastic element change	Shock absorbers change	Antiroll bars change

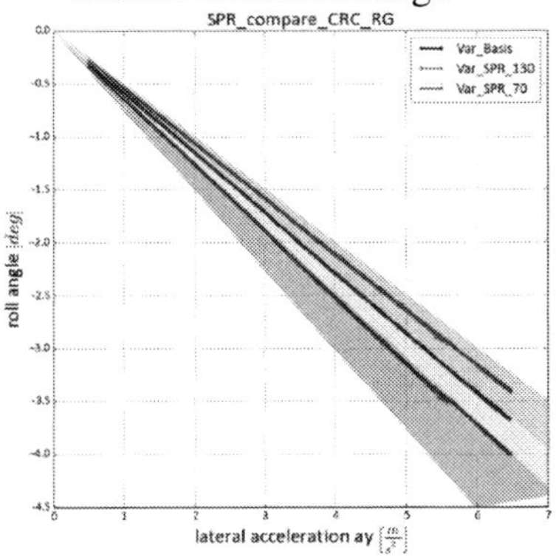

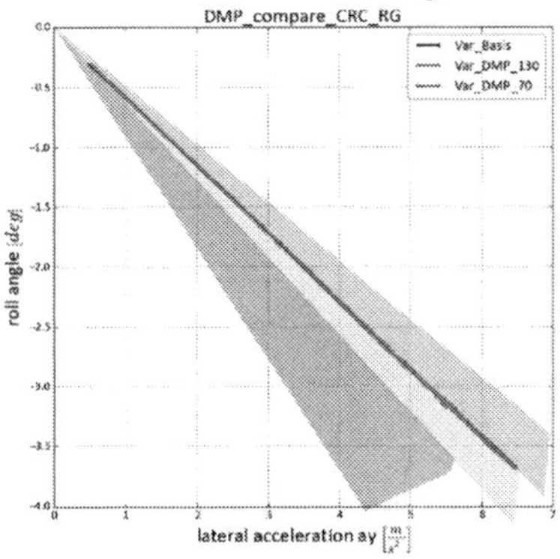

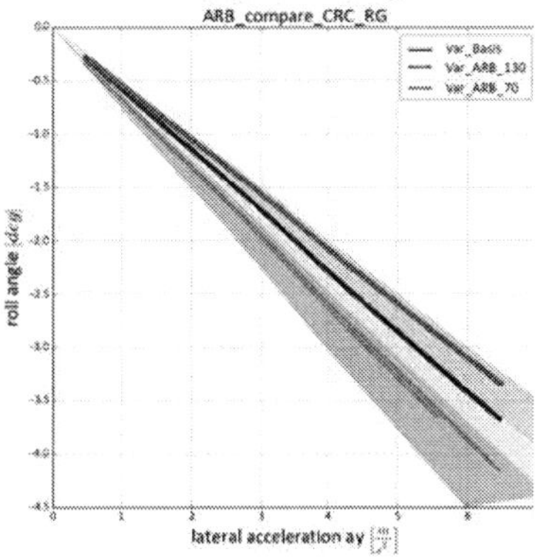

Considerable effect, increase of springs stiffness reduces vehicle roll	The effect is insignificant, change of shock absorbers' properties has almost no effect on lateral roll during long cornering	Considerable effect, increase of antiroll bar stiffness reduces vehicle roll

+30%	Stability is higher (roll is lower)	+30%	No effect	+30%	Stability is higher (roll is lower)
Basic	At the limit of normal	Basic	At the limit of normal	Basic	At the limit of normal
-30%	Stability is lower (roll is higher)	-30%	No effect	-30%	Stability is lower (roll is higher)

These diagrams show that in order to increase vehicle stability during the long cornering, it is necessary to increase the stiffness of the antiroll bar or elastic elements.

4. Summary

The paper presents a description of the main recommendations for control of the adjustable vehicle suspension. The presented materials show the feasibility of the proposed solutions, in particular basing on the computer simulation results. The implementation of the proposed approaches can significantly increase consumer properties of a vehicle without considerable expenses for algorithms development, calibration and setup. At the same time, it is necessary to take into account three important theses:

1 *The control system operates depending on the driving conditions.*
2 *The control system is designed on the basis of the suspension components.*
3 *The control system shall be determined by parameters and target requirements.*

This means that implementation of any logic of suspension control, including the one presented in this paper, does not exclude the necessity of suspension adjustment, its adaptation to the certain vehicle and check of vehicle safety and stability.

The list of necessary sensors and equipment for the adjustable suspension operation is presented in the paper separately, but whether each of the components should be installed on-board shall be checked in terms of economic and functional feasibility of a certain vehicle.

5. References

[1] Bakhmutov S V and Yurlin D V 2017 Simulation of vehicles' active suspension systems by means of a complex model with an external description of control systems *NAMI works* **2**(269), pp 6–15 (In Russian)

[2] Bazurov A A and Monahov V P 2016 Different types of suspensions of cars of various classes *Automotive Engineers Association Journal* **3**(98) pp 18–20 (In Russian)

[3] Torsten Nitschke and Andreas Nessel 2010 The closed air suspension system of the Porsche Panamera (Stuttgart : Vehicle Dynamics Expo 22.06.10) p 27

[4] Pevzner Ya M and Gorelik A M 1963 Pneumatic and hydro-pneumatic suspensions Ed. Gelfgat D B (Moscow: Leningradskaya printing house of Gosgortechizdat Publ) p 321 (In Russian)

[5] Afanasiev B A, Zheglov L F, Zuzov V N et al. 2008 *Design of all-wheel-drive wheeled vehicles* **2** Ed. Polungyan A A (Moscow: Bauman Moscow State University) Publ., p 528, ill. pp 26, 28. (In Russian)

IOP Conf. Series: Materials Science and Engineering **534** (2019) 012015 doi:10.1088/1757-899X/534/1/012015

Logical artificial intelligence Mivar technologies for autonomous road vehicles

O O Varlamov[1,2], D A Chuvikov[2,3], D V Aladin[1,2], L E Adamova[2] and V G Osipov[2]

[1] Bauman Moscow State Technical University, 2-ya Baumanskaya Street 5/1, Moscow 105005, Russian Federation
[2] MIVAR, 72 Oktyabrskaya str., Moscow 127521, Russian Federation
[3] Department of Automated control systems, Moscow Automobile and Road Construction State Technical University (MADI), Leningradskiy Pr. 64, Moscow 125319, Russian Federation

E-mail: ovar@yandex.ru

Abstract. The main field of our research is artificial intelligence already implemented in the experimental autonomous road vehicle. Artificial intelligence now is a really huge area of studies. This investigation is focused more on a development of intelligent systems. Such as expert systems, semantic text processing systems, semantic image understanding systems, robotics systems. Wi!Mi is a solution for developing mivar knowledge models. This tool belongs to the class of so-called ES shells – programs that allow significantly simplify and accelerate the ES development process. It uses a novel knowledge representation mechanism based on bipartite graphs and the significantly improved but simple inference engine, which is able to process large systems involving millions of variables in just a few seconds. A special representation model which is called mivar nets with linear computational complexity using if-then rules is used in Wi!Mi. The idea is that the subject domain is divided into objects and connections between them. Scientific investigation allows to implement new decision-making system called "RoboRazum" as a form of an embedded software platform with flexible capabilities to adapt to control by any robotic suites and systems. RoboRazum is a platform for designing autonomous robotic control systems. Provided solutions don't require powerful computational resources and could be used in small systems-on-chip like microcontrollers. That is why this solution is perfectly fit with conception of Internet of Things (IoT). Reasoned and fast logical decisions based on mivar technologies play important role for autonomous road vehicles.

1. Research areas of artificial intelligence

The main field of our research is artificial intelligence. Artificial intelligence (AI) now is a really huge area of studies and lot of research is going on there [1]. At present time research is being done more at a development of intelligent systems [1]. Such as expert systems [1-2], semantic text processing systems [3], semantic image understanding systems [3], robotics systems [4–10].

So our main task is to develop four basic directions.

- First direction is expert systems, systems intended to help experts to make decisions – how to make the developing process cheaper and faster, in other words ES shell [2].

- The second one is the semantic natural language processing – development of semantic platform for solving tasks of semantic text analysis and natural Russian language understanding [3].
- The third one is an Intelligent image processing (smart image recognition with the text description of the image as a result) [3].
- And the fourth one is robotic control systems. The question is how to force robotic systems to behave intelligently [4–12].

All these directions can be used for autonomous road vehicles.

2. About mivar technology

When you build intelligent systems, one of the main questions is how to represent knowledge about subject domain we are working on in a "good" way. What do it means? The question here is how to represent knowledge in a form, which will be understandable for a computer and for the intelligent system developer as well. Usually in scientific literature they called this feature "expressiveness of the model". But this is not the only question. We have to work with knowledge. That is why this knowledge representation mechanism should also be effective. It means that we need to be capable to effectively process this knowledge.

Universal modelling system. Knowledge representation (KR) is one of the fundamental problems of AI, which occupies developers to this day. Currently, there are many approaches to knowledge representation including predicate description, semantic networks, production rules, neural networks, evolutionary, agent-oriented, stochastic and many others. In our products we use our own knowledge representation technology.

Principles of mivar-based approach. The idea is that the whole world can be represented in a form of triples (object-property-relation). This concept in different modification served us as a foundation of mivar knowledge representation models in our products. Moreover this idea made it possible to develop Wi!Mi, Tel!Mi and Robo!Razum and to construct out own inference engine.

Wi!Mi is an expert system shell. This shell gave a possibility to develop expert systems in totally different subject domain while significantly reducing time and costs. The Wi!Mi shell uses its own mivar model of knowledge representation developed by the authors. The basis of mivar approach to knowledge representation is adaptive, dynamic description of the modeled domain.

The idea is that the subject domain is divided into objects and connections between them. And all this entities form the bipartite directed graph – mivar net. Mivar logical inference with linear computational complexity using if-then rules. On the base of this representation it became possible to construct the inference engine which works really fast. If we have some task for the system, this engine builds a solution algorithm how to solve this task. After you receive the algorithm you can run it and receive desired results. And this algorithm is constructed in just few seconds in really huge subject domains. On the base of the idea described above, a representation model giving a possibility to create an inference engine has been constructed. This preinstalled inference engine will be used in all systems created with Wi!Mi. The developer only has to fill the knowledge base without paying attention to inference engine.

Mivar knowledge *representation model.* The Wi!Mi shell uses its own MIVAR model of knowledge representation developed by the authors [12]. The basis of MIVAR approach to knowledge representation is adaptive, dynamic description of the modeled domain. The key concept in MIVAR approach to knowledge representation is the notion of MIVAR net. This net provides formalization and representation of human knowledge when creating expert systems using the Wi!Mi shell.

The mivar net is a way of representing the information part of MIVAR space (objects and rules for processing them) as a bipartite directed graph, consisting of objects P and rules R (Fig. 1). Together, these objects and rules form the domain model.

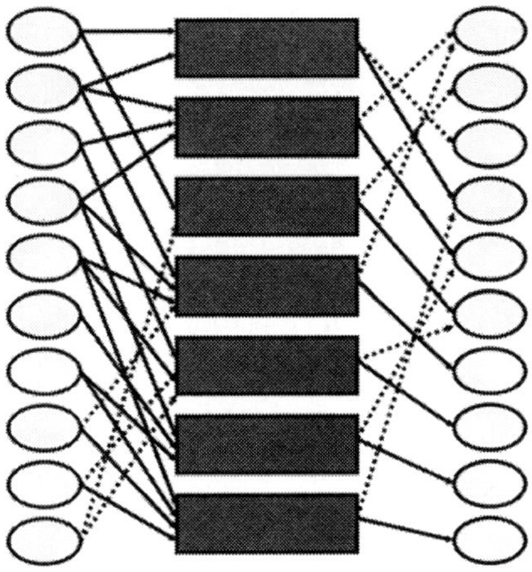

Figure 1. Bipartite graph of MIVAR net. Here objects are represented by yellow ovals and rules are represented by blue rectangles.

The mivar knowledge representation, described above gave us a possibility to construct an algorithm (method) that allows to build the solution scheme of different problems which occurred in considered subject domains. Using the above representation of domain knowledge, it is possible to construct an algorithm that allows searching for information within the mivar net.
The mivar net data model allows determining the open and hidden relationships between objects within the network, and building algorithms for inference path calculation for any given task within relevant subject area.

3. Robot Intellectualization

Integration of service robots is hindered by many factors: poor reliability, high cost of mechanisms, small battery capacity, high cost of sensors and low level of intellectualization. Due to production development, engineers have managed to solve problems associated with components of robotic systems, however the problem of service robot intellectualization is still urgent. Service civil robots should be smoothly integrated into human environment: thus, the robot should interact with infrastructure of the modern world and have capability to communicate with a human being. Therefore, the service robot should be controlled by the on-board autonomous artificial intelligence system. We propose our own approach to robot intellectualization. Implemented decision support system is called "Roborazum". It's a form of an embedded software platform with flexible capabilities to adapt to control by any robotic suites and systems. Provided solutions don't require powerful computational resources and could be used in small systems-on-chip (SoC) like microcontrollers and etc. That is why this solution is perfectly fit with conception of Internet of things (IoT).

4. An example of intellectualization platform integration

We consider autonomous road vehicles and intellectual transport environment elements as a particular case of robots. Robotic suite control system as an example of intellectualization platform integration collects information from sensors and sends it to the decision-making system. System processes the information and finds possible solution transmitted then to the robotic suite control system, and the last one sends control signals to engines. The task could be set in natural language, then using the software product Tel!Mi, system understands, what to do. After that system runs a search of possible solution in knowledge database. Also system gets information from sensors. This knowledge fills the solution models. After all information is processed, system generates control commands and sends them through hardware platform interfaces to the robot. Some projects have already been implemented using our robotics software intellectualization platform.

Expert systems are used in different models taking into account the following: road signs, road users, pedestrians, road surface marking. For example, the first model of autonomous car passing through an intersection contains 338 parameters and 320 rules. The second model of robot motion in the streets with heavy traffic consists of 240 parameters and 250 rules. As a result of model processing, the car obtains action algorithm for carrying out further actions.

5. Summary

The methodology of creating the control systems of autonomous road vehicles, integrated in intelligent transportation environment has been developed. The experimental autonomous road vehicle representing a platform with open protocol for development and testing of autonomous road vehicles technologies is described in the paper [6–7]. The problem of tactical level is to provide fast and correct decision-making with further logging of each reasoned step. In this case mivar technology implementation allows to find the way of solving the request and then applies logical solution. So all the responses will be logically approved and the response rate will be much higher than in standard systems. Reasoned and fast logical decisions play important role for autonomous road vehicles.

6. References

[1] Varlamov O O 2002 Evolutionary databases and knowledge for adaptive synthesis of intelligent systems. The Mivar information space *Radio & Communication* 286

[2] Varlamov O, Antonov P, Cibirova M, Khadiev A and Sergushin G 2015 MIVAR: machine-implemented method for automated construction of logical inference path in the knowledge base *Radio industry* **3** 28–43

[3] Varlamov O O 2017 The Mivar approach as a basis for a qualitative transition to a new level in the field of artificial intelligence *Radio industry* **4** 13–25

[4] Varlamov O O 2018 About metrics of self-sufficiency and intelligence of robotic complexes and cyberphysical systems *Radio industry* **1** 74–86

[5] Varlamov O O 2015 Prospects of developing mivar management systems for autonomous intelligent robots. *Radio industry* **3** 210–225

[6] Shadrin S S and Ivanov A M 2016 Technical aspects of external devices into vehicles networks integration *International journal of applied engineering research* **11** 7003–7006

[7] Shadrin S S, Varlamov O O and Ivanov A M 2017 Experimental autonomous road vehicle with logical artificial intelligence *Journal of advanced transportation* ID 2492765

[8] Chuvikov D A 2017 *Models and algorithms of reconstruction and examination of emergency events of road accidents* Abstract of PhD Thesis Moscow 24 p

[9] Chuvikov D A 2017 *Models and algorithms of reconstruction and examination of emergency events of road accidents* PhD Thesis Moscow 318 p

[10] Chuvikov D A 2017 About the expert system "Analysis MVA", based on the concept of the mivar approach *Problems of artificial intelligence* **2** 78–88

[11] Varlamov O O and Aladin D V 2018 About the creation of mivar control systems for monitoring the keeping of traffic rules on the basis of logical kernel (razumator) and expert systems *Radio industry* **2** 25–35

[12] Varlamov O O 2018 Wi!Mi Expert System Shell A Novel Tool To Build Knowledge-Based Systems With Linear Computational Complexity *International Review of Automatic Control* **6**

IOP Conf. Series: Materials Science and Engineering **534** (2019) 012016 doi:10.1088/1757-899X/534/1/012016

Systems of the self-driving vehicle

D S Vdovin and I O Khrenov

Bauman Moscow State Technical University, 105005 Baumanskaya 2th st., 5, Moscow, Russian Federation

E-mail: vdovin@bmstu.ru

Abstract. The article describes the basic systems of a self-driving vehicle distinguishing it from a conventional car under the control of the driver. The requirements for such systems are formulated, a number of tasks to be solved for each system is described, the interaction of these systems as a part of the design of a self-driving vehicle is described. A classification of types of self-driving control systems is proposed.

1. Introduction
In modern designs of self-driving vehicles, in addition to the conventional automobile systems, the following systems can be specified:

1. Perception of the environment system (sensor network);
2. Complex positioning system of the vehicle;
3. Navigation system (local and global route planning);
4. Internal control systems of the vehicle.

Further, each system is overviewed separately.

2. Perception of the environment system (sensor network)
Perception system of the environment can be divided into external and internal sensor network.

2.1 External sensor network
The purpose of the external sensor network is to use onboard measuring devices to recognize and classify objects of the road scene (Figure 1) and to deliver information about them to the virtual road scene assembler.

The virtual road scene is a combination of external traffic conditions, which include the following objects:

- Road scene objects:
  - Vehicles of categories M, N, O, L;
  - Pedestrians;
  - Cyclists/motor cyclists;
  - Horse drawn carts;
  - Traffic lights;
  - Signs;
  - Obstacles of a general type (moving and motionless);

  - Predefined events (traffic accidents, road closures, road repairs, traffic police officer's directions);
  - Vehicles with special signals (ambulance, police, etc.).
- Road scene environment:
  - Type and condition of the roadway or ground;
  - Borders of the road and zones of the ground, appropriate for driving;
  - Road surface marking;
  - Types of intersections and traffic rules and laws;
  - Damage to the road bed;
  - Weather.

Figure 1. Example of recognized objects "car" and "motorcyclist" on the image (left) and the assembled virtual road scene with suitable zones for driving (right).

Today, measurement and sensor equipment of an external sensor network usually consists of the following components (Figure 2):

- Optical and infrared sensors (video cameras, stereo cameras);
- Radar;
- Light sensors (LIDAR / LEDAR - 3D laser and 2D scanners);
- Ultrasonic sensors ("parktronic").

It is also possible to classify the external sensor network into a passive, non-radiant (for example, video cameras) and active (radar / lidar, video cameras with infrared illumination to build a depth map).

The main functional requirements for the external sensor network of a self-driving vehicle are:

- determining the distance to objects and their coordinates on the road scene;
- classification of road scene objects according to the list, presented above (the list may be enlarged);
- determination of the velocity vector of a moving object and predicting its movement;
- work in real time. Requirements on the reaction speed of the system are dictated, usually, by the maximum speed of the vehicle in the self-driving mode;
- reliability of recognition and accuracy of putting objects on the virtual road scene with an assigned probability of incorrect recognition per kilometer of path and operating time;
- work in "all-weather" conditions (day, night, rain, snow, fog, sun exposure and headlights of oncoming traffic, etc.);
- integration into the overall self-test system of the self-driving vehicle (system health check).

When recognizing the surrounding objects and environment using data from the video cameras, video processing algorithms are widely used (video streaming pre-processing - stabilization, color and brightness correction, etc.), 3D depth map reconstruction in the direction of the stereo camera view, algorithms for classifying and recognizing objects, tracking objects and prediction of their movement by extrapolation over several video frames. A depth map in the direction of the stereo camera view is then processed into a projection plane road scene using three-dimensional geometric transformations.

Scanning data from 2D/3D lidars and radars allows to build a three-dimensional cloud of points around a self-driving vehicle and use more sophisticated object recognition and tracking algorithms than using pairs of 2D images in a stereo camera (Figure 3).

Figure 2. On the left is a stereo video camera, in the center is an infrared 3D lidar, on the right is a car radar.

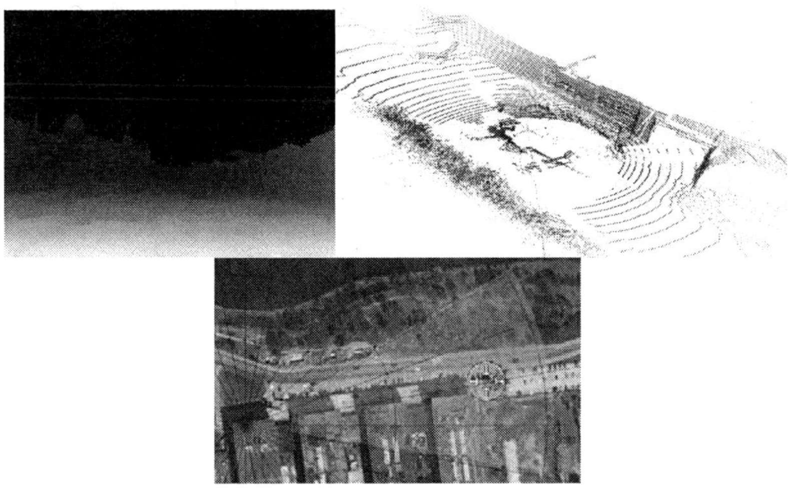

Figure 3. On the left - a depth map obtained by processing a stereo image, in the center - a point cloud obtained by scanning with a 3D lidar, on the right - the result of scanning by a radar (section of a three-dimensional figure).

Today, none of the above sensor systems provide the necessary reliability of recognition of the road scene and its objects. Therefore, when assembling the road scene, the integration of recognition data from all of this systems is used: compare the coordinates and dimensions of the obstacles; object speeds; different classification and recognition algorithms results. Then objects and their recognized states are assigned with recognition confidence factors and objects are placed on the virtual road scene. Decision-making algorithms in the top-level control system should take into account the factors of confidence for recognized objects while calculating the route for the self-driving vehicle (slow down, choose detour trajectories, assign security zones around poorly recognized objects with an unclear prediction of their movement, etc.).

2.2 Internal sensor network

The purpose of the internal sensor network is to track the state of the systems and aggregates of the vehicle. Information about the state of vehicle systems influences the route, speed planning and all behavior strategy of the self-driving vehicle. A simple example – fuel level or battery charge sensor determines the possibility of further movement of a self-driving vehicle without refueling / recharging.

IOP Conf. Series: Materials Science and Engineering **534** (2019) 012016 doi:10.1088/1757-899X/534/1/012016

This data is also very useful to other purposes: vehicle strategic motion planner, optimization tasks of the vehicle movement, scheduled maintenance and preventive maintenance, etc.

Sensors of the car internal sensor network (Figure 4) today are quite developed and usually consist of the sensors integrated into the vehicle's aggregates (wheel speed sensors, steering angle sensor, engine speed sensor, electronic brake and gas sensors, cooling system temperature sensors, pressure in wheels, etc.).

Additional requirement for internal sensor network mounted on a self-driving vehicle is to provide remote access to its measurement results for the general self-diagnosis system and the top-level control system that could be located outside the vehicle.

Figure 4. Bosch induction-type wheel speed sensor (left) and BMW potentiometer steering wheel sensor (right).

High-developed internal sensor network, integrated into a single on-board or cloud-based database, makes it possible to predict work condition of vehicle and its components, plan maintaining based on condition, to provide the high-level control system with alternative ways to distribute the power onboard and much more.

3. Vehicle positioning system

Global positioning satellite systems are widely used both as part of the onboard software and hardware of modern cars, as well as applications and GPS / GLONASS receivers on driver's smartphones (popular applications are TomTom, Garmin, etc.). Using a global positioning system on a self-driving vehicle is no different from using it in a conventional car and serves the following purposes:

- determine the coordinates of the location of the vehicle in real time on a global map;
- determine motion parameters: velocity vector and velocity value, acceleration;
- support the use of external databases of maps and online event maps;
- support of data exchange with external services;
- issuing operative restrictions on the route to the system of global route planning (traffic is blocked, the road is closed for this type of vehicles, etc.).

The main problem of using a global positioning system based on GPS / GLONASS / GALILEO satellites receiving signal is signal instability and low accuracy in determining coordinates (up to 30 meters).

For precise coordination on the map, Differential Global Positioning Systems (DGPS) having operation range 1 – 2 km and on-board inertial navigation system (limited to time of use for dozens of minutes due to accumulation of errors) are often used.

Integration of on-board inertial positioning system and global satellite system, as well as the use of previously known reference objects on the map (for example, vehicle could be coordinated with respect to the road's edge or marking lanes), allows precise positioning of the self-driving vehicle with sufficient accuracy in predefined operating conditions: with clear visibility, markup, etc. The task of such positioning is called SLAM-task (Simultaneous Localization and Mapping – definition of own coordinates on the map and the simultaneous construction of this map around a self-driving vehicle).

4. Navigation system (local and global route planning)

The navigation systems of a self-driving vehicle can be divided into a global system of route planning and path working operations performing and a local system of trajectory planning.

4.1 Global route planning system

The main tasks of the global route planning system are:

- Receipt and launch of the travel task from the dispatcher (operator) of the self-driving vehicles;
- Construction and selection of the general route of the self-driving vehicle on the map / in the area of operation (for example, Garmin Navigator route plan in the city), including taking into account the available online information on the condition on the roads;
- Planning and implementation of working operations along path (loading and unloading, sowing, combat missions, cleaning the territory, patrolling, etc.) and invoking the vehicle working equipment, installed on the self-driving chassis;
- Reporting actions on the completion of the travel task, events that happened during the operation.

It makes sense to equip the Global Intelligent Transport System with a system of vehicle global route planning and route working operations implementation.

Global route planning system requirements may include:

- Support online state of maps and routes:
  - on-board database of maps and routes;
  - updating the onboard database of routes from the cloud.
- Update databases of maps and states of roads according to the results of recognition (data supply):
  - signs (permanent, temporary), traffic lights, pedestrian crossings, bus stops, etc.;
  - width of the carriageway, the number of lanes;
  - condition and damages of the road surface;
  - types of intersections and rules of their passage;
  - temporary and permanent obstacles (repairs, accidents, open hatches).

4.2 Local route planning system

Currently, the driver is planning the local path of the vehicle by acting on its controls, based on the received information about the road scene in a visual way, by ear (car's audible warnings), in a tactile way (power response on the steering wheel, pedals, etc.).

In a self-driving vehicle, this task is solved by the vehicle high-level control system based on the road virtual scene and the planned global route.

Basic requirements for local route planning:

- Working in real time. Requirements on the reaction speed of the system are dictated, usually, by the maximum speed of the vehicle in the self-driving mode;
- Prediction of the road scene dynamics;
- Building a local trajectory
- Calculation of target states of a self-driving vehicle on a trajectory (speed, direction of movement, emergency stop, detour of an obstacle, state of installed working equipment and its operation needs on a local trajectory).

Systems of local movement planning can be divided into movement in a deterministic environment and a non-deterministic environment:

- highway (Traffic Laws are valid);
- urban environment (Traffic Laws are valid);
- quarry (Special Rules of Traffic are valid);
- closed territory (Special Rules of Traffic are valid).

Requirements for movement in a deterministic environment may be:

- Compliance with the Rules of the Traffic using typical scenarios of behavior:
  - Speed limit

- Compliance with road signs
- Parking and stopping rules
- Movement between lanes
- Start and end of movement
- Crossing / roundabout
- Exit and departure on highways
- Order of overtaking / avoiding obstacles

Route planning systems working in a non-deterministic environment have to deal with:

- Rough terrain (Traffic Laws are not relevant);
- Urban environment in case of emergency situations and combat actions (Traffic Laws are not relevant).

Requirements to the functions of the local trajectory planning system in a non-deterministic environment may include:

- Automatic evaluation of the feasible moving across terrain and ground traction capacity for the vehicle;
- Classification of borders / obstacles and ground surface type (house wall, curb, ditch / pit, marshland, steep slope, grass, asphalt, sand, snow, etc.).

5. Vehicle internal systems control

The main purpose of the internal systems control in the self-driving vehicle units is to calculate and apply control actions in order to maintain the local trajectory of movement and the target states of the vehicle, including its installed working equipment. At the same time, a control system can collect, store information about events on the route, the history of the vehicle's systems work, and calculate, together with the local movement planner, the most optimal vehicle modes of operation (minimum fuel consumption / charge, minimum travel time, minimum wear of machine parts, etc.).

Requirements for the internal control system of a self-driving vehicle may be:

- Real-time health-check and state assessment of the vehicle systems;
- Calculation of restrictions on the possibility of a vehicle moving ability to the global and local route planning system (fuel level, systems condition, available power, prediction of available traction force, etc.);
- Control signals calculation for vehicle systems and their execution;
- Diagnostics of vehicle units control systems, overall health check.

Today, most of the vehicle systems have their own electronic control units (electronic engine management system, electronic brake system, active steering electronic system, electronic transmission control system, see figure 5). Therefore, another requirement for the top-level control system of the self-driving vehicle is the integration of all electronic control units into a supervisory system, ensuring the transfer and processing of sensor readings into a seamless internal vehicle sensor network.

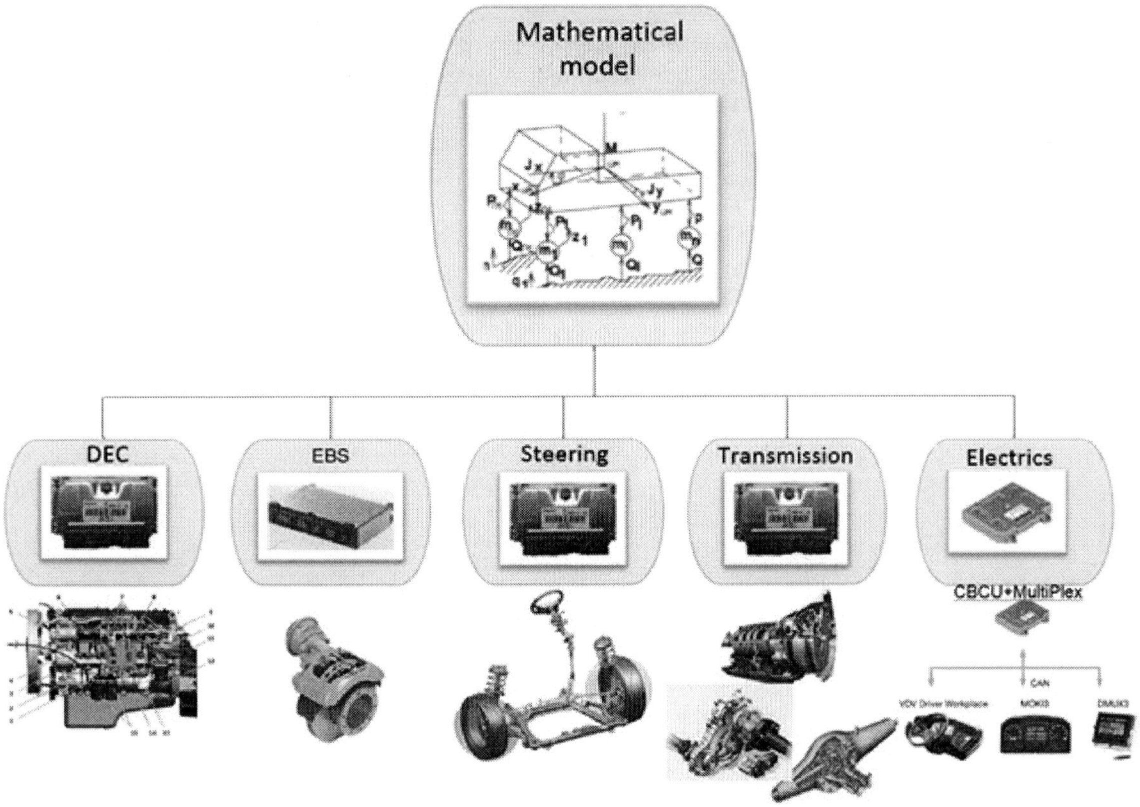

Figure 5. Electronic control units of car systems.

The most appropriate solution in meeting the latter requirement is building the control system of the self-driving vehicle based on a complex real-time mathematical model of the vehicle movement. This model should capture and predict vehicle behavior in motion, provide necessary signals and information for its individual on-board control systems.

6. Conclusions

Today development of each of the above-described systems of a self-driving vehicle is a separate research area and, usually, these developments are carried out by different scientific and engineering teams. However, these systems, closely intersecting in the design of the car, are highly dependent on each other and producing requirements to each other, often contradictory. The complex mathematical model (digital twin) of a self-driving vehicle on the virtual road, which is usually developed in the product design stage and then "works" onboard the vehicle, combining all workflows, can solve the problem of the inconsistency of these requirements, as well as identify possible functional problems at the junction of different control systems of the self-driving vehicle.

7. References

[1] Dyakov A S 2017 Analysis of domestic and foreign experience in the creation of unmanned special-purpose ground vehicles *NAMI Works* **4**
[2] Kotiev G O and Dyakov A S 2016 Method for developing undercarriage systems for high mobility unmanned ground vehicles *Izvestiya SFedU. Engineering sciences* **1**
[3] Keller A, Aliukov S 2015 Methodology of System Analysis of Power Distribution among Drive Wheels of an All-wheel-drive Truck *SAE Technical Paper* 2015-01-2788
[4] Tarasik V P 1997 *Mathematical modeling of technical systems: Proc. for high school* (Minsk: Design PRO)

IOP Conf. Series: Materials Science and Engineering **534** (2019) 012016 doi:10.1088/1757-899X/534/1/012016

[5] Albus James S, Huang Hui-Min, Messina Elena R, Murphy Karl, Juberts Maris, Lacaze Alberto, Balakirsky Stephen B, Shneier Michael O, Hong Tsai H, Scott Harry A, Proctor Frederick M, Shackleford William P, Michaloski John L, Wavering Albert J, Kramer Thomas R, Dagalakis Nicholas G, Rippey William G, Stouffer Keith A and Legowik Steven 2002 *4D/RCS Version 2.0: A Reference Model Architecture for Unmanned Vehicle Systems* **169**

[6] Fitzpatrick K, Hebert M, Pomerleau D and Schempf H 1992 Autonomous Cargo Transport System *Proceedings of the United States Postal Service 5th Advanced Technology Conference*

[7] LaValle S M 2006 Planning Algorithms *Cambridge University Press* (https://doi.org/10.1017/CBO9780511546877)

[8] Heutger M, Kückelhaus M and others 2014 Self-driving vehicles in logistics A DHL perspective on implications and use cases for the logistics industry *DHL Trend Research* (http://www.dhl.com/content/dam/downloads/g0/about_us/logistics_insights/dhl_self_driving_vehicles.pdf)

[9] Zhankaziyev S V 2016 *Intelligent transport systems: studies allowance* (Moscow: MADI)

[10] Keller A V, Gorelov V A, Vdovin D S, Taranenko P A and Anchukov V V 2015 Mathematical model of all-terrain truck *Proceedings of the ECCOMAS Thematic Conference on Multibody Dynamics*, Multibody Dynamics 1285-96

[11] Keller A and Aliukov S 2016 Efficient Power Distribution in an All-Wheel Ground Vehicles *SAE Technical Paper* 2016-01-1105

[12] Ivanov A M and Shadrin S S 2017 Development of autonomous vehicles' testing system *International automobile scientific forum, IASF-2017*

[13] Klebanov D A 2015 Development of technical and technological solutions for the creation and application of robotic cargo transportation systems in open pit mining *Thesis for the degree of candidate of technical sciences* Institute for Problems of Integrated Mineral Resources Development, Russian Academy of Sciences

[14] Levinson J S 2011 Automatic laser calibration, mapping, and localization for autonomous vehicles *PhD dissertation* Stanford University (http://purl.stanford.edu/zx701jr9713)

[15] Levinson J, Montemerlo M and Thrun S 2007 *Map-Based Precision Vehicle Localization in Urban Environments Robotics: Science and Systems* (Atlanta GA USA)

[16] Teichman A and Thrun S 2011 Practical object recognition in autonomous driving and beyond *Advanced Robotics and its Social Impacts Conference* (DOI: 10.1109/ARSO.2011.6301978)

[17] Levinson J, Askeland J, Dolson J and Thrun S 2011 Traffic Light Mapping, Localization, and State Detection for Autonomous Vehicles *2011 IEEE International Conference on Robotics and Automation* (DOI: 10.1109/ICRA.2011.5979714)

[18] Levinson J, Askeland J, Thrun S and others Towards Fully Autonomous Driving: Systems and Algorithms *2011 IEEE Intelligent Vehicles Symposium (IV)* (Baden-Baden, Germany)

IOP Conf. Series: Materials Science and Engineering **534** (2019) 012017 doi:10.1088/1757-899X/534/1/012017

Experimental system of safe distances and vehicle speed dynamic stabilization

S E Buznikov, D S Elkin, A M Saikin and V O Strukov

FSUE "NAMI", Moscow, Russian Federation

E-mail: sergey.buznikov@nami.ru

Abstract. One of the most promising direction for improving the level of traffic safety is various active safety systems creation. This area is intensively developed by leading foreign automakers and it is not complete. The relevance of researches aimed at effective active safety systems creating is determined by the world importance level of the traffic safety problem. Current study aims at the scientific substantiation of new technical solutions for the integrated vehicle active safety intelligent system. To conduct the research system analysis methods and modern control theory were used. The problem of structured set typical collisions prevention is reduced to the problem of vehicle state variables dynamic stabilization. The obtained result is the new technical solution scientific substantiation for the integrated intelligent system that combines the functions from several active safety systems. The high limit of the mass center velocity is defined as the minimum speed of the obstacle, wheels understeer or oversteer, turnover, slippage of the driving wheels and speed limits set by road signs, etc. To assess the control quality a modified quadratic functional is used that takes into account the operation costs as well as the software hardware initial cost. Analysis of the study results allows us to conclude that the intellectual properties developed by the system are acquired through the use of mathematical models and algorithms for indirect measurements performed in a minimal configuration of primary information sensors and actuators.

1. Introduction

One of the most promising direction for traffic safety improvement is various active safety systems creation [1]. This area is being intensively developed by leading foreign automakers. Leading foreign companies include the stabilization systems for longitudinal (ABS) and transverse (ESP) wheel slip [2, 3], turnover prevention, advanced emergency braking (AEBS), adaptive cruise control (ACC) [4], tire pressure monitoring (TPMS), position control on the lane (LDWS), etc. in the standard equipment of modern cars.

The purpose of the study is the scientific substantiation for new technical solutions of the vehicle active safety integrated intelligent system.

Based on the analysis of cars collision conditions it is possible to construct a structured set of typical collisions [5]. The problem of structured set collisions prevention is reduced to the problem of the vehicle state variables dynamic stabilization. An essential feature of the dynamic stabilization problem is the incomplete observability of the object state vector and the uncertainty of its dynamic boundaries. This feature determines the need to include the driver-operator in the control of traction loop, brakes and course.

The operator replacing at completing the unobserved state coordinates and their dynamic boundaries requires a large number of physical quantity sensors, which leads to a decrease in reliability, an increase in cost and other negative consequences [6].

The evolutionary process of automotive active safety systems improvement is built on a modular principle. This principle implies the expansion of additional functions by adding to the control system new blocks-modules that are connected to the common information bus [7], which creates a certain redundancy of technical facilities.

The integrated system involves combining the information and control functions of various active safety systems in a single control device at minimal configuration of primary information sensors and actuators.

2. Task statement

At the conceptual level, the task of adaptive cruise control is reduced to the safe distances and vehicle speeds dynamic stabilization in an automatic mode. Safe speed should be limited by the speed of oversteer and understeer, car turnover, driving wheels slippage, road signs, tire condition, etc.

The safe distances to obstacles in the front and rear hemispheres should be sufficient to prevent collisions with the stationary and passing vehicle when braking taking into account the condition of the road surface without blocking the wheels [8].

At the content level, the dynamic stabilization problem reduces to minimizing the modified quadratic control quality functional on a finite interval $[t_2 - t_1]$ in the form:

$$Q(t_2) = Q_1 + Q_2 \rightarrow min \qquad (1)$$

at $U \in U_{per}, R \in R_{per}$, where

$$Q_1 = \int_{t_1}^{t_2} C_0[V_{mz}(\tau) - V_m(\tau)]^2 d\tau + \int_{t_1}^{t_2} \sum_{i=1}^{2} C_1[\Delta L_i(\tau) - \Delta L_{bound\ i}(\tau)]^2 d\tau\,;$$

$$Q_2 = \int_{t_1}^{t_2} \Delta C(R,\tau) d\tau + Q_2(R,t_1);$$

$U = (U_1, U_2, U_3)^T$ – control actions vector on the gearbox, traction and brakes;
U_{per} – permissible range of control actions;
$R = (R_H, R_S)^T$ – technical solutions vector in the hardware (R_H) and software (R_S) facilities area;
$V_{mz}(t)$ – safe vehicle speed;
$V_m(t)$ – vehicle mass center speed;
$\Delta L_1(t)$ and $\Delta L_2(t)$ – distance to obstacles in the front and rear hemispheres on the lane;
$\Delta L_{bound\ 1}(t)$ and $\Delta L_{bound\ 2}(t)$ – boundary values of safe distances to obstacles in the front and rear hemispheres;
$\Delta C(R,t)$ – intensity of the operating system costs;
$Q_2(R,t_1)$ – cost of software and hardware systems set.

The best solution of the problem (1) in the proposed statement is the traction and brakes control algorithm providing stabilization of safe distances and speeds realized in the software and hardware environment with the minimum acquisition and operation cost.

3. Research results

To implement the stabilization safe speed algorithm, a single-step traction and braking control is used.

The specified speed V_{mz} is defined as the low limit of safe speeds determined from the conditions for typical collisions of a structured set prevention.

The components of the constraint vector includes:
V_{mz0} – specified speed of movement;
V_{mz1} – given speed of obstacle approaching;
$V_{bound\ 0}$ – boundary rollover speed;

$V_{bound\ 1}$ – boundary speed of the front wheels understeer;
$V_{bound\ 2}$ – boundary speed of the rear wheels oversteer;
$V_{bound\ 3}$ – boundary speed of driving wheels slipping;
$V_{bound\ 4}$ – boundary speed of tire cords rupture;
$V_{bound\ 5}$ – boundary speed of movement with spare wheel;
$V_{bound\ 6}$ – boundary speed specified by road signs in the their action zone;
$V_{bound\ 7}$ – boundary speed of the permissible vibrations level.

The resulting value V_{mz} is defined as:

$$V_{mz} = \min[V_{mz0}, V_{mz1}, V_{bound\ 0}, \dots, V_{bound\ 7}] \quad (2)$$

$$V_{mz1} = \begin{cases} 0,\ if\ V_{ob} = 0\ and\ \Delta L < \Delta L_{bound}; \\ V_{ob} + k_0(\Delta L - \Delta L_{bound}),\ if\ V_{ob} > 0\ and\ \Delta L > \Delta L_{bound}; \\ V_{mz0},\ if\ V_{ob} = 0\ and\ \Delta L \geq \Delta L_{bound}, \end{cases}$$

V_{ob} – obstacle speed;
ΔL – distance to the obstacle;
ΔL_{bound} – safe distance boundary value to the obstacle in the front hemisphere;
k_0 – speed controller gain factor of approaching to the obstacle.

The quantity ΔL_{bound} is defined as the maximum boundary distances of approaching to the obstacles during the delay time τ_m of the deceleration $\Delta L_{bound\ 1}$ and the limiting distance of the deceleration to the full stop $\Delta L_{bound\ 2}$:

$$\Delta L_{bound} = \max[\Delta L_{bound\ 1}, \Delta L_{bound\ 2}] \quad (3)$$

$\Delta L_{bound\ 1} = 0.5\hat{a}_{ob1}\tau_m^2;$

$\Delta L_{bound2} = V_m \cdot \tau_m + 0.5V_m^2\hat{a}_T^{-1} - 0.5V_{ob}^2\hat{a}_{ob}^{-1}.$

Speed identification of the passing obstacle V_{ob} is performed from the distance measurement to the nearest obstacle on the lane ΔL and the solution of the approach speed differential equation in discrete time with step ΔT:

$$V_{ob}(k) = [\Delta L(k) - \Delta L(k-1)] \cdot \Delta T^{-1} + V_m(k) \quad (4)$$

Safe distances stabilization in the front and rear hemispheres on the lane is performed by controlling the value of the braking deceleration a_T. Its value is determined from the equations of the boundary distances to the front and rear obstacles in the form:

$$a_T = \min[a_{T0}, a_{T1}, a_{T2}] \quad (5)$$

$a_{T0} = 0.5V_m^2[\Delta L_1 - \Delta L_{del} + 0.5V_{ob1}^2\hat{a}_{ob1}^{-1}]^{-1};$

$a_{T1} = 2\Delta L_2 \cdot \tau_{ob2}^{-2};$

$a_{T2} = 0.5V_m^2[V_{ob2}\tau_{ob2} + 0.5V_{ob2}^2\hat{a}_{ob2}^{-1} - \Delta L_{ob2}]^{-1};$

τ_{ob1} and τ_{ob2} – estimates of the brake systems operation lag of the front and rear obstacles;
V_{ob1} and V_{ob2} – front and rear obstacles speed;
$\hat{a}_{ob1}$ and $\hat{a}_{ob2}$ – estimates of the front and rear obstacles maximum decelerations.

Brake deceleration value is limited by the maximum value a_{Tmax}, which depends on the braking system parameters and the condition of the road surface.

The experimental intelligent active safety system includes:

- standard wheel speed sensors;
- front and rear cameras;
- video processing unit;
- three-axis solid-state accelerometers (MEMS);

- brake actuator;
- control board with software-controlled keys.

The experimental system is mounted on the "Lada Kalina" car with an electric drive and is being tested at the NAMI testing ground in Dmitrov.

The preliminary tests results confirm rather high efficiency of adaptive cruise control, automatic braking, driving stability, tire pressure and braking overheating temperatures monitoring function.

Information functions for estimating the vehicle state coordinates, such as mass center velocity, steering angle, tire pressure, brake overheating temperature, wear of tire cords, etc. are carried out using virtual sensors implemented programmatically.

Thus, the proposed technical solution is characterized by minimum hardware configuration, minimum operation and acquisition costs.

4. Conclusion

The analysis of conducted researches allows to formulate the following conclusions:

- The developed models and control algorithms for traction and brake channels are based on deterministic descriptions of traction, braking acceleration and course angle determined by control actions on the controls;
- Mathematical models are adequate in the operating range of velocities and accelerations and can be tuned to a particular model of the vehicle by introducing tuning data;
- The above control algorithms provide the solution of the vehicle state coordinates dynamic stabilization problem taking into account the extended velocity constraint vector, replacing the functions of the individual systems: ABS, ASR, ESP, ACC, AEBS and TPMS in minimal hardware configuration.

Acknowledgments

This scientific article was prepared based on the results of applied scientific research and experimental development, which is carried out with the financial support of the Russian Education and Science Ministry, Agreement No. 14.625.21.0042, the unique identifier of the project is RFMEFI62517X0042.

References

[1] Jeong E and Oh C 2017 Evaluating the effectiveness of active vehicle safety systems *Accident Analysis & Prevention* **100** pp 85–96

[2] Ni L, Gupta A, Falcone P and Johannesson L 2016 Vehicle lateral motion control with performance and safety guarantees *IFAC Proceedings Volumes* **49** pp 285–290

[3] Liang C, Xinzhao G, Jianhua G, Hongwei L, Libo C and Mingli S 2012 Coordinated control of electronic stability program and active front steering *Procedia Environmental Sciences* **12** pp 1379-86

[4] Davis L C 2012 Stability of adaptive cruise control systems taking account of vehicle response time and delay *Physics Letters A* **376** pp 2658–2662

[5] Buznikov S E 2018 System analysis of vehicle active safety problem *IOP Conf. Series: Materials Science and Engineering* **315**

[6] Saikin A, Buznikov S and Karpukhin K 2016 The Analysis of technical vision problems typical for driverless vehicles *Research Journal of Pharmaceutical, Biological and Chemical Sciences* **7** 2053-9

[7] Shadrin S S, Ivanov A M and Karpukhin K E 2016 Using data from multiplex networks on vehicles in road tests in intelligent transportation systems and in self-driving cars *Russian Engineering Research* **36**

[8] Buznikov S E, Elkin D S, Shabanov N S and Strukov V O 2016 The task of unmanned vehicle safe automatic braking *NAMI Proceedings* **265** pp 44–52

Thermostating system with intelligent management for electric vehicles

V S Struchkov, R H Kurmaev, D M Iakunov and I A Lubimov

FSUE "NAMI", Moscow, Russian Federation

E-mail: v.struchkov@nami.ru

Abstract. The article considers an optimum temperature regime maintainment method of the electric vehicles aggregates. Technical solutions for the creation of an intelligent thermostatic system are proposed. Vehicle equipped with the optimum temperature maintenance system developed during current research was placed under tests.

1. Introduction

The experience of electric vehicles operation in the Russian Federation in recent years reveals not only problems associated with the essential charging infrastructure, but also with low performance indicators, such as the mileage and the time necessary for charging [1, 2]. However, these indicators are the criterion of an electric vehicle applicability in everyday life, when it will not be only a concept car, but also mass-market car. Weather factors alone do not greatly constrain the use of an electric vehicle. Much more important circumstance is associated with the need to use power-consuming systems and devices designed to provide a comfortable and safe movement in winter. Exactly due to activation of many consumers the vehicle's mileage reduces to 50 km at negative temperatures.

It is obvious that the Russian versions of electric vehicles created by domestic and foreign manufacturers need special equipment. Even serial cars with internal combustion engines are often equipped with special "Russian packages" which include, for example, other characteristics of the wheels suspension and less sensitive neutralization systems to fuel quality.

2. Main part

Taking into account the specific operation conditions, a modular thermostating system of power electrical components was developed [3, 6, 7]. The general view of the developed liquid thermostating system is shown in Figure 1.

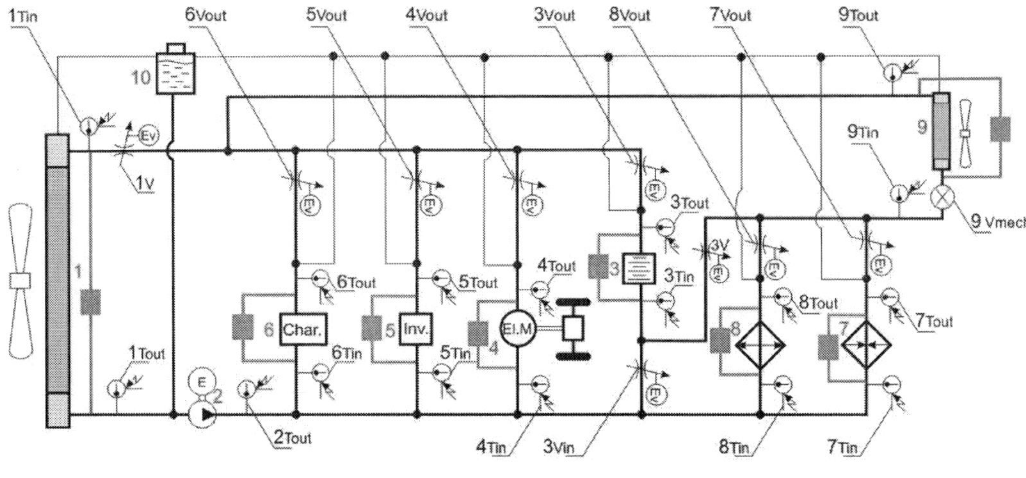

Figure 1. Schematic diagram of the thermostating system for electric vehicle components.

The coolant – unit 1 carries out maintenance of the temperature. This unit is a liquid-air radiator comprising an electric fan. It receives the cooling liquid from the common (main) discharge circuit of the coolant discharge from all aggregates of the vehicle. The coolant temperature at the inlet to the unit is regulated by a liquid electric valve (1V) in the open-close mode according to the temperature sensor values (1 T_{out}) based on the coolant temperature at the cooling radiator exit.

In the charger circuit 6, the regulation is carried out by a temperature sensor (6 T_{out}) and an electric valve (6 V_{out}) in the open-close mode.

In the inverter circuit 5, control realized by a temperature sensor (5 T_{out}) and an electric valve (5 V_{out}) in open-close mode.

In the motor circuit 4 the temperature is maintained by a temperature sensor (4 T_{out}) and an electric valve (4 V_{out}) in open-close mode.

Thermostating of the traction battery (TB) 3 is carried out using a liquid-air heat exchanger with a fan, which is located in the battery case and works constantly when the ignition is on. This cooling method is optimal for such design [3, 4, 5].

When the electric vehicle ignition is switched on, the electric pump of the cooling system is switched on before starting the movement. Circulation of the coolant is carried out in the full mode of the electric pump, through all pressure pipelines of all vehicle aggregates and constant discharge into the common bypass circuit through the expansion tank. In this circuit, the coolant discharge is carried out continuously regardless the temperature control regimes in each circuit of all units to ensure removal of the steam-air mixture from the used units.

Regulation of the temperature state in each unit is provided by a change in the flow rate of the heat carrier, which is discharged into the common liquid circuit of the main vehicle radiator. At the same time, when the specified coolant temperature is reached at the radiator unit input, the total heat release is provided by using an electric fan.

Further adjustment of the each unit temperature regime is carried out in an automatic mode, using temperature sensors, by changing the flow rate of the each unit coolant separately, by opening and closing the regulating electric valves.

To ensure the set temperatures in the winter and summer period, the system uses the electrical heater and the electrical cooler of liquid. The developed intelligent thermostating control system allows automatic monitoring of the entire system and makes optimal decisions to maintain the specified parameters (Figure 2).

IOP Conf. Series: Materials Science and Engineering **534** (2019) 012018 doi:10.1088/1757-899X/534/1/012018

Figure 2. Intelligent thermostating control system.

The most dependent climatic conditions component in vehicle is the TB. Let's consider the basic modes of maintaining the optimum temperature.

a) At the TB average temperature from 15 °C to 35 °C an electric valve (3 V_{in}) is constantly open; the valve 3 V_{out} operates in the open-close mode from the temperature sensor 3 $T_{battery}$. The valves 3 V, 8 V_{out}, 7 V_{out} and 9 V_{mech} are closed. Heat carrier circulation is taking place through the main radiator and expansion tank 10 (Figure 3).

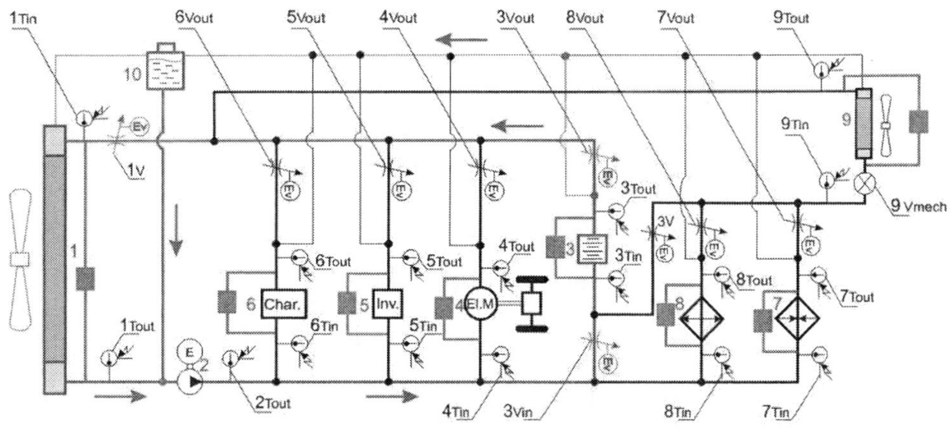

Figure 3. Coolant circulation at the TB cooling mode through the main radiator.

b) At the TB temperature below 15 °C a liquid preheater 7 is turned on; and simultaneously opens the electric valve 7 V_{out}, 3 V and 3 V_{out}; the electric valve 3 V_{in} is closed. In this case, the valve 8 V_{out}, 9 V_{mech} should be closed. The liquid temperature at the heater outlet is maintained by a temperature sensor 7 V_{out} in the heater operating mode switch on-switch off. The heating mode is switched off by the sensor 3 $T_{battery}$ when the temperature reaches 15 °C. Then the system is regulated according the algorithm "a" (Figure 4).

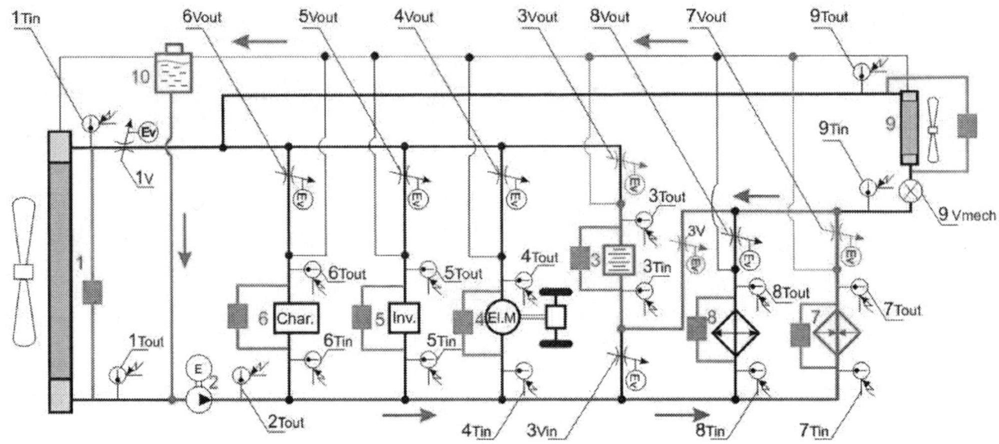

Figure 4. Circulation of the coolant through the expansion tank at the TB heating mode using a preheater.

c) At the TB temperature above 35 °C a liquid cooler 8 is switched on; and simultaneously the valve 8 V_{out} and 3 V opens, valve 3 V_{in}, 7 V_{out}, 9 V_{mech} closes. The liquid temperature in the liquid cooler is maintained by the coolant in the range –10 ± 5 ° C. When the required TB temperature is reached the sensor 3 $T_{battery}$ turns off the cooling system. Then the system is regulated according the algorithm "a" (Figure 5).

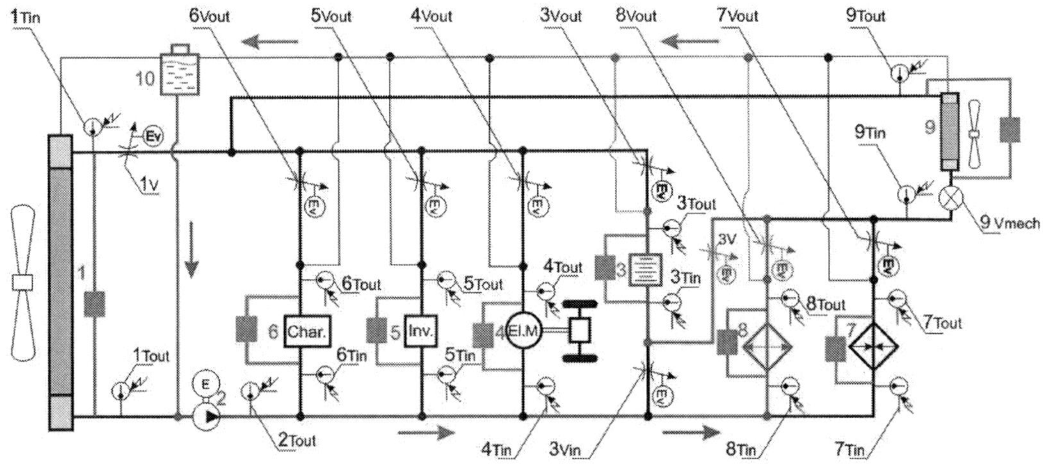

Figure 5. Circulation of the coolant through the expansion tank at the TB cooling mode with using an additional cooler.

To ensure the microclimate in the passenger compartment, it is possible to use "b" and "c" modes with the heating and cooling of the TB. In this case, it is necessary to open the valve (9 V_{mech}). At the separate interior heating mode a liquid electric heater (7) must be switched on separately, an electric valve (3 V) must be closed and the mechanical crane 9 V_{mech} must be opened.

The developed thermostating system was implemented on an electric vehicle prototype based on the "Lada Kalina". The reequipped vehicle is shown in Figure 6.

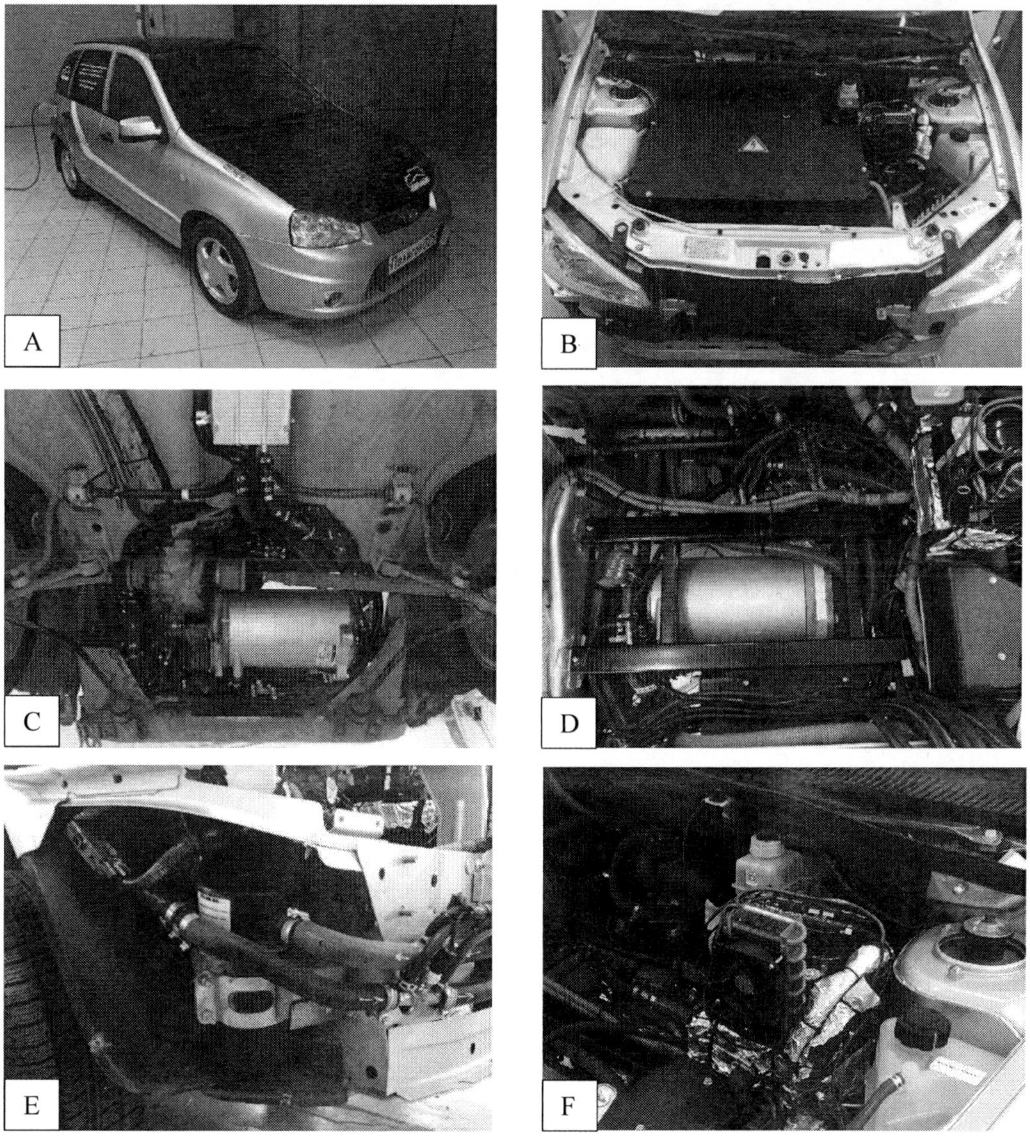

Figure 6. Vehicle equipped with the developed thermostating system.
(A – general vehicle view, B – under hood space, C – vehicle front part bottom view, D – vehicle front part top view, D – electric heater of liquid, E – cooler of liquid).

The TB is located at the top of the engine compartment, attached to the power frame. The electric motor and inverter are installed in the bottom part, under the TB and fixed to the same frame. The charger is installed in the floor tunnel. The electric heater of liquid is located on the right side of the bumper, in front of the front wheel and the coolant cooler on the left under the hood. Such arrangement of the thermostating system elements is rational and operable for a given vehicle.

After assembling and setting of all necessary systems, laboratory road tests were conducted at the Dmitrov Test Center of FSUE "NAMI" (Figure 7).

Figure 7. Tests at the Dmitrov Test Center FSUE "NAMI" (A, B – laboratory tests; C, D – road tests).

Main tests and operability measurements of the thermostating system were carried out on the chassis dynamometer. Load parameters for the stand were obtained on the test center roads. On the chassis dynamometer, two main tests of the thermostating system were carried out:

- tests of the thermostating system when the vehicle is moving according to the mixed cycle;
- tests of the thermostating system with uniform movement of the vehicle.

At the mixed cycle tests the vehicle movement chart is shown in Figure 8.

IOP Conf. Series: Materials Science and Engineering **534** (2019) 012018 doi:10.1088/1757-899X/534/1/012018

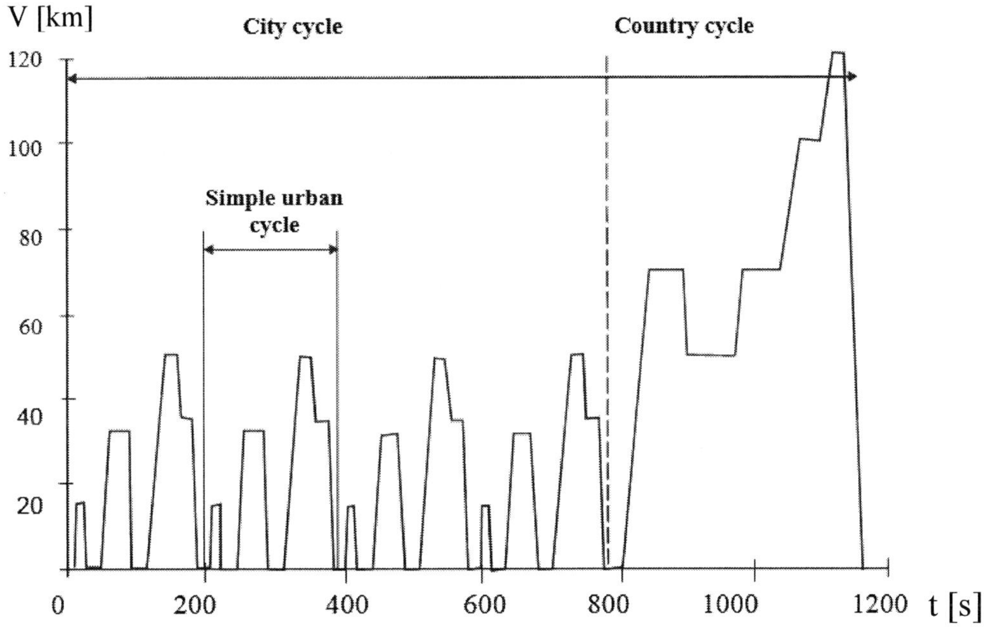

Figure 8. The vehicle movement chart in the mixed cycle.

The vehicle was installed on the chassis dynamometer, monitoring and recording systems were switched on. Three mixed cycles were conducted in a row, all received data were recording on the PC. Figures 9 – 12 show the data on the coolant temperature drop at the inlet and outlet of the vehicle aggregates.

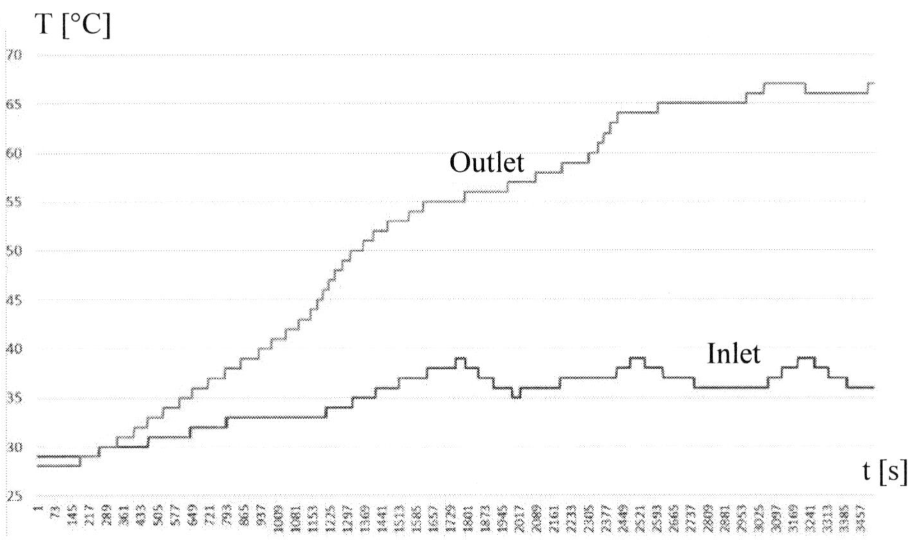

Figure 9. Temperature drop at the inlet and outlet of the motor.

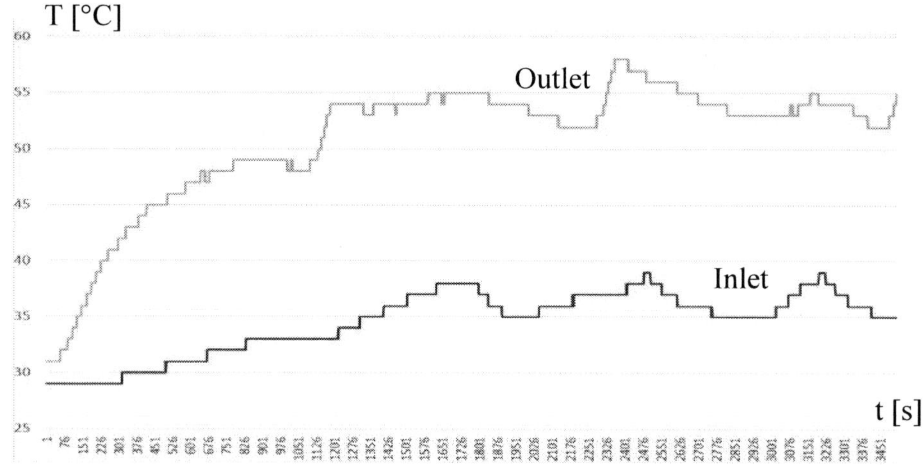

Figure 10. Temperature drop at the inlet and outlet of the inverter.

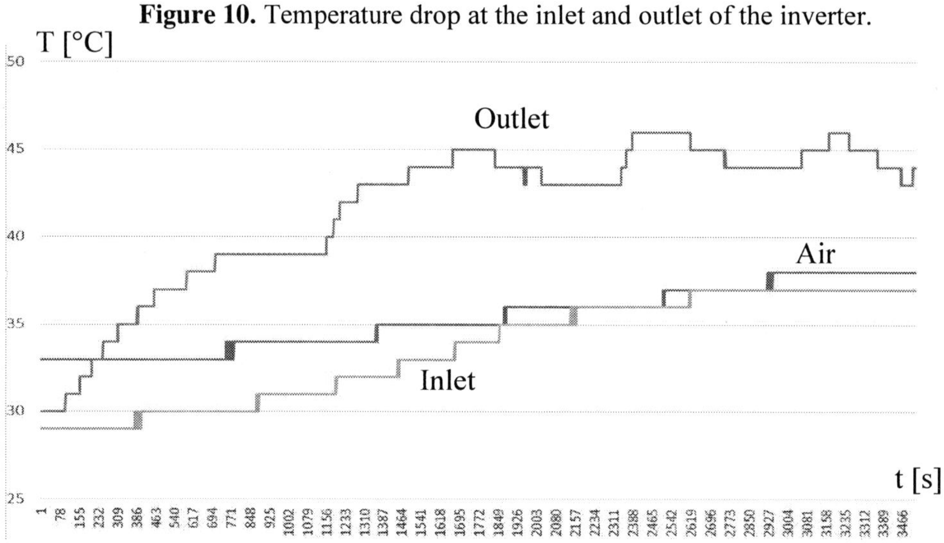

Figure 11. Temperature drop at the inlet and outlet of the TB.

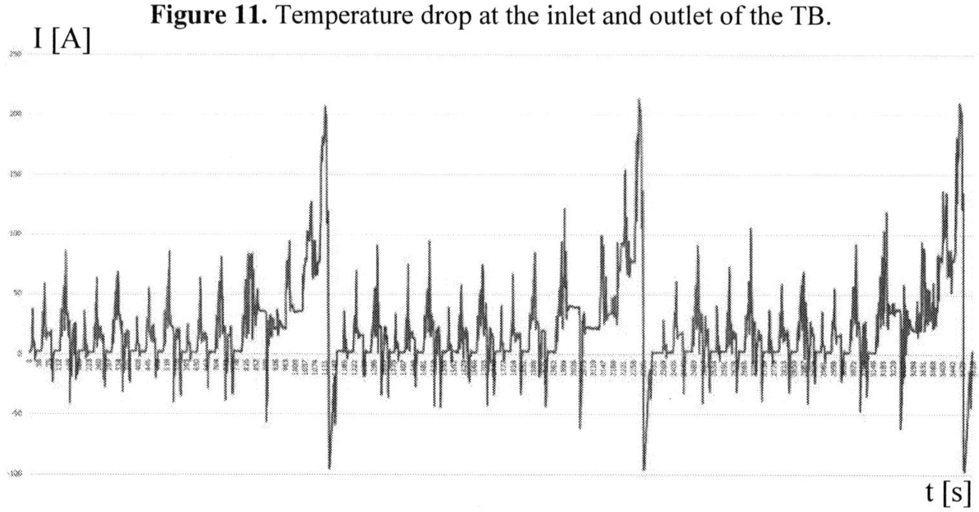

Figure 12. Energy consumption for moving in mixed cycles.

After analyzing of the obtained data, at vehicle movement in the mixed cycles, the following conclusions can be made:

- thermostating system ensures stable vehicle operation;
- aggregates temperature satisfies the manufacturers requirements;
- thermostating system effectively removes heat to the environment.

After conducting the series of drives in the mixed cycle, the tests continued at constant speeds 40 km/h and 80 km/h for 10 minutes each. The obtained data on the coolant temperature drop at the inlet and outlet of the motor, inverter and TB are presented in Figures 13 – 16 respectively.

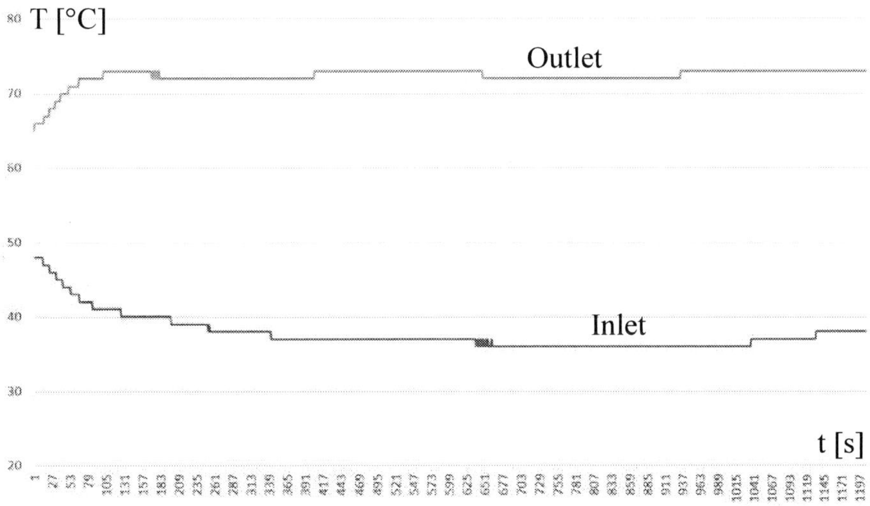

Figure 13. Temperature drop at the inlet and outlet of the motor.

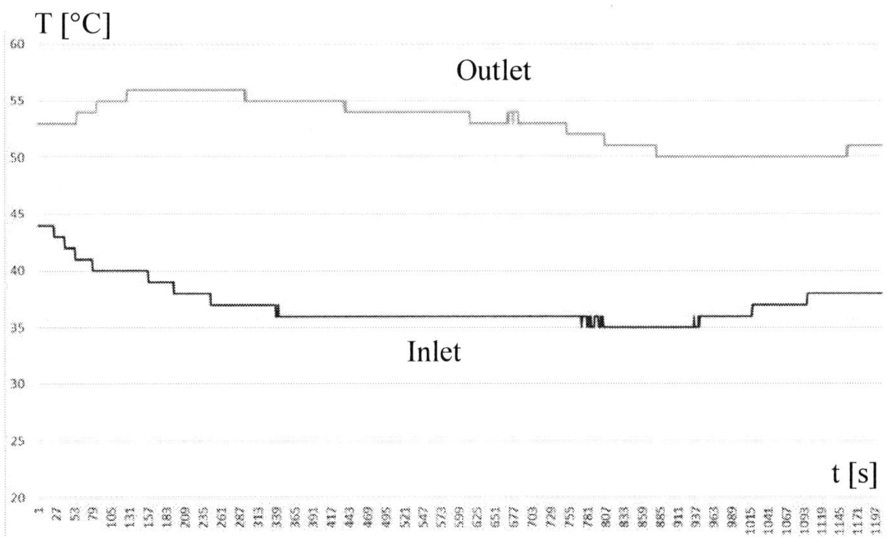

Figure 14. The temperature difference at the inlet and outlet of the inverter.

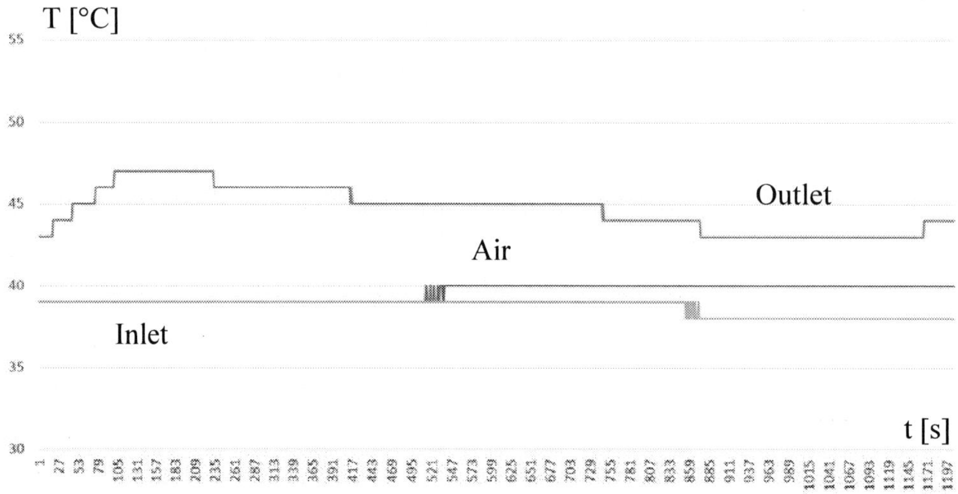

Figure 15. Temperature drop at the inlet and outlet of the TB.

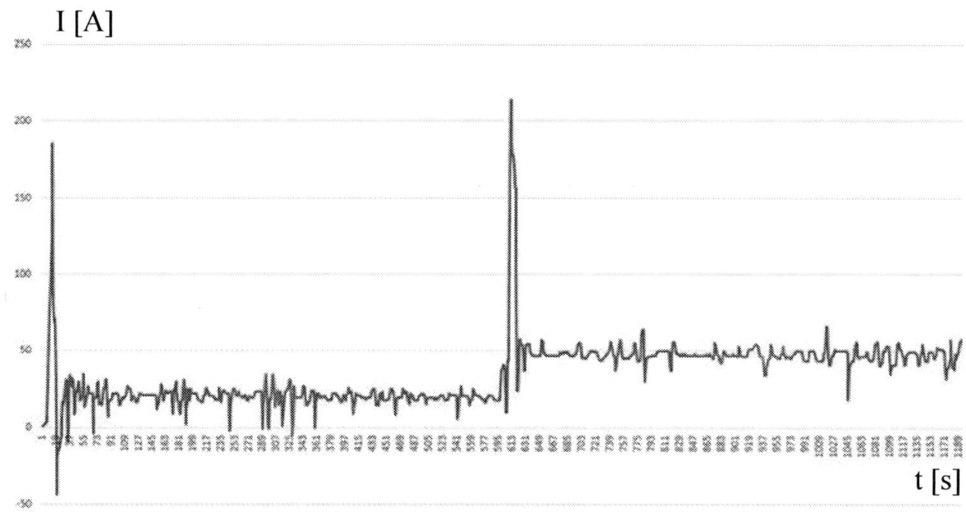

Figure 16. Power consumption for driving at constant speed.

After analyzing the obtained data at constant speed of the vehicle movement, the following conclusions can be made:

- thermostating system ensures stable vehicle operation;
- aggregates temperature satisfies the manufacturers requirements;
- thermostating system effectively removes heat to the environment.

3. Conclusions

The ability to adapt the thermostating system to the external operation conditions and the vehicle components operating conditions allows speaking about "intellectual" control. Complex tests of the vehicle equipped with a new thermostating system confirmed the developed circuit operability for connecting and regulating the components temperature. The laid down principles should be used at creating thermostating systems for electric vehicles, since it allows to redistribute heat fluxes more rationally. Modularity of the design implies full or partial use on vehicles, operating in various climatic and road conditions.

Acknowledgments
The article was prepared with financial support of the Russian Education and Science Ministry, the agreement No. 14.624.21.0047 from October 26, 2017. Unique identifier of the agreement is RFMEFI62417X0047. The theme is "Development and experimental testing of efficient energy distribution technology for domestic electric vehicles with additional power sources based on photoelectric converters".

References

[1] Kurmaev R H, Terenchenko A S, Karpukhin K E, Struchkov V S and Zinov'ev E V 2015 Maintaining the required temperature of high-voltage batteries in electric cars and hybrid vehicles *Russian engineering research* **35** pp 666 – 669
[2] Karpukhin K E, Kurmaev R K, Terenchenko A S, Struchkov V S and Tsimbaluk M A 2016 Aspects of construction of combined thermo-statics system for electric vehicle ARPN *Journal of engineering and applied sciences* **11** pp 13674 – 80
[3] Karpukhin K E, Shorin A A, Terenchenko A S, Umnitsyn A A and Kondrashov V N 2016 Research of effectiveness of accumulator systems of hybrid motorcars and electromobiles in conditions of negative temperatures *Russian engineering research* **8** pp 26 – 29
[4] Xia G, Cao L and Bi G 2017 A review on battery thermal management in electric vehicle application *Journal of power sources* **367** pp 90 – 105
[5] Al-Zareer M, Dincer I and A.Rosen M 2017 Novel thermal management system using boiling cooling for high-powered lithium-ion battery packs for hybrid electric vehicles *Journal of power sources* **363** pp 291 – 303
[6] Karpukhin K E, Terenchenko A S, Kolbasov A F and Kondrashov V N 2017 Using photoelectric converters in road transport in order to improve energy efficiency in Russia *International journal of mechanical engineering and technology* **8** pp 529 – 534
[7] Kolbasov A, Karpukhin K, Terenchenko A and Kavalchuk I 2018 *Concept of intellectual charging system for electrical and plug-in hybrid vehicles in Russian Federation* IOP Conference Series: Materials Science and Engineering

IOP Conf. Series: Materials Science and Engineering **534** (2019) 012019 doi:10.1088/1757-899X/534/1/012019

Driver assistance systems: analysis, tests and the safety case. ISO 26262 and ISO PAS 21448

O M Kirovskii and V A Gorelov

Department of Tracked Vehicles and Mobile Robots, Bauman Moscow State Technical University, 2nd Baumanskaya st. 5, 105005 Moscow, Russian Federation

Abstract. Modern automotive system must comply with strict safety requirements. This paper focuses on two aspects of safety: functional safety per ISO 26262 (FS) and the safety of the intended functionality (SOTIF) per ISO PAS 21448. The FS encompasses a lifecycle ensuring the absence of unreasonable risks due to internal failures of the system. SOTIF concentrates on non-deterministic parts and algorithms (e.g. neural networks), as full specification of their performance is out of reach now. At the same time, FS and its lifecycle are better known to the community and have better history of implementation. This paper presents an approach to integration of FS and SOTIF requirements based on FS lifecycle.

1. Introduction

Increasing autonomy of road vehicles is one of the most important trends shaping today's automotive industry. One of the challenges that developers and OEMs face is the requirement to guarantee safety and security to people in and around the autonomous vehicle (AV) under all circumstances. The methods for safety assurance are constantly evolving for many years already, without any final stop in sight. This introduction gives a brief definition of the terms related to automotive safety and main strategies the engineers pursue in assuring it.

According to the Technical Regulation of the Eurasian Economic Union TR TS 018/2011 "Safety of Wheeled Vehicles" [1], "safety of the vehicle" is defined as "state (...) where the risk of harm to life, health of people as well as to the property (...) is eliminated or minimized". The lawmakers acknowledge the existence of conditions under which complete elimination of risks related to the vehicle is impossible. The intent of the law is to motivate the vehicle designers and manufacturers and other vehicle lifecycle participants to take measures against those risks through the entire lifecycle.

Norms on functional safety (e.g. ISO 26262 [2]) share this paradigm. Functional safety (FS) addresses faults resulting from malfunctions of systems (i.e. failures of its components). Modern autonomous driving (AD) systems and advanced driver assistance systems (ADAS) highly depend on the data from external sources provided by non-deterministic, often AI-based algorithms (e.g. object recognition, path planning using machine learning techniques). In those cases, even a fault-free system may exhibit hazardous behavior due to the limitations of the system's performance. The new paradigm of Safety of the Intended Functionality (SOTIF) as formulated in ISO PAS 21448 [3] addresses the issue.

Both FS and SOTIF prescribe a lifecycle process for identification and control of risks. The FS lifecycle is better known and more widely implemented because of its longer history, while SOTIF measures are becoming necessary with increasing autonomy of vehicles. The two lifecycles can be combined in order to achieve throughout traceability of requirements and facilitate creation of the Safety Case.

2. Scope of FS and SOTIF

It is known that one may find more than 40 definitions of safety only among different ISO standards. Surprisingly, both ISO 26262 and ISO PAS 21448 agree on this issue and define safety as absence of unreasonable risks. Based on the definition, the scope of each standard is defined:

- Functional safety: absence of unreasonable risk due to hazards caused by malfunctioning behavior of E/E Systems (ISO 26262-1, Def. 1.51)
  - Malfunctioning behavior is defined as a failure or unintended behavior of an item with respect to its design intent (which in turn is undefined).
- Safety of the Intended Functionality: Absence of unreasonable risk due to hazards caused by performance limitations of the intended behavior or by reasonable foreseeable misuse by the user (ISO PAS 21448, Def. 3.5)
  - Performance limitation: insufficiencies of the function itself (example: incorrect perception of the situation or caused by model assumptions that are not feasible) – ISO PAS 21448, Def. 3.4

The scopes defined in both standards are not separated clear enough, as there is no clear distinction provided between "malfunction" and "performance limitation". Both are defined as deviations from desired behavior. It can be said that there is an overlap between the scope of the FS and SOTIF. Instead of refining the definitions in order to draw a clearer distinction, a more useful approach would be to define the aspects of safety which are best treated with FS methodology and those where SOTIF methodology is more beneficial.

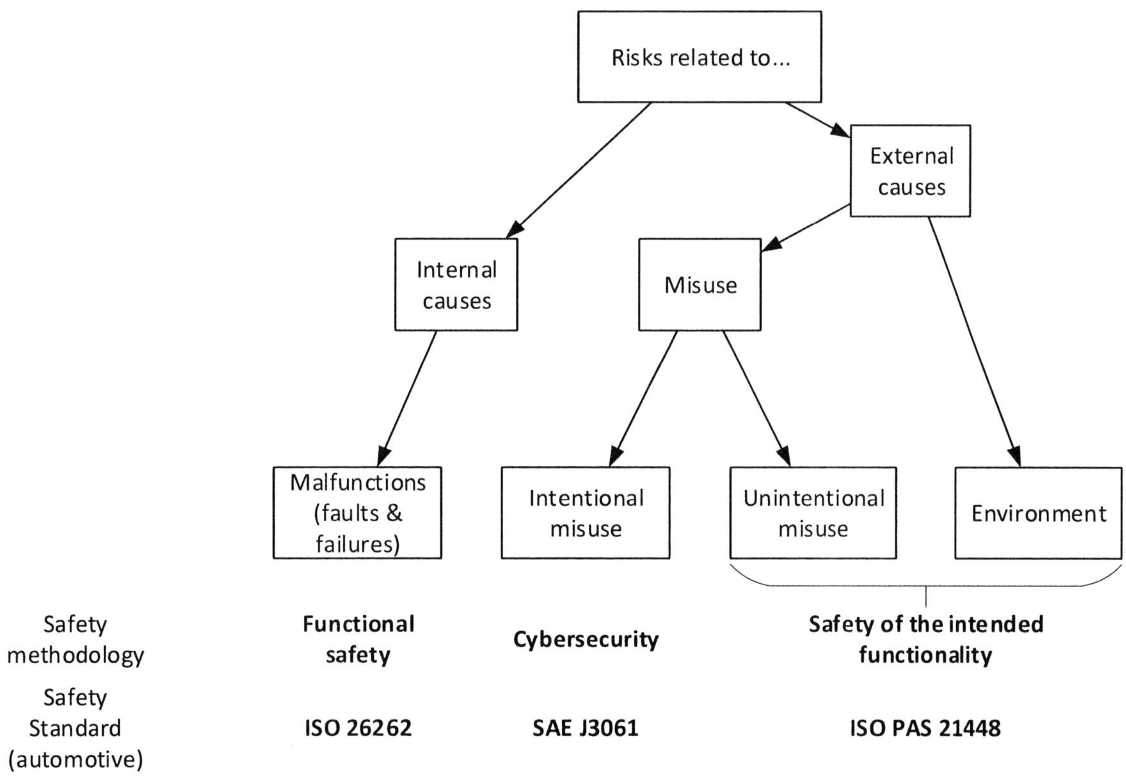

Figure 1. Safety methodology: use cases.

Safety methodologies are employed to control (i.e. systematically identify, reduce, mitigate and assess the residuals) risks. Different methodologies here strive for the same goal. Meaningful use of various methodologies together may be achieved only if methods related to different methodologies are put together into one lifecycle.

Figure 1 comes without a claim for completeness. Besides that, cross-dependencies exist, e.g. functional safety addresses unintentional misuse if a misuse case becomes hazardous in case of malfunctions; cybersecurity also handles the misuse cases which are not entirely intentional, e.g. in the area of human-machine interface (HMI). This paper will address FS and SOTIF lifecycle; cybersecurity is not addressed. However, the approach defined here is capable of adapting cybersecurity and possibly other safety methodologies (i.e. safety of mechanical components, chemical safety etc.)

3. Generalized Safety Lifecycle (FS + SOTIF)

3.1. ISO 26262

The lifecycle of functional safety is presented on the well-known "3-V diagram" in the ISO 26262.

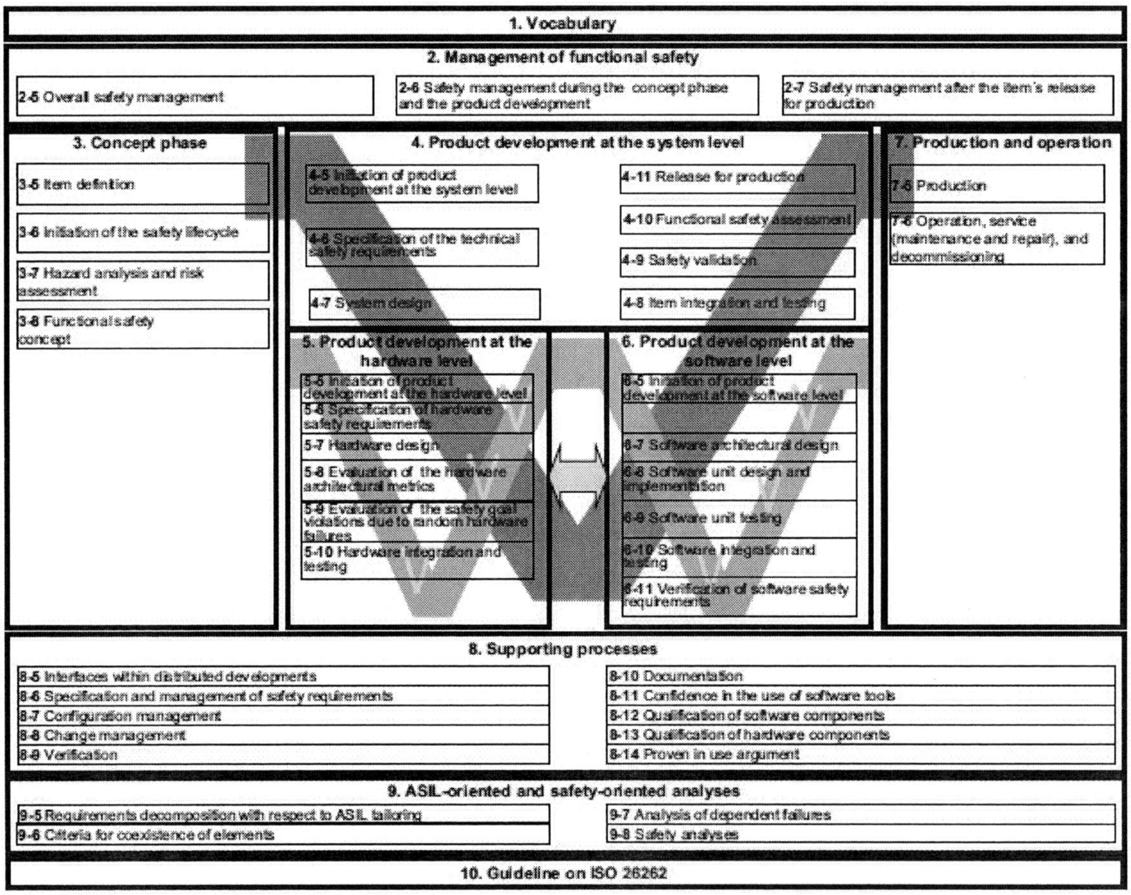

Figure 2. Safety lifecycle according to ISO 26262:2011 [2].

The numbers 1 to 10 mark the parts of the standard relevant to each lifecycle item. The lifecycle consists of five phases: concept phase (Part 3 of the standard), product development on the system (Part 4), HW (Part 5) and SW (Part 6) levels as well as production and operation (Part 7). For now, SOTIF is limited to the concept phase and system-level activities. Technical details of ensuring the safety of the intended functionality on SW level are part of what may be called "Safety of Machine Learning (ML)". Currently those aspects are not addressed within the ISO PAS 21448.

IOP Conf. Series: Materials Science and Engineering **534** (2019) 012019 doi:10.1088/1757-899X/534/1/012019

3.2. ISO PAS 21448

SOTIF lifecycle consists of three major steps:

1. "Evaluate by analysis": identification and evaluation of *triggering events* (i.e. specific conditions of a driving scenario that serve as an initiator for a subsequent system reaction possibly leading to a hazardous event, see [3], def. 3.15).
2. "Evaluate known hazardous scenarios": identification of relevant hazardous scenarios from the set of known hazardous scenarios, explicit hazard testing
3. "Evaluate unknown hazardous scenarios": "statistical testing", i.e. search validation via exposure of the system to the statistically distributed conditions of real world or simulation.

ISO PAS 21448 consists of 12 chapters which run along the V-lifecycle close to the lifecycle of the ISO 26262. Fig. 10 in [3] provides a link from SOTIF lifecycle to FS lifecycle. However, this link is too general and therefore not useful. Table 1 provides a view of a lifecycle integrating FS and SOTIF requirements (generalized safety lifecycle).

4. Generalized Safety Lifecycle (a suggestion)

In Table 1, *italicized* items show work products (WP) existing in FS lifecycle and getting additional requirements from SOTIF lifecycle. Triggering Event Analysis (TEA), the WP highlighted **bold**, is the only SOTIF-specific WP.

Of 14 WPs in Generalized Safety Lifecycle, one (TEA) is newly introduced due to SOTIF and 3 do not get any additional requirements. The residual 10 WPs contain both FS and SOTIF requirements.

Table 1. Generalized safety lifecycle: FS + SOTIF.

Functional Safety		Safety of the Intended Functionality		Generalized Safety Lifecycle (work products)
Chapter	Work product	Chapter	Work product	
3-5	Item Definition	5	Functional and System Specification	*Item Definition*
3-7	Hazard Analysis and Risk Assessment	6	Hazard Identification Hazard Analysis Risk Evaluation Specification of Validation Target	*Hazard Analysis and Risk Assessment*
		7	Identification of Triggering Events Assessment of the Acceptability of Triggering Events	**Triggering Event Analysis (TEA)**
		8	Functional modifications to reduce SOTIF risks	*Functional Safety Concept*
3-8	Functional Safety Concept			
		9	Definition of V&V Strategy	*SOTIF V&V Strategy*

4-6.4.6	System Verification Report	10	SOTIF Verification Report	*System Verification Report*
4-6.4.6.2	Validation Plan			*Validation Plan*
4-7.4	Technical Safety Concept			Technical Safety Concept
4-7.4	System Design Specification			System Design Specification
4-7.4.3	Safety Analysis Report			*Safety Analysis Report*
4-8.4	Integration Testing Plan & Report			Integration Testing Plan and Report
4-9.4	Validation Plan & Report	11	SOTIF Validation Report	*Validation Plan & Report*
4-10.4	Functional Safety Assessment Report			*Safety Analysis Report*
4-11.4	Release for Production Report	12	SOTIF Release Report	*Safety Analysis Report*

The next chapter of this paper goes through specific work products of the Generalized Safety Lifecycle and presents the integration of the FS and SOTIF lifecycles on the level of work products.

5. Integration of the Lifecycles

Now we have defined the work products which should be changed or where additional information needs to be added in order to integrate the requirements of ISO PAS 21448. The next question is: which information needs to be added to each of those WPs and to the safety lifecycle as a whole? What are the requirements of the SOTIF lifecycle?

The paper [4] presents the general safety case (SC) for ML. Even though SOTIF is not equal to "automotive ML applications", it is possible to use the safety case model of the paper to cover SOTIF-related safety cases as well.

The goal of the ML safety case (according to [4]): to prove that the residual risk associated with functional insufficiencies in the object detection and classification function is acceptable. The argument here shall be provided over the space of assumptions on the operational profile of the system, inputs to the ML functions and performance limitations of its algorithms.

The assumptions contain data which are not project-specific and therefore not part of the project's safety case. They may be fully or partially shared between projects. Each project may use the whole set of available data or a sub-set. The data may be found via safety analyses which are part of standard safety lifecycle as well as via SOTIF-specific steps of the safety lifecycle. The only SOTIF-specific WP identified in Table 1 is the Triggering Event Analysis (TEA). However, another allocation of SOTIF requirements may result in other WPs characterizing the SOTIF-specific part of the lifecycle.

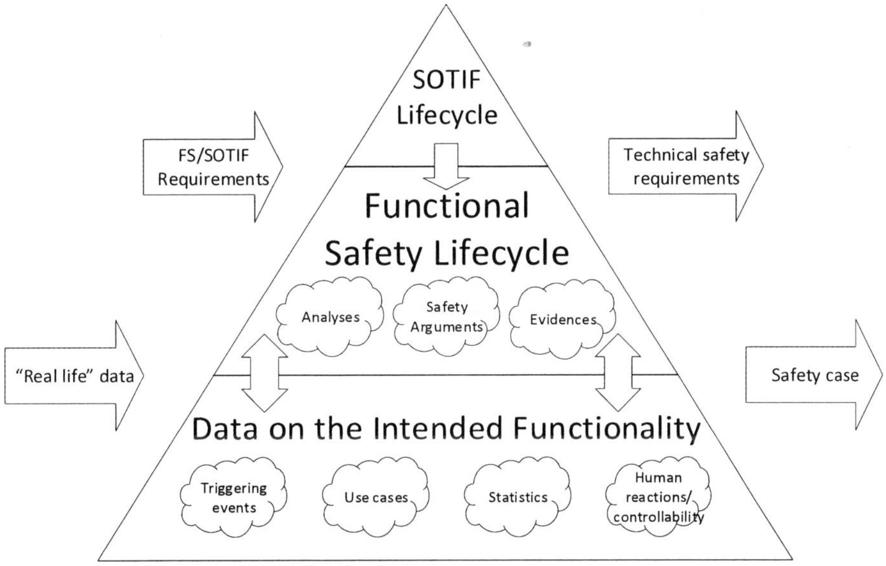

Figure 3. Structure of the Generalized Safety Lifecycle.

The Generalized Safety Lifecycle as defined in §4 is represented graphically on the Figure 3. It consists of functional safety lifecycle according to ISO 26262 with some WPs receiving requirements from ISO PAS 21448 alongside with normal safety requirements. The safety lifecycle is extended via SOTIF lifecycle, currently containing only the procedure of the Triggering Event Analysis. Both FS and SOTIF parts of the Generalized Safety Lifecycle are based on the data on the intended functionality.

The data on the intended functionality include the information currently known about the functionality to be implemented in the project. The data are project-independent; therefore, it may be stored centrally in a databank within an organization or an organizational division.

The necessary data on the intended functionality include:

- Triggering events
- Known relevant use cases, both safe and hazardous, with references to the relevant triggering events
- Statistics on the occurrence of the known relevant use cases
- Expected human reactions in hazardous cases. This is specifically needed for the systems featuring SAE automation levels of 1 to 3, as the safety case for those systems typically includes human driver as a safety measure.

The databank may include more data, depending on the data model implemented within the organization.

5.1. Hazard Analysis and Risk Assessment

The HARA consists of hazard identification and hazard evaluation.

Hazard identification is an inductive process aimed at the identification of possible hazards as well as risk related to them. Hazard identification always takes into account the situation (use case, driver's and road conditions etc.), which may be considered external to the system and the deviation of the actual system behavior from some expected behavior. Generalized Safety Lifecycle follows the FS lifecycle in consideration of functional failures (malfunctions) as deviations for hazard identification.

Hazard evaluation happens differently within FS and SOTIF frameworks. This is caused by the different understanding of the "reasonable" or target levels of risk in the two standards.

ISO 26262 features four Automotive Safety Integrity Levels (ASILs) as well as "QM" safety level for the cases where the residual risk is considered acceptable to be minimized by quality management approaches without implementation of any specific safety measures. ASIL levels play a dual role: they measure risks associated with a specific malfunction in a given situation (with ASIL A for the lowest risk and ASIL D for the highest), and at the same time ASIL allocation defines the requirements of the standard applicable to the system, i.e. the amount of effort to be invested in order to minimize the existing risk to a tolerable level. Thus, both possible levels of risk to be found in the system as well as the tolerable level are given by the standard itself, even though this information is implicit (unlike e.g. IEC 61508 where quantitative risk assessment is possible).

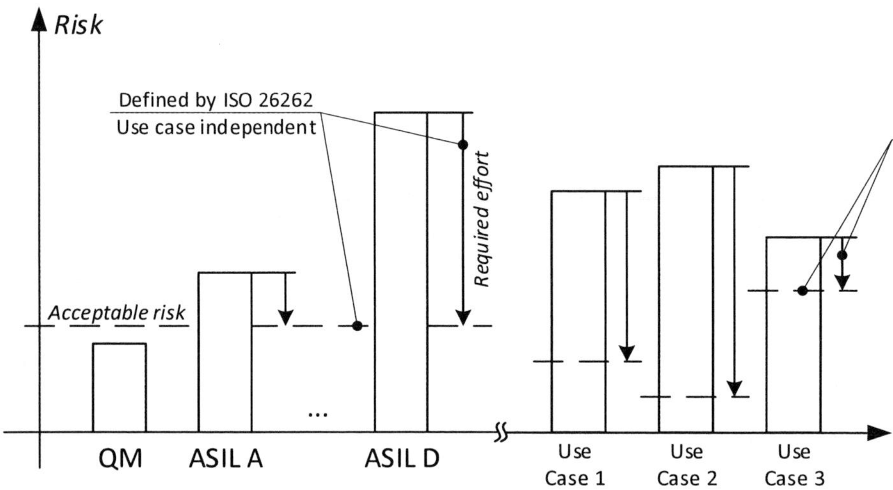

Figure 4. Acceptable risk and required effort: ISO 26262 and SOTIF.

According to ISO 26262, only the most hazardous use case is relevant for the definition of the maximum ASIL for the relevant safety goal. Then, the defined safety goal is used as a proxy for all technical activities. At the same time, SOTIF standard prescribes neither the acceptable level of risk nor the effort required to reach it, neither implicitly nor explicitly. Each use case shall be considered separately. Although the required technical measures can be shared between use cases (e.g. using safety goals as proxy similar to ISO 26262), safety of each use case shall be eventually proven separately, unless the use cases considered belonging to one equivalence class (see §5.3 below).

5.1.1. Use-cases for hazard identification

A use-case consists of the functional range, desired behavior, functional system boundaries, and the scenario (cf. [4]). Scene is the static part of scenario. The class diagram of the use case according to [3] is presented on the Figure 5.

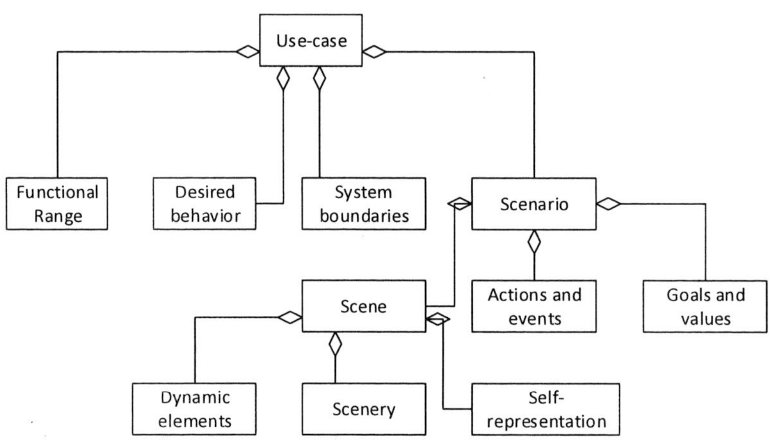

Figure 5. Use-case definition according to [3].

Example of a fully-defined use-case is given below:

1. Functional range: Automated Emergency Braking (AEB)
2. Desired behavior: The system activates braking if a collision with an obstacle is considered highly probable
3. System boundaries: Camera / radar system, AEB ECU, braking ECU, braking system.
4. Scenario
   - 4.1. Actions and events:
     - i) The ego vehicle (i.e. the vehicle equipped with the AEB functionality considered in the analysis) is moving forward
     - ii) No obstacle exists in the path of the ego vehicle within the hazardous distance (2 sec)
   - 4.2. Goals & values
     - i) Goal: keep travelling down the road, no lane change, no direction change
     - ii) Value: preferred speed, lateral distance, longitudinal distance
   - 4.3. Scene
     - 4.3.1. Dynamic elements: other vehicles on the road (e.g. a bus in front of the ego vehicle, at distance of 3 sec and another car behind the ego vehicle, distance 5 sec)
     - 4.3.2. Scenery
       - i) Intra-urban
       - ii) Right-hand traffic
       - iii) Two-lane road: right lane and left lane
       - iv) Bicycle lane on the right of the right lane
       - v) Pedestrian pass on the right of the bicycle lane
       - vi) Road is symmetric (i.e. it has two lanes, bicycle lane and pedestrian pass in the opposite direction)
     - 4.3.3. Self-representation: ego vehicle is a passenger car.

New use-cases are obtained via iterating the changeable parameters. Changeable parameters include actions and events, values, and parameters of the scene.

In order to prevent the explosion of the number of considered use cases, creativity is required in the iteration process. Only parameters relevant to the chosen functionality shall be iterated. The list above considers the AEB functionality, which influences the longitudinal dynamics of the vehicle. Therefore, lateral conditions (e.g. cars travelling in other lanes) are irrelevant for the hazards related to the AEB. They should not be iterated.

5.1.2. Hazard evaluation

The hazards are evaluated on the basis of their probability. SOTIF considers non-probabilistic models for system limitations, i.e. the system will experience a malfunction whenever a hazardous use case shows up. Therefore, the probability of a hazard is fully defined by the probabilistic characteristics of the hazardous use case.

Probabilities of the realization of a use case are expressed per hour of driving or per kilometer. HARA shall define this probability, which is later used for the definition of required validation mileage (see §4.3). Obviously, the probability cannot be defined on the level of detail required for the use case definition (see list in §5.1.1). Use cases may be grouped into sets where probability of exposure may be calculated for the whole set. This strategy will require the validation to be based on the whole set of use cases.

Probabilistic modelling yields relatively precise data for determination of probabilities of hazardous use cases (e.g. [5]). An example of underlying risk-generating process is shown on Figure 6. The probability of the absorbing states on the right (both "Safe driving" and "Accidents") is known from the traffic and accident statistics. Quality of the assessment of the probabilities of driving situations depends on the precision of the modelling of the transitional process (e.g. transition rates for Markov chain modelling).

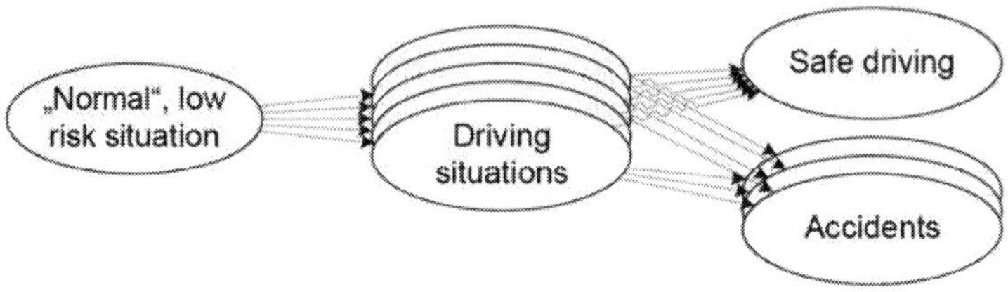

Figure 6. Risk-generating process as Markov chain (e.g. [5].

Quality-of-fit parameters (confidence intervals, likelihood etc.) may be used to understand how good the probability assessment is. The likelihood can be increased using data from previous validations. Models yield quality-of-fit estimations directly, while for other methods of the assessment those estimations shall be drawn based on assumptions.

Known use cases (i.e. use cases identified as hazardous for a given malfunction) are part of the "Data on the Intended Functionality" (see Figure 3). They are mostly not project-specific and belong to a centralized data-store.

5.2. Triggering Events Analysis

Triggering Event Analysis (TEA) is the only SOTIF-specific WP of the Generalized Safety Lifecycle. It is important to understand that a triggering event is not a use case, it is a characteristic of multiple use cases.

The TEA consists of the identification of the triggering events (TEs) and the evaluation of their acceptability (similarly to HARA).

The suggested process includes analysis of SOTIF-critical characteristics (SOTIF-CCs), similar to critical characteristics as defined in the FMEA process (e.g. [6]), related to the functionality potentially affected by hazards related to functional insufficiencies.

5.2.1. Identification of relevant TEs

The inclusion of SOTIF-CCs harmonizes the TEA process with FMEA process and allows the implementation of the TEA via any common FMEA tool. Example of SOTIF-CCs related to camera are sharpness, brightness and classification quality. The failures are identified based on the SOTIF-CCs using a HAZOP-like procedure. Those failures may be linked directly to malfunctions previously defined within the project (normally malfunctions are defined at the step of Item Definition).

The list of triggering events shall be prepared, based on the groups described in [3]: environment and location, road infrastructure, urban infrastructure, highway infrastructure, driver behavior (including reasonably foreseeable driver misuse), expected behavior of other drivers/road users, etc. The list may be pre-defined and uploaded into the FMEA tool. Then, pre-defined TEs are linked with failures, which are in turn linked to malfunctions. A failure net linking TEs to malfunctions appears.

5.2.2. Evaluation of TE acceptability

According to [3], a TE is acceptable if all use cases related to it have low exposure probability, high controllability, or risks associated to them bear low severity. Using the definitions from ISO 26262-3, in order to qualify as acceptable a TE shall be either linked only to use cases occurring far less than 1% of all driving time (e.g. an airplane landing on a highway), or the use cases easily controllable (i.e. more than 99% of drivers may resolve the hazardous situations without risks), or the use cases where there is no risk of any bodily injury.

Having the TEA and HARA prepared in a systematic way allows easy identification of acceptable and non-acceptable TEs via database tools.

Relevant triggering events are part of the "Data on the Intended Functionality" (see Figure 3). They are not project-specific and belong to a centralized data-store. The list of TEs relevant to a given project is a subset of that general list.

5.3. Validation of the Unknown Use Cases
The goal of the validation of the unknown use cases is to show that the system under development does not pose any unacceptable risk even for the use cases which were not directly considered during the development process.

The selection of the use cases to be picked for testing shall follow the distribution of the use cases in the general population. This does not necessarily include specific use cases, especially those related to the triggering events, as they should be already sorted out during verification phase.

Basic probabilities of the use cases are given in the HARA (see §5.1). The probabilities are given per hour or per kilometer. However, just inverting the number will give under 50% of confidence in safety of this use case. The full validation mileage shall be calculated considering the probabilities of the use cases to be binomially distributed [7]. Typical validation distance for a level 2 function is measured by many hundreds of thousands of kilometers, up to one million. However, the proper planning of the validation distance is more important than its length. The conditions of driving (e.g. weather, locality, intra/extra-urban driving etc.) shall be weighted according to the distribution experienced by the target vehicles. Statistics related to the road traffic gathered by local traffic authorities (e.g. so-called "Green Book" by AASHTO [8]) is the main source of information related to this.

Driving and accidents statistics are part of the "Data on the Intended Functionality" (see Figure 3). They are not project-specific and belong to a centralized data-store.

5.4. Joint Safety Assessment Report

The Safety Assessment Report (SAR) summarizes safety evidences and provides a safety argument, i.e. explanation of the findings of the safety lifecycle. The SAR is normally written in a semi-formal way. For Generalized Safety Lifecycle, the SAR is basically created according to the requirements of ISO 26262, but includes SOTIF-related evidences as well.

No specific changes in the process of creation of the SAR are required.

6. Conclusions

The main contribution of this paper is the Generalized Safety Lifecycle, integrating the lifecycles of functional safety and the safety of the intended functionality. Distribution of the requirements between WPs is shown (Table 1). Concept of a separate databank for storage of the non-project specific data on the intended functionality is introduced (§5, Figure 3). Integration of both lifecycles makes SOTIF clearer and easier to implement.

Functional safety is often perceived as "just another requirement" from the authorities or the customer. However, if implemented based on system engineering methodology, both FS and SOTIF help improving the understanding of the system under development and making it safer. They facilitate future updates and modifications, thus reducing the cost of development and aftermarket activities.

Acknowledgments

The authors want to express their gratitude to Mr. Joseph Miller (J D Miller Consulting, formerly ZF Friedrichshafen), Mr. Konstantin Byakov, PhD (Bauman MSTU), Dr. Brian Murray, Dr. Stefan Rinkenauer and Dr. Sari Buelent (all ZF Friedrichshafen), and many other colleagues who contributed to the development of this concept.

References

[1] EAEC Technical Regulation TR TS 018/2011 *Safety of Wheeled Vehicles* (Moscow)
[2] ISO 26262-1:2011. *Road vehicles – Functional safety* (Geneva)
[3] ISO/PRF PAS 21448 2018 *Road vehicles – Safety of the intended functionality* (Geneva)

[4] Burton S Gauerhof L and Heinzemann C 2017 *Making the Case for Safety of Machine Learning in Highly Automated Driving* SAFECOMP Workshops LNCS 10489 pp 5–16
[5] Lurie O and Miller J 2018 Hazard Analysis and Risk Assessment beyond ISO 26262 Management of Complexity Via Parametrization *SAE Technical Paper 2018-01-1067* I information on https://doi.org/10.4271/2018-01-1067
[6] AIAG VDA 2017 *Failure Modes and Effect Analysis Manual* (Berlin)
[7] Koester F Form T Lemmer T Plaettner J 2016 Wie gut müssen automatisierte Fahrzeuge fahren – PEGASUS *Pegasus Project*
[8] American Association of State Highway and Transportation Officials 2001 *A Policy on Geometric Design of Highways and Streets* (Washington)

IOP Conf. Series: Materials Science and Engineering **534** (2019) 012020 doi:10.1088/1757-899X/534/1/012020

Test bench development for signal registration in millimeter wave automotive radars

D V Valuyskiy, A A Panarina, S V Vityazev and V V Vityazev

Ryazan State Radio Engineering University (RSREU), Gagarina str., 59/1, Ryazan, Russia, 390005

Abstract. A development of a bench for automotive radar signals registration and analysis is considered in this paper. The bench is based on the AWR1642boost evaluation module with 77 GHz carrying frequency. Supporting software is developed to help signal analysis and reflection model creation. The bench aims an investigation of signal reflection properties from some specific types of obstacles and the development of the reflection mathematical models.

1. Introduction

Millimeter wave radars are widely used in different types of modern remote sensing systems including automotive radar imaging. Millimeter wave radars have some important advantages over centimeter waves including less antenna size and system form factor; much simpler technical design of wavelength dependent components; high transmission power and wide frequency band at the same time; cheapness of target products [1]. Radars make advanced driver assistance systems (ADAS) more functional and reliable in poor weather conditions, in low light, at long distances and in some other cases [2]. One of functions ADAS radar can offer is a detection of obstacles at the road in front of a vehicle at a distance enough to prevent an accident. Some fallen trees and other unwanted objects, pothole, icing can become very dangerous obstacles. The development of ADAS radar which would be able to solve the described problem efficiently requires the knowledge of the mathematical model of radio wave reflection from different obstacle types. Specific properties of road obstacle reflections are of a great interest.

In [3] the analysis of radar reflections from the asphalt surfaces in millimeter wave range is carried out. The reflections from dry, wet and iced asphalt are considered and compared. An ability to detect a road surface condition with icing and snowing detection is analyzed in [4]. Wide set of experiments on scattering cross section measurement for different types of typical road objects is carried out in [5].

In this paper the development of hardware and software facilities for the registration of test signals in millimeter wave radar system (77 GHz frequency band) is considered. Unlike the works mentioned above the main purpose of the test bench development is to understand the properties of reflections from obstacles of a specific kind: unwanted objects and potholes; and to form their mathematical model.

2. Hardware components of the bench

An AWR1642boost evaluation module from Texas Instruments is used as a basic hardware component of the bench. This is a highly integrated system-on-a-chip which includes ARM-family master core, C674x digital signal processor (DSP) core and the radar subsystem with analog and radio frequency transceiver path and analog to digital conversion [6]. Frequency modulated continuous wave (FMCW) signals are used to get good range and velocity resolution. Frequency band is centered at 77 GHz. Azimuth resolution is achieved with an antenna array, including four receiving and two transmitting antennas with an ability to use multiple input multiple output (MIMO) approach to improve the resolution.

The AWR1642boost module is installed on a front bumper of a car to carry out field experiments.

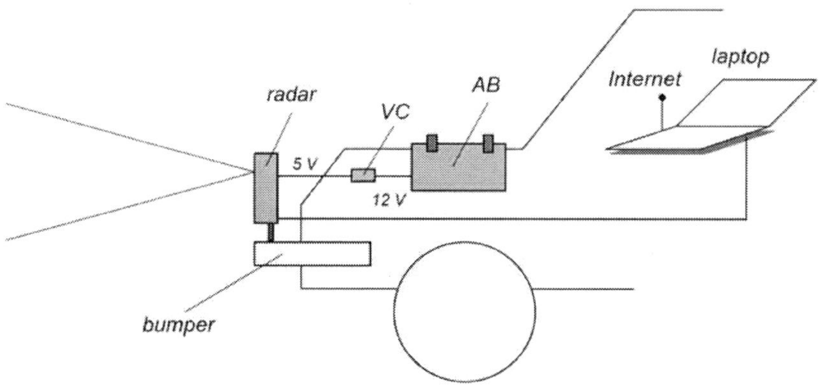

Figure 1. A diagram of AWR1642boost based test bench, installed on a car.

The AWR1642boost board is placed in a radio transparent case to protect it from dirt, water or mechanical damages. Easy installation and removal of the module is provided by bolting. The module is powered by the automotive battery (AB). The battery voltage of 12 V is converted to the supply voltage of 5 V required by the module. Voltage converter (VC), made on the basis of the stabilizer SCV0042-5V-0.9 A, is used for this purpose. The structure of the bench is shown in figure 1.

The module is connected to the laptop via the debug port, allowing reading signal records. Laptop connection to the Internet should be provided to allow the use of the software shell Demo Visualizer from Texas Instruments. This tool provides user interface when working with the module.

The functionality of the bench is as follows. The car is sent to the place of the experiments. In place the module in the case is installed onto the front bumper of the car. Easy installation is performed with two nuts. Power and laptop interface cables are connected to the radar. These cables are pre-installed and are not removed from the vehicle. On the laptop in the Demo Visualizer environment, the required radar operation mode is set, including the selection of signal parameters and the desired radar characteristics. An object, the reflecting properties of which are studied is placed in front of the car at the required distance, or the car is placed so that the object of interest is in the field of view of the radar. Demo Visualizer on the laptop screen builds radar image of the observed scene in real time. When the scene is set up, the signal recording is started for further processing and analysis in the laboratory. The signal can be recorded when the vehicle is stationary or in motion.

3. Software components of the bench

The software used for signal recording and analysis can be divided into two components. The first one is the software used during signal recording. The second component is a specially developed environment, which combines the capabilities of signal modeling, real records analysis, reflection model development and testing.

The software used during signal registration process includes the tools which are provided together with evaluation module. This includes Demo Visualizer, which has already been mentioned above, and a set of program codes loaded on the processor. The processor codes are open source and allow modification and reload of the executable files. The codes are divided into the programs for the ARM control core, and the programs executed by the DSP signal processing core. The Code Composer

IOP Conf. Series: Materials Science and Engineering **534** (2019) 012020 doi:10.1088/1757-899X/534/1/012020

Studio integrated development environment is used to modify these programs and reload the processor.

DSP software implements a signal processing algorithm typical for ADAS FMCW-radars. Fast Fourier transform (FFT) with pre-weight processing provides range resolution. The second FFT provides the resolution for velocity. Averaging over the receiving channels from four antennas and the use of two-time threshold processing (before and after velocity resolution) increases the reliability of object detection. The third FFT is performed across the receiving channels and provides azimuth resolution. The azimuth FFT can be 4 (according to the number of real antennas) or 8 point if MIMO technology is used (according to the number of transmitting and receiving antenna pairs). A 1-point discrete Fourier transform (DFT) is performed before the azimuth FFT. In fact, it just repeats the velocity FFT but only for those ranges and velocities which were previously detected, so this DFT is very low complexity. This approach is used because only limited internal DSP memory can be available in practice. It becomes possible not to store the results of the second FFT, but to perform it again, but in much simpler conditions.

The DSP software includes general purpose library functions and also the codes written specifically for this task. The software can be easily extended or modified. In particular, in the case of the development of the described test bench DSP software is modified to give the ability to record "raw data", that is, fixing the signal samples coming directly from the analog-to-digital converter (ADC) without any digital processing. The volume of recording in the current implementation of software is limited by the DSP memory and corresponds to the reflection of one frame of radar pulses.

ARM-core software performs system initialization and organizes chip components interconnections.

The Demo Visualizer environment executed at the laptop receives the data calculated by the DSP-core and visualizes it as a radar imagery.

A specially developed software shell targeting signal modeling, real data analysis and reflection model formulation and testing is another part of test bench software. This shell includes three main components which address the three parts of graphical user interface located at the top, middle and bottom of the screen, respectively. The first component is responsible for selecting the mode of operation: signal model or real recording. If we are working with a signal model, the type and parameters of the obstacle, such as material, size, distance, location geometry and some other parameters are selected. In the case of real record operation, the parameters at which this record was made are entered.

The second component performs the tasks of calculating the signal model, visualizing the characteristics of the signal, analyzing its properties.

The third component performs signal processing and visualizes the results. A user can set the processing mode, select the preferred modification of the algorithm, analyze the quality of the image. The samples of the signal generated during mathematical modeling for the case of four point reflectors are shown in figure 2(a). The model is calculated for the case of four receiving antennas and a frame of radar pulses. The result of the processing is shown in figure 2(b). It represents the output of 2-dimentional FFT procedure across range and velocity bins.

IOP Conf. Series: Materials Science and Engineering **534** (2019) 012020 doi:10.1088/1757-899X/534/1/012020

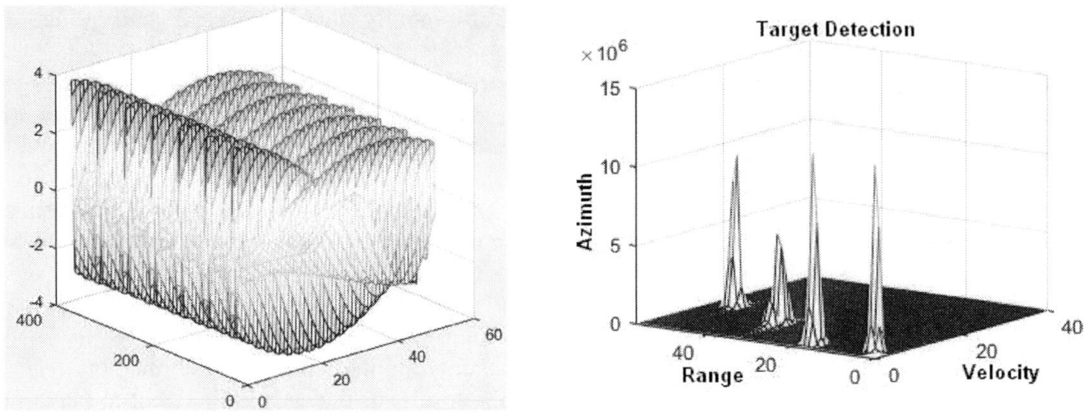

(a) (b)

Figure 2. A model of a reflected signal (a) and the results of its processing in the developed software shell (b).

4. Conclusion

The paper describes the development of software and hardware for registration of test signals to study the properties of reflection of electromagnetic waves in the millimeter range from objects that can influence much the safety on the roadway. Hardware components of the test bench and software tools that perform analysis of test records and their processing are considered. Further work is related to the development of a mathematical model of the reflections, which is waited to be formed on the basis of the test records. Then the model should be integrated in the software environment of ADAS radar modeling and investigation, which finally is waited to result in a development of the improved algorithms for the detection of obstacles of the given type.

Acknowledgments

The reported study was funded by RFBR according to the research project № 18-07-00419 A.

References

[1] Hasch J, Topak E, Schnabel R, Zwick T, et al. 2012 Millimeter-Wave Technology for Automotive Radar Sensors in the 77 GHz Frequency Band *IEEE Transactions on microwave theory and techniques* **60**(3) pp 845-860 (DOI: 10.1109/TMTT.2011.2178427)

[2] Patole S M, Torlak M, Wang D and Ali M 2017 Automotive radars: A review of signal processing techniques *IEEE Signal Processing Magazine* **34**(2) pp 22-35 (DOI: 10.1109/MSP.2016.2628914)

[3] Sarabandi K, Li E S and Nashashibi A 1997 Modeling and Measurements of Scattering from Road Surfaces at Millimeter-Wave Frequencies *IEEE transactions on antennas and propagation* **45**(11) pp 1679–88 (DOI: 10.1109/8.650080)

[4] Bystrov A, Abbas M, Hoare E, Tran T Y, et al. 2014 Remote road surface identification using radar and ultrasonic sensors *Proceedings of 11th European Radar Conference* pp 185-188 (DOI: 10.1109/EuRAD.2014.6991238)

[5] Shnajder V B 2014 Radar system for the traffic safety of ground vehicles *PhD dissertation* (Moscow: MAI) p 109

[6] 2017 *AWR1642 Evaluation Module (AWR1642BOOST) Single-Chip mmWave Sensing Solution* User's Guide – Texas Instruments Incorporated p 24

Wheel vehicle dynamics real-time simulation for on-board stand-alone moving control system realization

N V Buzunov[1], G O Kotiev[2] and B V Padalkin[3]

Bauman Moscow State Technical University, 105005 Baumanskaya 2th st., 5, Moscow, Russian Federation

E-mail: [1] buzunovnv@bmstu.ru, [2] kotievgo@yandex.ru, [3] padalkin@bmstu.ru

Abstract. The main problem of design on-board stand-alone moving control system is necessity of prototype object during all design stages. The real-time simulation models of wheel vehicle dynamics help to cope with such difficulties. These models must be similar to real wheel vehicle object sufficiently. The research describes the order of creation real-time simulation models of wheel vehicle dynamics using implicit numerical implementation method with application of higher order derivatives for solving differential equations system. Also the verification and proof of adequacy method is presented. It is noticed that the created real-time model corresponds to physical wheel vehicle object if the solving step of numerical implementation method no more than 0.001 second. Such real-time models can be used for design on-board stand-alone moving control systems.

Current development trends for vehicle on-board network are responsible for advanced level of automation of both separate systems and entire object control and diagnostic processes. On-board devices conjunction into single hardware-software system represents on-board information-control system. The presented approach into on-board network construction is a basis for stand-along moving control system construction.

One of the most prospective approaches to the construction of on-board control systems is application of real-time operating controlled objects at stages of debugging and calibration of simulation mathematical models. Virtual prototypes significantly simplify stand-alone moving control system construction process, they make it possible to avoid a number of errors at stages of prototyping, which leads to significant reduction of control system development time and significant cost reduction in the future.

Study object real-time dynamic simulation mathematical model performance is characterized by the following criterion: performance quality depends not only on logical validity of calculations, but also on calculations time [1]. Model functions in real-time, if operating speed corresponds to physic process speed on controlled objects.

Real-time simulation process can be described with the use of diagram, which is illustrated on Figure 1 [2].

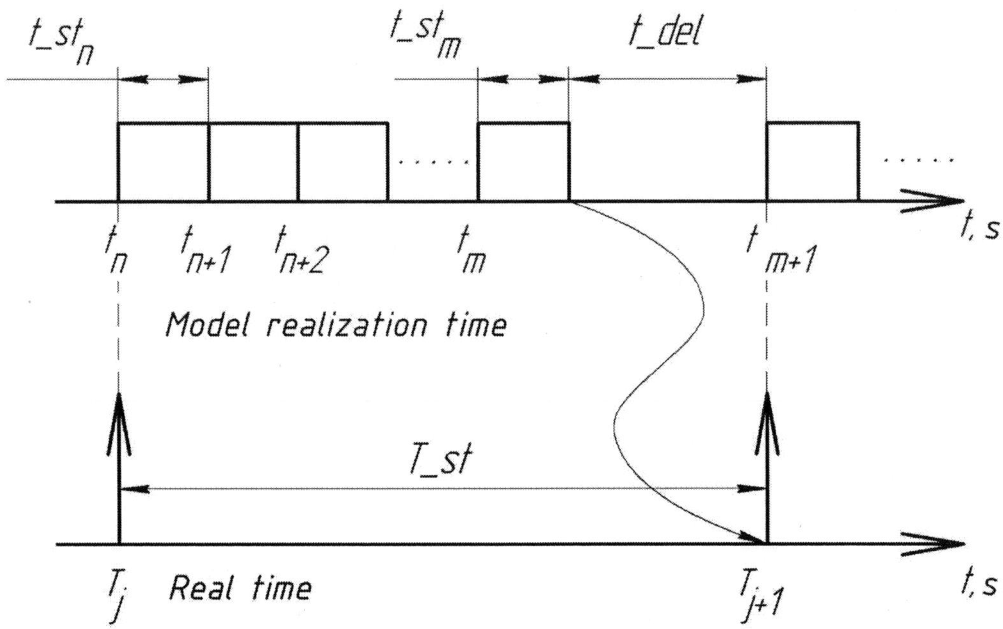

Figure 1. Real-time implementation of mathematical model:
T_j, T_{j+1} – current values (samples) of "real-time", s;
T_st – time step, s;
$t_n...t_{m+1}$ – current value of simulation time, s;
$t\_st_n...t\_st_m$ – numerical integration steps, s;
t_del – duration of delay during numerical integration, s;
n . . . m+1 – sequence numbers of numerical integration steps;
j . . j+1 – sequence numbers of "real-time" samples.

Implementation of fixed numerical integration step in cases of non-linear systems study is not possible [3], [4]. This fact defines requirements for selection of non-linear system mathematical model realization numerical method: method-operating algorithm has to provide integration step changing.

During stand-alone moving control system and wheeled vehicle real-time dynamic model co-operative work implementation, one of the main tasks in study of object respond on different control actions. For mathematical model adequacy to real object in defined area, transient process has to be simulated more accurately, then stationary process. Therefore, integration numeric method has to provide dynamic changing of implementation step during simulation. Runge-Kutta method, Miln method and implicit integration method with the use of higher derivatives meet the given requirement [5].

Another requirement to the numerical integration methods is based on the fact that given simulation mathematical model of wheeled vehicle has to operate in real-time mode to make its interoperability with physical control system possible. In order to meet this requirement, numerical integration method has to provide sufficient computational performance, considering solution stability check for recurrent iteration. Approximate solution stability check with the use of Runge-Kutta method significantly reduces computational performance of numerical integration.

Miln method and implicit integration method with the use of higher derivatives are free from this disadvantage. While using Miln method, it is necessary to preliminary estimate four first function samples and corresponding derivatives («starting section»), where after received data is put into approximation formulas for fifth and following samples of the required function. Estimation of initial values can be performed with any other numerical method. When it is needed to reduce given solution step in order to evaluate first required function sample, following after "starting section", "starting section" have also to be recalculated with the modified integration step.

In this context, implicit integration method with the use of higher derivatives is represented as an optimal one for realization of real-time dynamic models, which are to be implemented in the framework of development of stand-alone moving control systems of wheeled vehicles.

Numerical simulation can be represented with the following scheme.

Let us assume, that displacement vector x_i and velocity vector $\dot{x}_i$, as projections on generalized coordinates, are known at the moment t_i. Position and velocity at the moment $t_{i+1}=t_i+h$, where h is forecast step, are forecasted. Herewith, vectors of time derivatives of miscellaneous orders are estimated in generalized coordinates for the moment t_i.

Process of numerical integration of differential equation is based on Taylor expansion of functions $x(t)$ and $\dot{x}(t)$ at a point t_i; in consists of three stages.

1. Tentative forecast of displacement and velocity with order of accuracy of h^4 and h^3, correspondingly:

$$x_{i+1} = x_i + h \cdot \dot{x}_i + \frac{h^2}{2} \cdot \ddot{x}_i + \frac{h^3}{3!} \cdot \dddot{x}_i, \tag{1}$$

$$\dot{x}_{i+1} = \dot{x}_i + h \cdot \ddot{x}_i + \frac{h^2}{2} \cdot \dddot{x}_i. \tag{2}$$

Following the estimation results, values of second and third derivatives are calculated at a point $i+1$. At this stage, received function values and also values of its first, second and third derivatives represent initial approximation.

2. Tentative forecast correction with accuracy to h^5. Based on results, received on first stage, displacement and velocity of object are estimated [5]:

$$x_{i+1} = x_i + h \cdot \frac{(\dot{x}_i + \dot{x}_{i+1})}{2} - \frac{h^2}{10} \cdot (\ddot{x}_{i+1} - \ddot{x}_i) + \frac{h^3}{120} \cdot (\dddot{x}_{i+1} + \dddot{x}_i) \ , \tag{3}$$

$$\dot{x}_{i+1} = \dot{x}_i + h \cdot \frac{(\ddot{x}_i + \ddot{x}_{i+1})}{2} + \frac{h^2}{12} \cdot (\dddot{x}_i - \dddot{x}_{i+1}). \tag{4}$$

Values, received according to equation (3) and (4), are implemented for modification and correction of second and first derivative values at a point $i+1$. Received at this stage function values and also values of its first, second and third derivatives represent second approximation.

3. Stability check of estimated values x_{i+1}, $\dot{x}_{i+1}$. According to equation (3) and (4), with the use of second approximation, displacement and velocity at a point $i+1$ are estimated. These values represent third approximation. Then absolute values of differences of second and third approximations are estimated correspondingly. The received difference values should not exceed accuracy, defined before computations had started. In case of fulfillment of this condition, third approximations are used for next simulation step. Otherwise, computation, starting from paragraph 1, should be repeated for a moment t_{i+1}, assuming for this iteration step $h/2$.

In the presented paper realization of real-time dynamic model of plane non-linear motion of multi-axle wheeled gear (MWG) with axle arrangement 6x6 is performed on C++ high-level programming language. Implicit integration method with the use of higher derivatives is implemented as an integration method. Presented in papers [6], model of non-linear vehicle motion is implemented as an initial ("reference") model. This model is realized with Simulink software, which is a component of program complex Matlab™.

Regarded virtual object is a base for development of stand-alone moving control system hardware and software.

IOP Conf. Series: Materials Science and Engineering **534** (2019) 012021 doi:10.1088/1757-899X/534/1/012021

In investigated real-time model, motion of vehicle, as of rigid body, is analyzed in horizontal plane of even non-deformable supporting surface and it consists of translational motion of mass center and of rotational motion around mass center (Figure 2) [11].

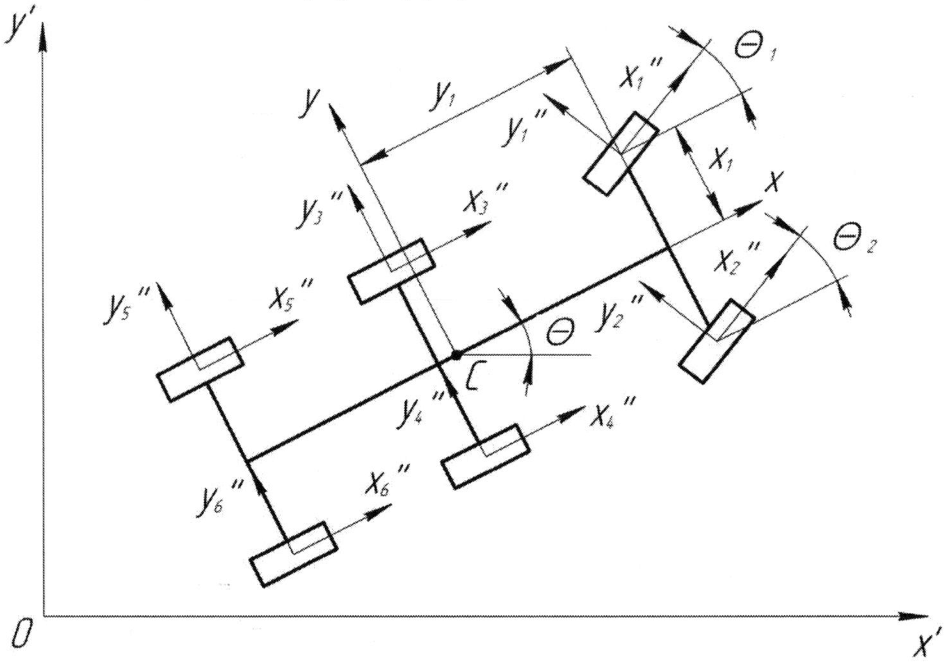

Figure 2. Systems of coordinates, used while simulating of non-linear motion of wheeled vehicle with axle arrangement 6x6:
x' - O - y' – fixed coordinate system (FCS);
x - C - y – moving coordinate system (MCS), attached to object mass center;
x_i'' - O_i - y_i'' – moving coordinate system (MCS), attached to axis of rotation of i-th wheel;
Θ – heading angle of wheeled vehicle model;
θ_i – angular displacement of i-th wheel.

System of equations (5), describing presented motion, makes it possible to calculate current accelerations, based on values of forces and moments, which are interacting a vehicle:

$$\begin{cases} a_x = \dfrac{dV_x}{dt} - \omega_z \cdot V_y = \dfrac{1}{m} \cdot \left(P_{wx} + \sum\limits_{i=1}^{6} R_{xi} \right) \\ a_y = \dfrac{dV_y}{dt} + \omega_z \cdot V_x = \dfrac{1}{m} \cdot \left(P_{wy} + \sum\limits_{i=1}^{6} R_{yi} \right) \\ J_z \cdot \dfrac{d\omega_z}{dt} = \sum\limits_{i=1}^{6} M_{tri} + \sum\limits_{i=1}^{6} M\left(\vec{R}_i\right) \\ V_{x'} = \dfrac{dx'}{dt} = V_x \cdot \cos\theta - V_y \cdot \sin\theta \\ V_{y'} = \dfrac{dy'}{dt} = V_x \cdot \sin\theta + V_y \cdot \cos\theta \\ \omega_z = \dfrac{d\theta}{dt} \end{cases}, \tag{5}$$

where m – vehicle mass; J_z – vehicle mass moment of inertia about an axis z; $\vec{V}$ - vehicle mass center velocity vector; $\vec{a}$ - vehicle mass center acceleration vector (absolute derivative of vehicle mass center velocity vector); $\frac{d\vec{V}}{dt}$ - relative derivative of vehicle mass center velocity vector; $\vec{\omega}$ - vehicle turning rate vector; θ - vehicle turning rate about an axis x'; x', y' - coordinates of vehicle mass center in fixed coordinate system; x - y - moving coordinate system, attached to vehicle; x_i'', y_i'' - coordinate system, attached to vehicle i-th wheel; $\vec{R}_i$ - vector of undercoat interaction force, affecting i-th wheel; $\vec{P}_w$ - vector of air resistance; M_{tri} – moment of resistance to i-th wheel turn.

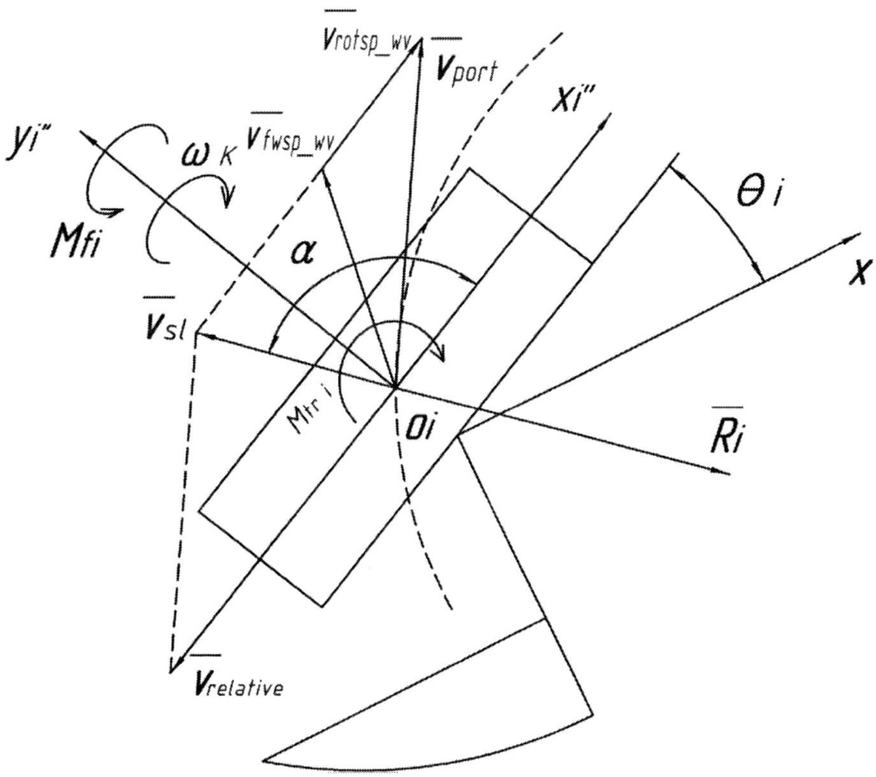

Figure 3. Wheel analytical model.

Figure 3 shows wheel analytical model [12], [15], which is implemented for the presented real-time model. The following conventional signs are used:

$\vec{V}sl$ – vector of sliding velocity of wheel contact point relative to supporting surface;

$\vec{V}relative$ – vector of wheel relative velocity at a point of contact area;

$\vec{V}port$ – vector of transport velocity of wheel rotational center O_i (vector of velocity of point MCS x - C - y, which is currently connected to O_i of MCS x_i'' - O_i - y_i'', relative to FCS x' - O - y');

$\vec{V}fwsp\_wv$ – vector of linear velocity of translational motion of mass center of wheeled vehicle in FCS x' - O - y';

$\vec{V}rotsp\_wv$ – vector of velocity of rotational motion of i-th wheel center around mass center of wheeled vehicle;

α – rotational displacement of vector of sliding velocity of wheel relatively to x_i'' axis of x_i'' - O_i -y_i'' system;

IOP Conf. Series: Materials Science and Engineering **534** (2019) 012021 doi:10.1088/1757-899X/534/1/012021

ω_{κ} – angular velocity of i-th wheel rotation;

$\vec{R}_i$ – force of i-th wheel interaction with supporting surface;

θ_i – rotational angle of i-th wheel;

M_{fi} – modulus of rolling resistance of i-th wheel;

M_{tri} – modulus of cornering resistance of i-th wheel.

Table 1 presents input data and model parameters.

Table 1. Input data and model parameters.

Parameter	Value
Vehicle mass, kg:	24500
Length, m:	5,9
Wheel track, m:	2,1
Mass center height, m:	1,66
Maximum velocity, km/h:	110
Wheel rolling radius, m:	0,615
Coefficient of engagement in X axis direction, *max*	0,6
Coefficient of engagement in Y axis direction, *max*	0,6
Coefficient, which is defining shape of φ(s) diagram:	0,005
Initial linear velocity of mass center v_x, m/s:	2
Displacement angle of drop arm (const), °	5
Minimum turn radius during accelerated motion, m	77,3
Minimum turn radius during slow down motion, m	42,3

Research and verification of the presented real-time model is realized in two stages. In the frame work of first stage vehicle moves with constant displacement angle of drop arm with acceleration, during the second stage – with slow down. Simulation is realized with different values of maximum decision step (0,001 s; 0,003 s; 0,005 s). Duration of each heat is 30 s. On completion of all the heats, the received results (of linear and angular model parameters) are compared with Simulink-model verified data.

Figure 4 shows assemblage of linear velocities V_x of vehicle mass center for different values of maximum simulation step and also velocity V_x of mass center of verified Simulink-model. Figure **5** shows simulation errors for different maximum steps. Tables 2 and 3 present maximum error values of all the analyzed parameters during accelerated and slowdown motion.

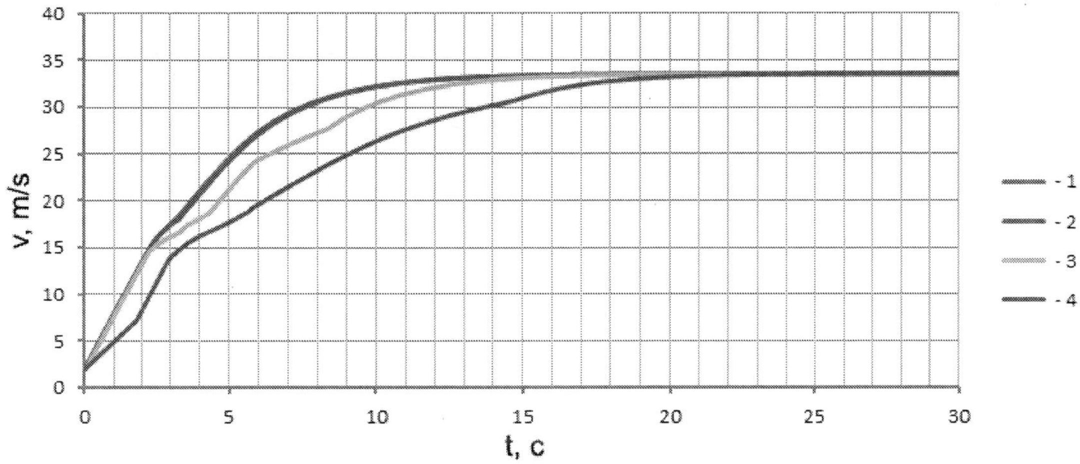

Figure 4. Time dependencies of linear velocity *V_x* of vehicle mass center:
1 – dependency, received with verified Simulink-model;
2 – dependency, received with analyzed real-time model with maximum simulation step 0,001 s;
3 – dependency, received with analyzed real-time model with maximum simulation step 0,003 s;
4 – dependency, received with analyzed real-time model with maximum simulation step 0,005 s.

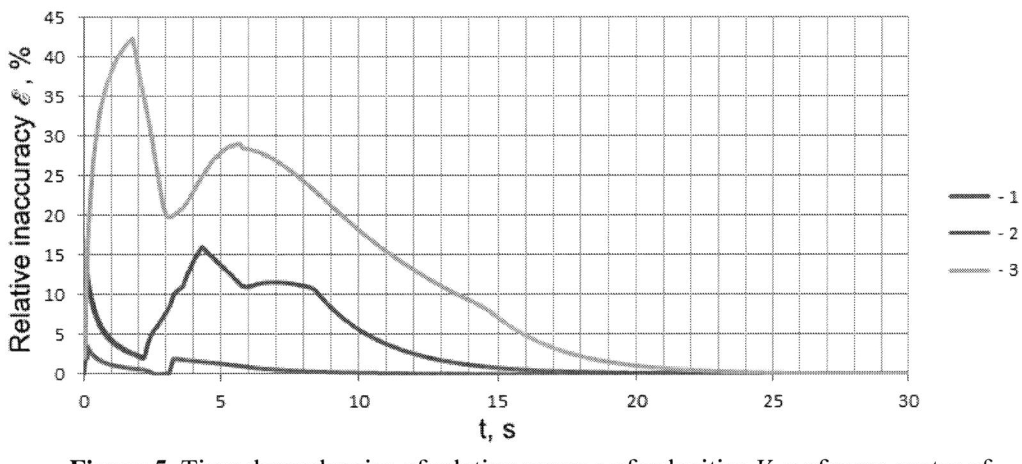

Figure 5. Time dependencies of relative errors ε of velocities *V_x* of mass center of analyzed model in conditions of different model parameters:
1 – linear velocity error for maximum simulation step 0,001 s;
2 – linear velocity error for maximum simulation step 0,003 s;
3 – linear velocity error for maximum simulation step 0,005 s.

Table 2. Maximum relative simulation error values for non-linear accelerated vehicle motion.

Model parameter	Error, *max* (max. simulation step 0,001 s)	Error, *max* (max. simulation step 0,003 s)	Error, *max* (max. simulation step 0,005 s)
Mass center velocity, v_x	4 %	16 %	42 %
Mass center velocity, v_y	5 %	23 %	90 %
Angular velocity, ω	6 %	45 %	57 %
Mass center X-coordinate	0,45 %	16 %	41 %
Mass center Y-coordinate	9,5 %	55 %	85 %
Angular displacement, Θ	3 %	19 %	54 %

Table 3. Maximum relative simulation error values for non-linear slowdown vehicle motion.

Model parameter	Error, *max* (max. simulation step 0,001 s)	Error, *max* (max. simulation step 0,003 s)	Error, *max* (max. simulation step 0,005 s)
Mass center velocity, v_x	0,65 %	5 %	40 %
Mass center velocity, v_y	3 %	15 %	53 %
Angular velocity, ω	5 %	26,5 %	50 %
Mass center X-coordinate	9,5 %	35 %	72 %
Mass center Y-coordinate	1 %	30 %	51 %
Angular displacement, Θ	2 %	31 %	54 %

Figure 6 shows vehicle mass center tracks, which are plotted as a result of slowdown heats of "reference" and analyzed objects in conditions of different maximum values of simulation step. It should be noticed, that increase in simulation step results in "delayed" motion of simulation object in relation to verified model, but for the most part tracks match in superimposition, except final vehicle position.

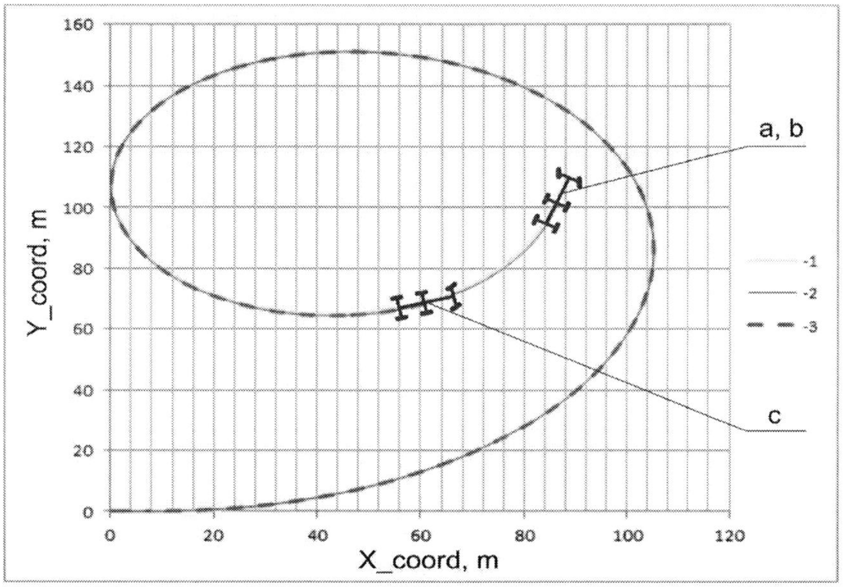

Figure 6. Vehicle motion tracks in conditions of different model parameters:
1 – motion track of «reference» model;
2 – motion track for maximum simulation step 0,001 s;
3 – motion track for maximum simulation step 0,005 s;
a – vehicle position of «reference» model simulation;
b – vehicle position of simulation with maximum simulation step 0,001 s;
c – vehicle position of simulation with maximum simulation step 0,005 s.

The computational experiments performed show that simulation real-time mathematical model perform and adequacy significantly depends on selected maximum solution step. Based on data, presented in tables 2 and 3, the conclusion is that analyzed model can be implemented as a vehicle emulator, while solution steps are no greater than 0,001 s. Such conclusions are compatible with native and foreign researches results, presented in [18], [19], [20] and [21].

The presented results show, that it is possible to apply real-time dynamic models of wheeled vehicle motion in development of on-board stand-alone moving control systems.

The article is written based on the results of work carried out with the financial support of the Ministry of Education and Science of the Russian Federation under the agreement No. 14.574.21.0178 (Unique work identifier: RFMEF57417X0178).

References

[1] Scherbin M A 2015 Modern on-board information and control system of automotive vehicles *J Automobile engineers association* **3** (92) pp 26-29
[2] Belanger J Venne P and Paquin J N 2009 The what, where and why of real-time simulation *J Transactions on Power Delivery* **24** pp 390-399
[3] Liu C S and Peng H 1998 A state and parameter identification scheme for linearly parameterized systems ASME *J of Dynamic Systems, Measurement and Control* **120** pp 524-528
[4] Tavoosi V, Kazemi R and Hosseini S M 2014 Vehicle handling improvement with steer-by-wire system using hardware in the loop method *J of Applied Research and Technology* **12** pp 769-781
[5] Demidovich B P, Maron I A and Shuvalova E Z 1967 *Numerical methods of analysis*

(Moscow: Science) p 368

[6] Gorelov V A, Kotiev G O and Beketov A A 2008 Movement mathematical model of all-wheeled vehicle *J Automobile engineers association* 1(48) pp 50-54

[7] Keller A, Aliukov S 2015 Effectiveness of Methods of Power Distribution in Transmissions of All-Wheel-Drive Trucks *SAE Technical Paper* 2015-01-2732

[8] Keller A, Aliukov S, Anchukov V, et al. 2016 Investigations of Power Distribution in Transmissions of Heavy Trucks *SAE Technical Paper* 2016-01-1100

[9] Zhileykin M M, Kotiev G O and Nagatsev M V 2018 Synthesis of the adaptive continuous system for the multi-axle wheeled vehicle body oscillation damping *IOP Conf Series: Materials Science and Engineering* **315 issue 1** article number 012031

[10] Novikov V V, Pozdeev A V and Diakov A S 2015 Research and testing complex for analysis of vehicle suspension units *International Conf on Industrial Engineering, ICIE* **129** pp 465-470

[11] Skotnikov G I, Jileykin M M and Komissarov A I 2017 Increasing the stability of the articulated lorry at braking by locking the fifth wheel coupling *IOP Conf Series: Materials Science and Engineering* **315 issue 1** article number 012027

[12] Gorelov V A and Komissarov A I 2016 Mathematical model of the straight-line rolling tire - Rigid terrain irregularities interaction *2nd International Conf on Industrial Engineering, ICIE 2016* **150** pp 1322-28

[13] Keller A V, Gorelov V A, Vdovin D S, et al. 2015 Mathematical model of all-terrain truck *Proceedings of the ECCOMAS Thematic Conference on Multibody Dynamics* pp 1285–96

[14] Keller A V, Gorelov V A and Anchukov V V 2015 Modeling truck driveline dynamic loads at differential locking unit engagement *Procedia Engineering* **129** pp 280–287

[15] Vol'skaya N S, Zhileykin M M and Zakharov A Y 2017 Mathematical model of rolling an elastic wheel over deformable support base *IOP Conf Series: Materials Science and Engineering* **315 issue 1** article number 012028

[16] Keller A and Aliukov S 2015 Efficient power distribution in an all-wheel drive truck *Lecture Notes in Engineering and Computer Science* **2218** pp 1201-1206

[17] Keller A and Aliukov S 2015 Methodology of System Analysis of Power Distribution among Drive Wheels of an All-wheel-drive Truck *SAE Technical Paper* 2015-01-2788

[18] Choi G J, Yoo Y M and Lees K P 2000 A real-time multibody vehicle dynamic analysis method using suspension composite joints *International J of Vehicle Design* **24** pp 259-273

[19] Gyoojae C 2001 Efficient solving methods exploiting sparsity of matrix in real-time multibody dynamic simulation with relative coordinate formulation *J of Mechanical Science and Technology* **15** pp 1090-1096

[20] Sung-Soo K and Jeong W 2007 Subsystem synthesis method with approximate function approach for a real-time multibody vehicle model *J Multibody System Dynamics* **17** pp 141-156

[21] Weidong P and Papelis Y E 2005 Real-time dynamic simulation of vehicles with electronic stability control: Modeling and validation International *J of Vehicle Systems Modelling and Testing* **1** pp 143-167

IOP Conf. Series: Materials Science and Engineering **534** (2019) 012022 doi:10.1088/1757-899X/534/1/012022

Dynamic model of the robotic vehicle motion on a deformable irregular terrain

N S Volskaya[1] and I V Chichekin[2]

Wheeled Vehicles Department, Bauman Moscow State Technical University, 105005 Baumanskaya 2th st., 5, Moscow, Russian Federation

E-mail: [1] volskaja52@mail.ru, [2] hiv2@mail.ru

Abstract. The paper presents a system approach to the estimation of the off-road performance of the wheeled vehicles operating on deformable terrains. A problem-solving approach to the development of the model for the wheeled vehicle – irregular deformable terrain system is presented. The model was used for the simulation of the vehicle motion on different terrains. The presented method for the calculation of the wheel sinkage (calculation of the resistance to the motion of a wheeled vehicle) is based on the Y.S. Ageykin model. The paper presents the analysis of the research results.

Introduction

It is known that deformation and adhesion properties of the terrain surface is one of the main factors defining the type of the running gear, arrangement of the axles, type and properties of the suspension of a robotic wheeled vehicle. Relative deformations of the tire and the terrain affect the main aspect of the off-road performance – the wheel sinkage. On the other hand, the wheel sinkage significantly depends on the vertical load on the wheel axis. As the evidence from practice shows, additional loads can exceed the static loads by more than 100 % due to the dynamic loads caused by the terrain irregularities and time-varying deformation properties of the terrain. There is no ready method, algorithm or software for the analysis of the off-road performance parameters of the vehicle with dynamic loads on its systems, primarily, of the “wheeled vehicle – deformable irregular terrain” system.

Therefore, the problem of development, selection and analysis of the dynamic models of the real load conditions of the components and units of the vehicle sprung and unsprung parts taking into account conditions of the vehicle motion over the irregular deformable terrain is of engineering and scientific interest.

Research objective: development of spatial dynamic models of the wheeled vehicles moving on the deformable irregular terrain defined by its physical and mechanical non-linear properties.

Statement of the problem

The following assumptions based on the research in the field of modeling and simulation of the loading modes of the vehicle chassis and on the analysis of the dynamical models of the all-wheel-drive vehicles [1, 2, 3, 4, 5, 6, 7, 8, 9,10, 11] can be made:

1. Terrain deformation properties are estimated with the use of the method developed by Y.S. Ageykin [12, 13]. Vertical and shear deformations of the terrain are defined by its physical and mechanical state and depend on its type, moisture W and density ρ_C. Depending on the physical state of the terrain,

the following parameters are calculated: stress-stain modulus E, angle φ_0 of internal shearing resistance, terrain cohesion c_0; the depth H_G of the soft layer is set by the researcher.

2. Tire interaction with the terrain is described by a simulation model presenting the terrain as an elastoviscoplastic or viscoplastic body (Shvedov and Bingam models).

3. The vehicle is presented as a dynamic model consisting of the oscillating sprung and unsprung parts M, m which are connected by means of elastic elements c_{Susp} , c_{tyre} and damping elements k_{susp} , k_{tyre} .

4. Stochastic road microprofile is defined by its spectral density $S_q(\theta)$. It is a random function based on the following assumptions: it is time independent and ergodic, microprofile ordinates q are distributed normally.

$$S_q(\theta) = A \cdot \theta^{-b}, \tag{1}$$

where A and b - spectral density coefficients defining the microprofile type, θ - track frequency. Ride quality criterion for the selection of the suspension concept is the level of the equivalent root mean square of the vibrational accelerations at the chassis points located above the wheels.

Research objectives:

1. Development of dynamical models of a two-axle vehicle. The models must represent irregularities and varying physical and mechanical state of the deformable terrain interacting with the tire.
2. Analysis of the mutual influence of the vehicle sprung and unsprung part oscillations and deformations of the irregular terrain.
3. Development of a calculation method for the evaluation of the parameters of the tire interaction with the deformable terrain. The method must take into account dynamic loading, terrain relaxation properties and cyclic loading.
4. To use the developed spatial dynamic model of the two-axle all-wheel-drive vehicle for the estimation of the optimality of the selected design parameters defining the dynamic deflection of the suspension and the size of the tire of the ZIL 432720 vehicle.

Research results

In the presented research, the structure of five dynamic models of a two-axle vehicle (fig. 1) on the deformable irregular terrain has been developed. In contrast to the known models (fig. 1, b, c, d, e), the developed models take into account relaxation of the terrain according to the Shvedov and Bingam models (fig. 2). The most realistic dynamical model of the vehicle is spatial model 1, a. The model has eight degrees of freedom. For every dynamic model, a system of differential equations is generated. The solution of these systems of non-linear equations is carried out with the use of Laplace transformations. In standard form, oscillation equation for a spatial system can be presented as:

$$\mathbf{A(p)} \cdot \begin{bmatrix} z_C(p) \\ \alpha(p) \\ \beta_1(p) \\ \beta_2(p) \\ \xi_{1,1}(p) \\ \xi_{1,2}(p) \\ \xi_{2,1}(p) \\ \xi_{2,2}(p) \end{bmatrix} \cdot \frac{1}{q_0} = \mathbf{B(p)} \cdot \begin{bmatrix} q_{1,1}(p) \\ q_{1,2}(p) \\ q_{2,1}(p) \\ q_{2,2}(p) \end{bmatrix}, \tag{2}$$

where: $\mathbf{A}(\mathbf{p})$ – matrix of vehicle parameters; $\mathbf{B}(\mathbf{p})$ – perturbation matrix; z_C – vertical displacement of the sprung part; α – pitch angle of the vehicle sprung part; β_1 and β_2 – roll angles of the vehicle sprung part under front and rear axels respectively; $\xi_{i,j}$ – displacement of the unsprung parts; q_0– half of the microprofile height; $q_{i,j}$ – displacement of the wheel contact patch relative to the microprofile at the *i*-th axle at the *j*-th side of the vehicle.

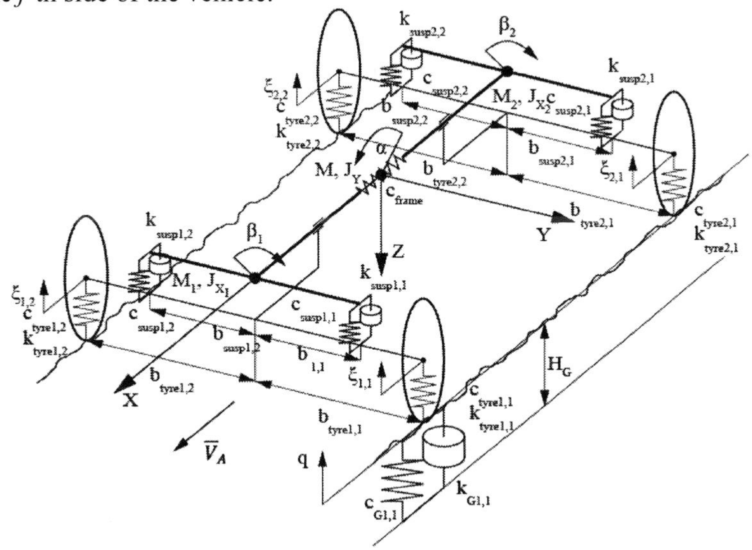

a) Model 5

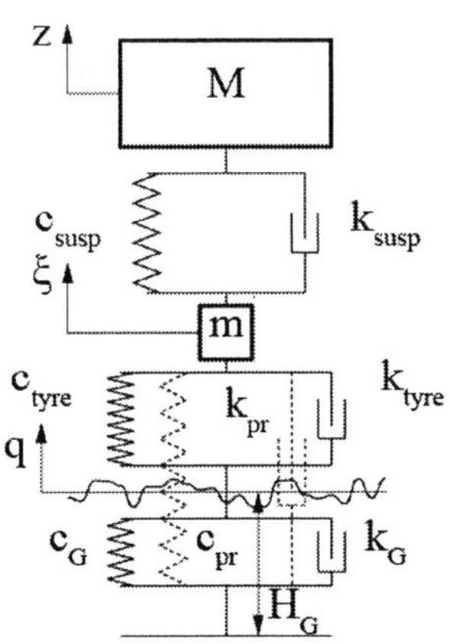

b) Model 1

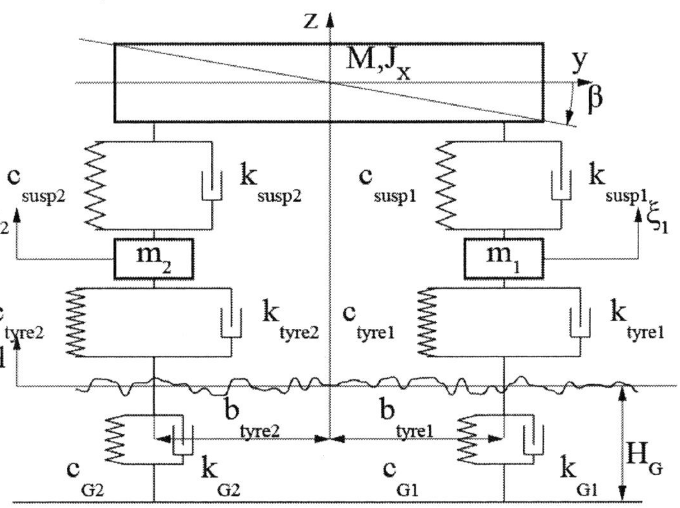

d) Model 3

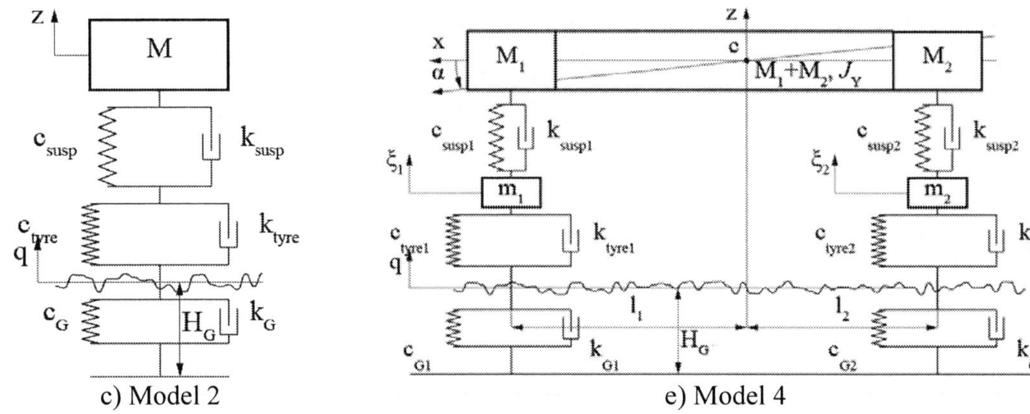

Figure 1. Dynamic models.

The developed models allow a researcher to:

1) find the influence of the ratio of the sprung and unsprung parts (M , m), suspension and tire stiffnesses, shock absorber damping (c_{susp}, c_{tyre}, k_{susp}), type and design of the suspension, tire type and size, tire inflation pressure (p_W), axle base, track width and lateral distance between the springs on the vehicle off-road performance;

2) evaluate numerical values of the vehicle off-road performance parameters: additional dynamic load caused by the vehicle oscillations propagated through the tires into the terrain (ΔP); wheel sinkage and tire deformation (z and h); terrain compaction after each wheel pass (ρ_{Ci}); coefficient of rolling resistance of the terrain and tire deflection (f); coefficient of the free tractive force (Ψ_T); probability of loss of contact between the wheels and the terrain; probability of contact between the vehicle body (vehicle axles) and the terrain and etc.;

3) evaluate numerical values of the criteria of the ride quality and component strength of the vehicle: root mean square vibrational accelerations and displacements at every point of the vehicle body ($\ddot{z}_{A_{i,j}}$ and $z_{A_{i,j}}$); ratio of the microprofile change after the first wheel pass ($S_q(\nu)_{2,j}$); lateral and longitudinal angular accelerations ($\ddot{\alpha}$, $\ddot{\beta}_1$ and $\ddot{\beta}_2$); deflection of the springs and the probability of the suspension maximum deflection (f_P); dynamic loads of vehicle frame (ΔP_{dyn}); crossing of the front and rear axels; frame torsion angle and frame rail distortion angle (β_{P_M} and α_{JI}).

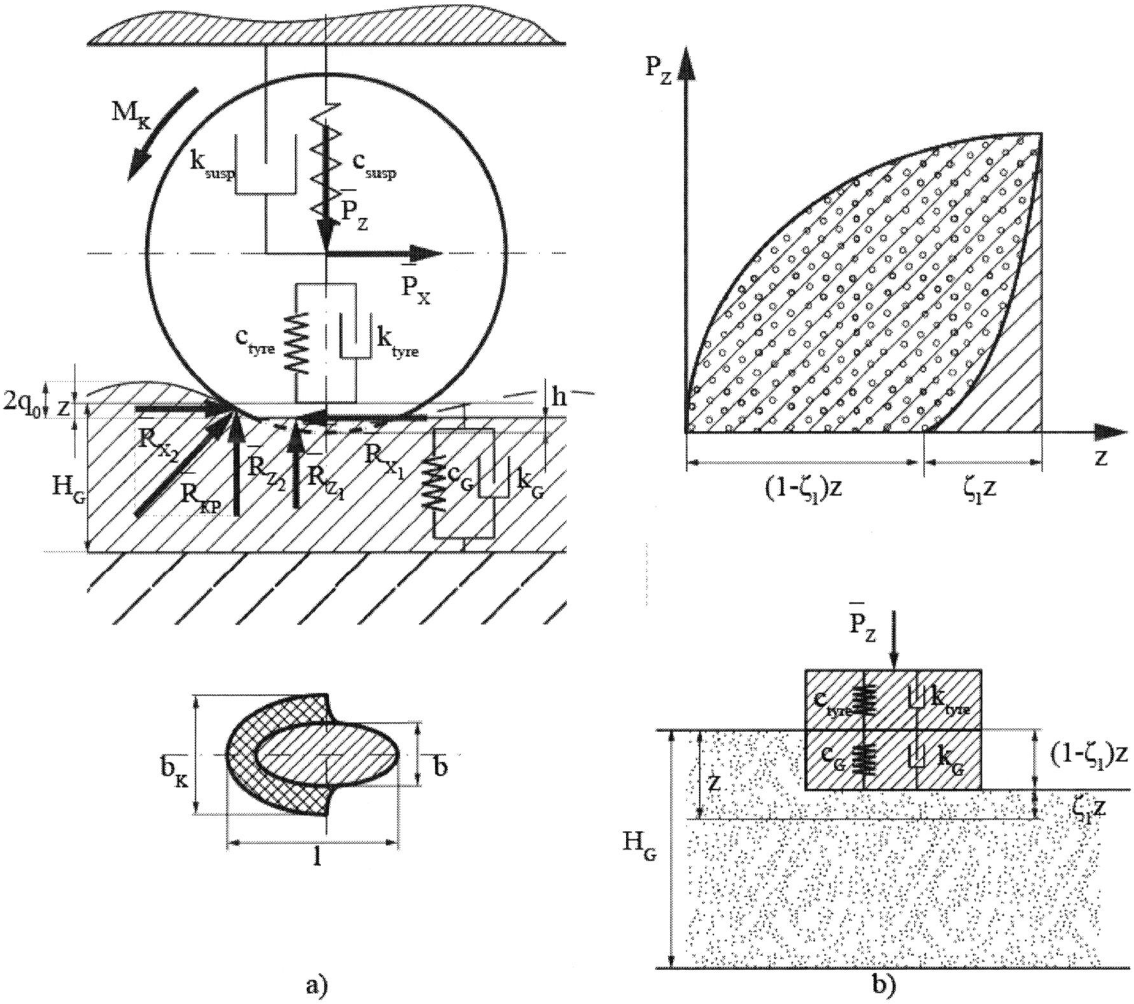

Figure 2. Dynamic model of the wheel interaction with a deformable irregular terrain.

Model efficiency and its scope in the tire – terrain interaction domain

Figures 3-9 show examples of calculation results for the five models. The following parameters were estimated:

1) influence of the vehicle velocity on the value of additional dynamic loads on the deformable irregular terrain ($A = 10^{-3}$) with different moisture (fig. 3). For the velocities within the range 10…50 km/h for the left and right vehicle wheels, minimal and maximal additional dynamic loads for models 2 and 5 differ by a factor of 4.5. At the same time, ΔP for model 2 increases by 25 % and 100 % in cases of the terrain moisture W=70 % and W=90 % respectively, when $V_a = 50$ km/h, the maximal values of ΔP for model 5 differ by about 13 %.

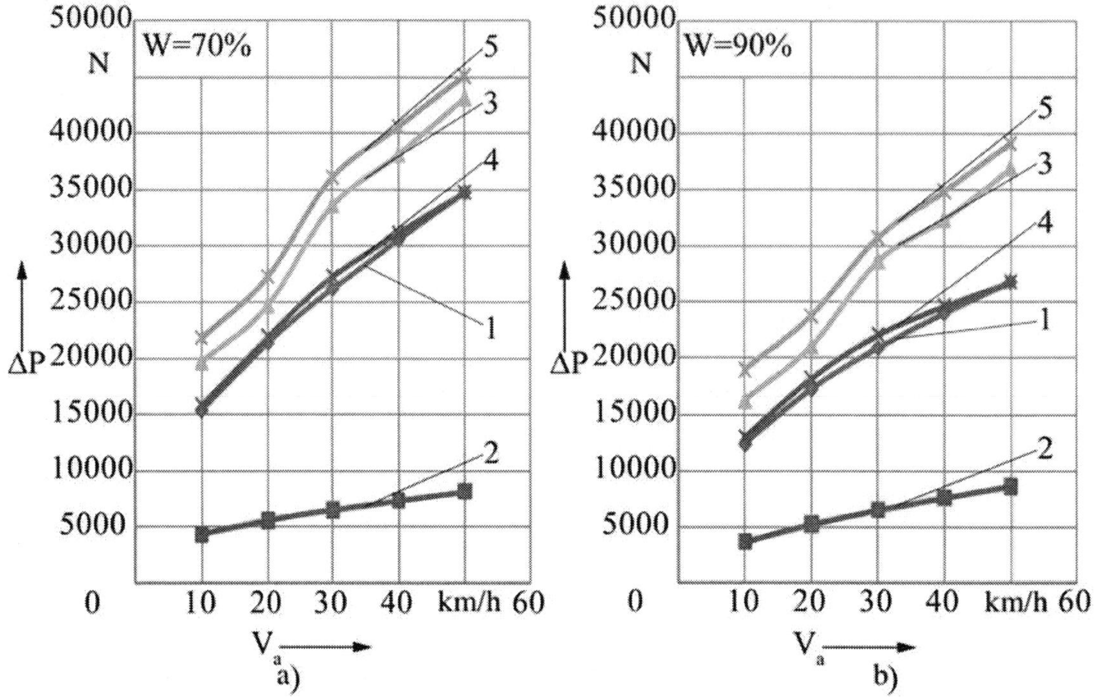

Figure 3. Influence of the vehicle velocity on the values of additional dynamic loads on the deformable irregular terrain with different moisture content ($A=10^{-3}$; 1…5 – numbers of the dynamic models).

2) figure 4 illustrates the influence of the terrain irregularities with $A=10^{-3}$ or $A=10^{-4}$ on the additional dynamic loads in the same interval of varying velocities of the vehicle. Depending on the choice of the dynamic model within the analyzed interval of the vehicle velocities on the deformable terrains with different degree of irregularity, additional dynamic loads obtained with simple model 2 and complex spatial model 5 can differ by a factor of 9, relative increase in the load being 230 % and 100 % respectively;

3) figure 5 shows influence of the dynamical model type on the wheel sinkage z which determines rolling resistance on the given terrain with different physical and mechanical properties depending on its moisture. When the soil moisture W=70 %, different dynamical models give conflicting results: in model 2, when the the wheel velocity increases, the wheel sinkage decreases down to zero. For model 5, in case of low velocity, sinkage $z = 0.15$ m and it decreases insignificantly (by 10 %) when the velocity increases.

4) figure 6 shows the wheel sinkage z for the front left wheel as a function of the vehicle velocity on plain and irregular deformable terrains with the same soil. The maximal value of z ($z = 0,32\ m$) is obtained on the overmoisted terrain with W = 90 % in case of minimal velocity. On a plain terrain, during motion without oscillations, the sinkage is less, than in case of dynamic loads (z=0.2 m on an average);

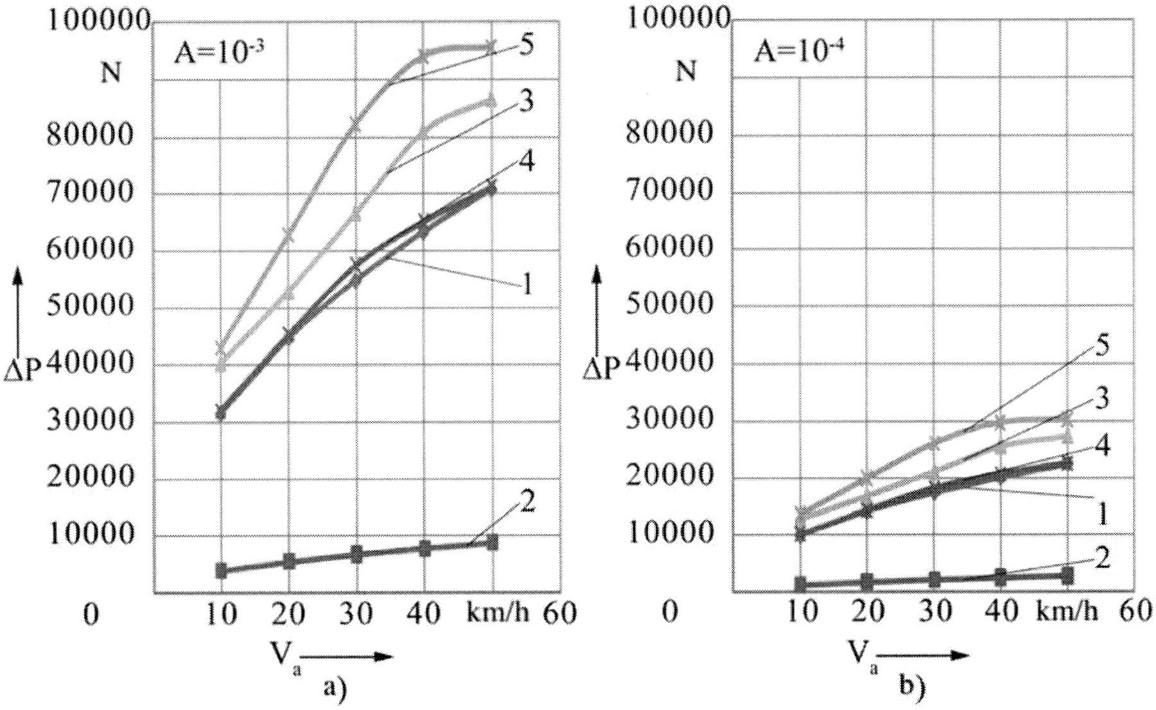

Figure 4. Influence of the terrain irregularity degree on the additional dynamic load.

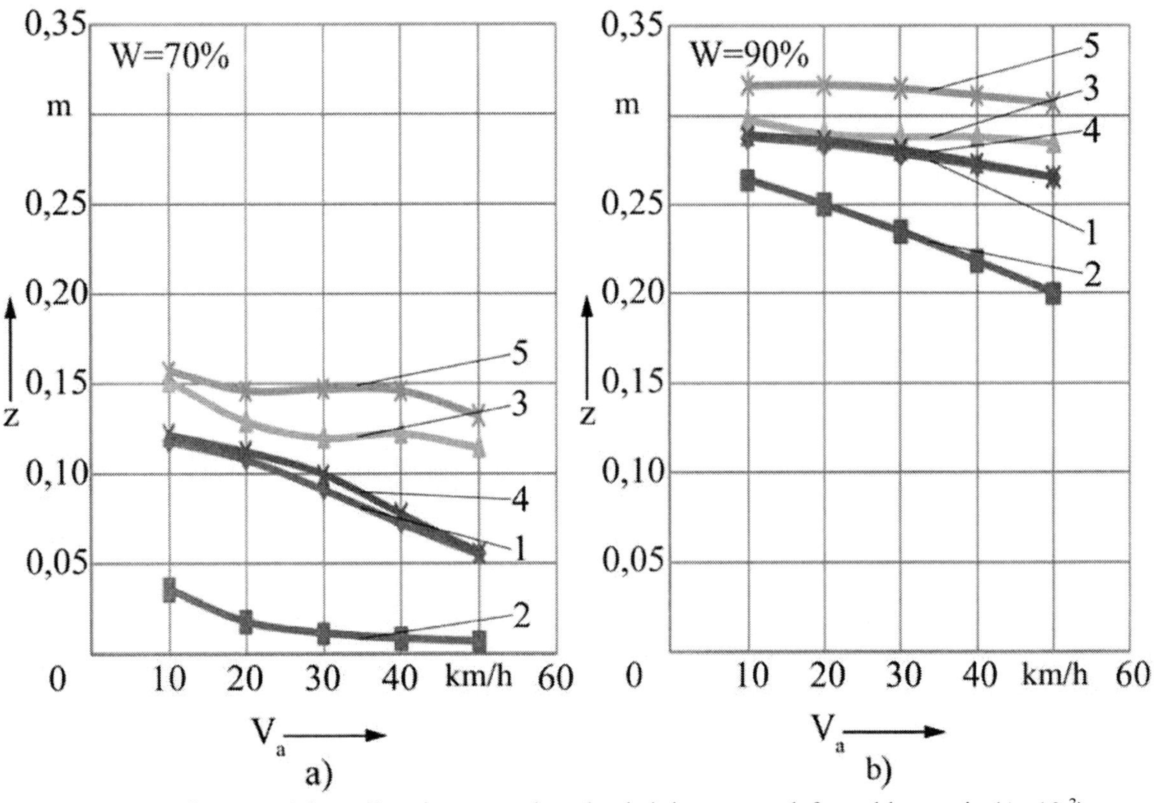

Figure 5. Influence of the soil moisture on the wheel sinkage on a deformable terrain ($A=10^{-3}$).

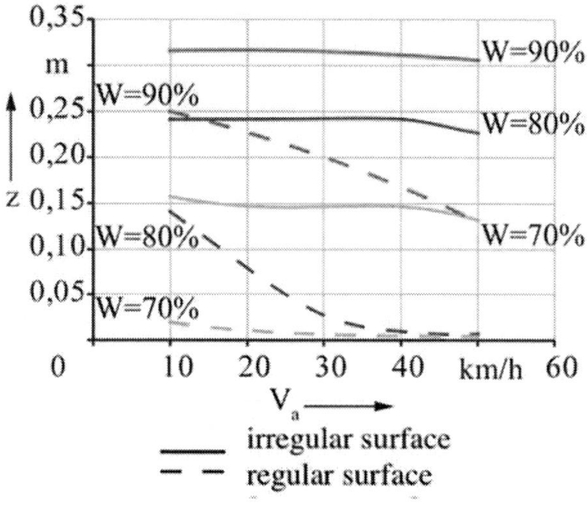

Figure 6. Dependency of the sinkage on the vehicle velocity on plain and irregular terrain.

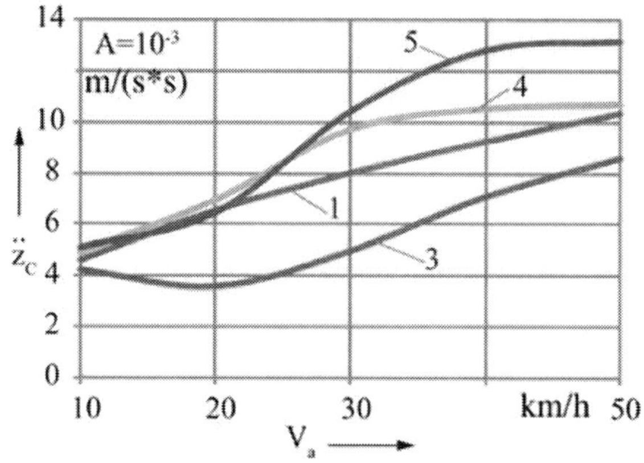

Figure 7. Ride quality parameters on an firm terrain.

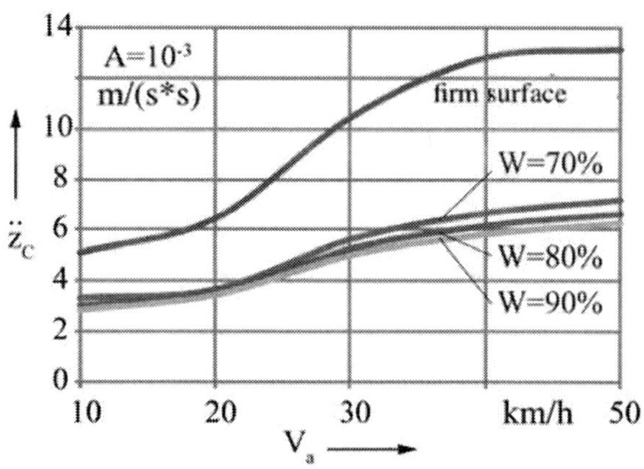

Figure 8. Ride quality parameters on firm and deformable terrains (spatial dynamic model 5).

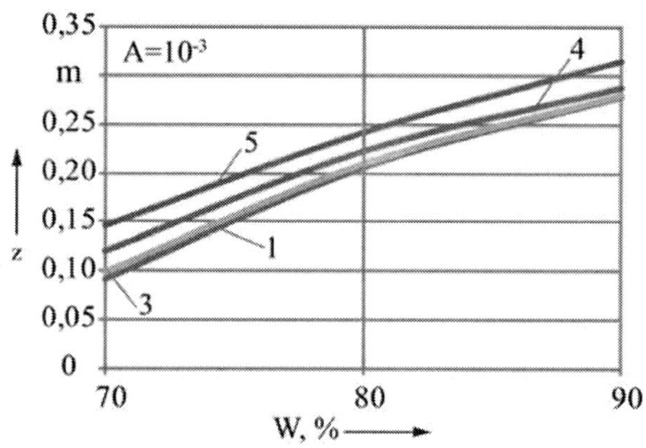

Figure 9. Dependency of the wheel sinkage on the terrain moisture for five dynamic models.

5) figure 7 shows ride quality parameters of the vehicle during motion on the irregular deformable terrain;

6) figure 8 shows ride quality parameters for firm terrain and deformable terrain with different moisture content. The parameters have been obtained with dynamic model 5. These parameters define maximum allowable velocities of the vehicle. Allowable root mean square values of the vibrational acceleration during short-period motion is 4 m/s^2, which means that the allowable velocity on firm irregularities with $A = 10^{-3}$ is $8\ km/h$ (fig. 8), on soft terrain – $10 \div 22\ km/h$ according to model 1, $17 \div 19\ km/h$ according to model 4 and $16 \div 18\ km/h$ according to model 5 (fig. 8);

7) for estimation of the cross-country capability, which is limited by the road clearance, figure 9 shows dependencies of sinkage z on terrain moisture, they are estimated for 4 models. The calculation results show, that motion is possible if the terrain moisture is no more then W=85 % and road clearance is 0,3 m. Models 1-4 have shown that the vehicle can move without any body – terrain contact on all the

analyzed deformable terrains. Model 5 shows that body –terrain contact might occur on a terrain with the limit moisture content W=90 %.

Analysis of the calculation results shows, that oscillations of unsprung parts have the most significant influence on the dynamic loads.

The analysis has shown that:

- it is necessary to take into account additional dynamic loads ΔP, while determining parameters of the tire interaction with irregular deformable terrain;
- transversal angular oscillations have a significant influence on ΔP values, so it is important to take them into account, especially if vehicle is moving on an overmoisted terrain;
- if spectral density coefficient A of the the terrain profile is larger then 10^{-3}, it is necessary to take into account transversal and lateral angular accelerations (model 3 and model 5);
- weight of unsprung parts has the most significant influence (more than 50 %) on wheel dynamic loads of the terrain and on the wheel sinkage.

The paper describes a method for estimation of the and off-road performance of an vehicle. A dynamic model of the wheel interaction with an irregular deformable terrain was used.

For estimation of the wheel sinkage *z* and tire deflection *h,* the following equations system should be solved:

$$q_{tyre} = \pi \cdot h \cdot \frac{(p_0 + p_W)}{2 \cdot b} \cdot \left(\frac{B}{H} + \frac{3 \cdot H}{2 \cdot B}\right) \cdot \left(1 - \frac{h}{B}\right); \quad q_{dyn} = \frac{q_{tyre}}{1 + \dfrac{V_a}{l \cdot 2n \cdot \phi_0}}; \qquad (3)$$

$$q_{dyn} = \frac{E \cdot z}{0,5 \cdot a \cdot (b + b_k) \cdot arctg\left[\dfrac{H_G - z}{0,5 \cdot a \cdot (b + b_k)}\right] + \dfrac{2 \cdot I \cdot E \cdot z \cdot arctg\left[\dfrac{\pi \cdot (H_G - z)}{b + b_k}\right]}{\pi \cdot (I_1 \cdot X_1 \cdot k_{\beta 1} \cdot b + I_2 X_2 \cdot k_{\beta 2} \cdot c_0 + X_3 \cdot z)}};$$

$$P_Z = 0,25 \cdot \pi \cdot q_{tyre} \cdot \left[(1 - 0,5\xi) \cdot 2 \cdot b \cdot \sqrt{D \cdot h - h^2} + \xi \cdot b_k \cdot \sqrt{D(h+z) - (h+z)^2}\right],$$

where $q_{ш}$ - contact patch pressure, expressed as a function of the tire deflection, MPa; D - wheel diameter, m; B and H - width and height of the tire profile, m; p_0 - terrain pressure caused by the tire carcass stiffness, MPa; p_W - tire inflation pressure, MPa; P_Z - wheel load, N; b_k - track width, m; I, I_1, I_2 - coefficients of influence of the contact surface size and shape on the terrain deformation; V_a - velocity of the vehicle, m/s.

First, sinkage and tire deflection for the vehicle front and rear axles are calculated. Then, on the assumption of the constant sinkage, the same parameters of the tire-terrain interaction are calculated for the new physical and mechanical properties of the terrain. The new terrain properties are recalculated for each axle passing over the already deformed terrain by the following dependencies: $\rho_n = f(n; q_{dyn}; \rho_т)$; $E, \varphi_0, c_0 = f(\rho_n; W_т; W)$.

After evaluation of h and z for each wheel of the two-axle vehicle we calculate the following parameters: tire rolling resistance, f_{tyre} ; terrain rolling resistance , f_{gr} ; total rolling resistance, f ; longitudinal slip, s ; free tractive force, ψ_T .

The above mentioned method has been used for the calculation of the relative deformations of the tire and terrain with provision for the additional dynamic load. Assessment of the additional dynamic load ΔP_C is based on the following dependency:

IOP Conf. Series: Materials Science and Engineering **534** (2019) 012022 doi:10.1088/1757-899X/534/1/012022

$$\Delta P_{dyn} = \ddot{z}_C \cdot M + \ddot{\xi}_C \cdot m, \qquad (4)$$

where $\ddot{z}_C$ and $\ddot{\xi}_C$ – root mean-square vibrational accelerations of the sprung and unsprung parts m and M of each wheel of the vehicle.

For estimation of $\ddot{z}_C$ and $\ddot{\xi}_C$, terrain damping and stiffness properties are taken into account (fig. 2). Calculations of the floatation and tractive effort for a single wheel and for the wheels of the front and rear axles of a two-axle vehicle (plain model) on loam terrain with different degree of irregularity and different physical and mechanical states have shown that the vehicle oscillation parameters have significant influence on the obtained results.

The experimental study of the vehicle ride quality and off-road performance on a deformable irregular terrain has been carried out.

For estimation of the adequacy of the developed model of the vehicle motion and for validation of the method for the terrain damping properties assessment, a series of experiments was performed on a test rig and on a real vehicle on a natural deformable terrain.

Difference between calculated and experimental root mean-square values of the vibrational accelerations does not exceed 6%.

The authors have used the results of the presented research for estimating optimality of the chosen parameters of the given two-axle vehicle on given deformable terrains.

The developed spatial dynamic model of the vehicle was used for estimation of the vehicle ride quality and off-road performance on a standard route. The route consists of 4 segments shown in table 1.

The parameters of the ride quality and off-road performance were calculated for the two-dimensional model and spatial models (4), (5) on the standard route. Parameters Z, ψ_T , which define ability to move in off-road conditions, for the spatial and two-dimensional model differ by 20 % on the average.

Table 1. Surface properties of the segments of the standard route.

Segment	Type of the segment surface	Initial data
1	Irregular road with firm surface	$H_G = 0$ м, $A = 10^{-4}$
2	Uneven soil road	$H_G = 0$ м, $A = 10^{-3}$
3	Smooth deformable terrain	$H_G = 0{,}4$ м, $A = 0$
4	Irregular deformable terrain	$H_G = 0{,}4$ м, $A = 10^{-3}$

The significant influence of the unsprung part on the vehicle ride quality and off-road performance has been confirmed. Unsprung part oscillation loads amount to 50-60 % of the additional dynamic loads on the irregular deformable terrain.

General results and conclusions

1. Dynamic models of the vehicle oscillation have been developed. Vehicle sprung and unsprung parts are connected with elastic and damping elements. Oscillations of this parts are related to the deformations of the deformable irregular terrain with non-linear mechanical properties depending on its physical state. Numerical disarrangement of theoretical and experimental data is within the range of 6-12 %.
2. Stiffness and damping properties of the deformable irregular terrain have significant influence on the oscillation parameters of the vehicle sprung and unsprung parts. Estimations of additional dynamic loads for firm and deformable irregular terrain ($A=10^{-3}$) have disarrangement up to 100 %. Increase in terrain moisture by 30 % relative to the basic value W=70 % leads to decrease in additional dynamic loads by 10 %.

3. Application areas of the developed dynamic models have been defined. If sprung and unsprung parts masses are in the following relation: $\frac{M}{m} > 10$, it is possible to disregard the unsprung part for simplicity and use dynamic model 2. It is recommended to select dynamic systems taking into account transversal angular oscillations (models 3 and 5) for estimation of additional dynamic loads on microprofile with irregularities higher than 50 mm. Calculations with the use of the spatial dynamic model (model 5) lead to a 25 % increase of dynamic loads in relation to the two-dimensional model (model 4).
4. As a performance criterion for the two-axle vehicle, its average velocity on the standard route is chosen. The route consists of segments with different deformation properties Spatial model 5 is recommended as the most realistic one for the simulation of the vehicle – deformable terrain system.

References

[1] Gorelov V A and Komissarov A I 2016 Mathematical model of the straight-line rolling tire - Rigid terrain irregularities interaction *Procedia Engineering* **150** pp 1322-28 (DOI: 10.1016/j.proeng.2016.07.309)
[2] Keller A V, Gorelov V A, Vdovin D S, et al. 2015 Mathematical model of all-terrain truck *Proceedings of the ECCOMAS Thematic Conference on Multibody Dynamics* pp 1285–1296
[3] Keller A, Aliukov S, Anchukov V, et al. 2016 Investigations of Power Distribution in Transmissions of Heavy Trucks *SAE Technical Paper* 2016-01-1100
[4] Gorelov V A, Komissarov A I and Miroshnichenko A V 2015 8×8 wheeled vehicle modeling in a multibody dynamics simulation software *Procedia Engineering* **129** pp 300-307 (DOI: 10.1016/j.proeng.2015.12.066)
[5] Keller A V, Gorelov V A and Anchukov V V 2015 Modeling truck driveline dynamic loads at differential locking unit engagement *Procedia Engineering* **129** pp 280-287 (DOI: 10.1016/j.proeng.2015.12.063)
[6] Kotiev G O, Padalkin B V, Kartashov A B and Diakov A S 2017 Designs and development of Russian scientific schools in the field of cross-country terrain vehicles building *ARPN Journal of Engineering and Applied Sciences* **12**(4) pp 1064-71
[7] Keller A and Aliukov S 2015 Efficient power distribution in an all-wheel drive truck *Lecture Notes in Engineering and Computer Science* **2218** pp 1201-06
[8] Keller A and Aliukov S 2015 Effectiveness of Methods of Power Distribution in Transmissions of All-Wheel-Drive Trucks *SAE Technical Paper* 2015-01-2732
[9] Volskaya N S, Zhileykin M M and Zakharov A Y 2018 Mathematical model of rolling an elastic wheel over deformable support base *IOP Conf. Series: Materials Science and Engineering* **315** 012028 (doi: 10.1088/1757-899X/315/1/012028)
[10] Keller A and Aliukov S 2015 Methodology of System Analysis of Power Distribution among Drive Wheels of an All-wheel-drive Truck *SAE Technical Paper* 2015-01-2788Vdovin D and Chichekin I 2016 Loads and Stress Analysis Cycle Automation in the Automotive Suspension Development Process *Procedia Engineering* **150** pp 1276-79 (doi: 10.1016/j.proeng.2016.07.285)
[11] Keller A and Alyukov S 2016 Power Distribution in Transmissions of Multi-Wheeled Vehicles *SAE Technical Paper* 2016-01-1103
[12] Ageykin Y S 1981 *Cross-country ability of cars* (Moscow: Mashinostroyeniye) p 232 (in Russian)
[13] Ageykin Y S 1972 *All-terrain wheels and composite running gear units. Theory and calculation* (Moscow: Mashinostroyeniye) p 184 (in Russian)

IOP Conf. Series: Materials Science and Engineering **534** (2019) 012023 doi:10.1088/1757-899X/534/1/012023

Simulation of the tire enveloping properties in case of the unmanned car wheel interaction with a rigid uneven surface

Y Y Levenkov[1], N S Vol'skaya[1] and O A Rusanov[2]

[1] Department of Wheeled Vehicles, Bauman Moscow State Technical University, 105005 Baumanskaya 2th st., 5, Moscow, Russian Federation
[2] Department of «Ground Vehicles», Moscow Polytechnic University, Moscow, 107023 Russian Federation

E-mail: levenkov_yy@bmstu.ru

Abstract. A computational and experimental method is presented for assessing the influence of the roughness of a rigid surface on the stiffness and damping properties of automobile tires. The method is based on the solution of tire-ground contact problem using finite-element analysis. In the paper, the influence of two types of irregularities (peaks and pits), their size and shape and tire air pressure on tire enveloping properties is analyzed.

1. Introduction

The disturbing effect of road irregularities on the sprung masses of the unmanned vehicle essentially depends on the tire enveloping properties, i.e. on the capability of the tire to reduce the displacement of the wheel axis in comparison with the height of the irregularity and to increase the interaction area of the wheel contact patch. To simulate the mechanical properties of a tire, it is necessary to use nonlinear mechanics methods that take into account a set of nonlinear factors (physical and geometric nonlinearities, hyperelastic properties of materials, contact interaction) [1,2,3,4,5,6,7,8].

The current level of the computing technology development allows considering this research problem in a more precise statement, using complex analytical models (taking into account the listed nonlinear factors), reproducing complex reinforced tire design and taking into account the interaction of the tire with the uneven road surface. Therefore, the presented research issue is relevant.

Research objective: creation of new, more accurate compared to the well-known, experimentally-confirmed methods of modeling of the car pneumatic tires, interacting with the uneven road surface; the methods should take into account all the tire general design features, mechanical properties of the materials, loading conditions of car wheels, geometrical parameters of irregularities.

Research tasks:

1. Development of a refined analytical 3D model of car pneumatic tire, taking into account its general dimensional design features (sidewalls and parts of the rubber tread, carcass, tire belt, bead bundle), reflecting nonlinear properties (rubber incompressibility and hyperelasticity, geometrically nonlinear deformation of the tire cord, contact of the tire with the rigid uneven surface), making it possible to estimate tire stiffness properties at the static and dynamic load (for various values of the vertical load on the wheel, different tire pressure values, different shapes of the road irregularities).

2. Development of the method for determination of stiffness and damping characteristics of the car pneumatic tire and for estimation of the tire enveloping properties at the static interaction of the wheel

with the rigid uneven surface and of the wheel movement along a rough road with irregularities of various sizes and shapes.

3. Determination of Mooney-Rivlin hyperelastic material properties (for rubber modeling in the framework of the method of estimation of the tire enveloping properties of the car pneumatic tires), based on a comparison of the results of calculations and tests of samples cut from the sidewall of the tire, as well as on the basis of a comparison of the results of calculations and tire deformation tests on a flat rigid surface during vertical loading.

4. Calculation of the tire deformation when it interacts with the rigid uneven surface; allowing to reveal non-linear dependencies of the tire stiffness and damping coefficient on its loading characteristics, shape and dimensions of irregularities of the road.

2. Modeling of the car pneumatic tire

In the framework of the presented research, there were analyzed the studies of Russian and foreign scientists, who carried out investigations of external mechanics of tires; papers on modern methods of the tire enveloping properties analysis were studied, as well as general methods of analysis of tire enveloping characteristics (experimental and theoretical) [9,10,11,12,13,14,15,16].

Based on the performed survey, the following conclusions were made:

1. Finite-element analysis (FEA) is the most effective method of assessment of the tire enveloping characteristics.

2. High accuracy of calculation can be provided by solid finite-element models of tires compared to shell (and other) models.

The new analytical tire model is represented by the equations which define mathematical model of interaction of the elastic car wheel with the road surface: 1) nonlinear theory of elasticity, 2) dynamic equilibrium in increments of displacement.

Equation of dynamic equilibrium of the mechanical system at the moment t:

$$\mathbf{M}\cdot\overset{\circ\circ}{\mathbf{U}}_t+\mathbf{C}\cdot\overset{\circ}{\mathbf{U}}_t+(\mathbf{K}_T+\mathbf{K}_G)\cdot\Delta\mathbf{U}_t=\Delta\mathbf{R}_t+\mathbf{R}_{Nt-\Delta t},$$

where: $\mathbf{M}$, $\mathbf{C}$ are the global mass matrix and the global damping matrix of FEA system; $\mathbf{U}$ is the vector of nodal displacements; $\overset{\circ}{\mathbf{U}}=\frac{\partial}{\partial t}\mathbf{U}$ is the vector of nodal velocities; $\overset{\circ\circ}{\mathbf{U}}=\frac{\partial^2}{\partial t^2}\mathbf{U}$ is the vector of nodal accelerations; $\mathbf{K}_T,\mathbf{K}_G$ are the matrix of tangential stiffness (taking into account nonlinear properties of material) and the matrix of geometric stiffness in reference configuration; $\Delta\mathbf{R}$ is the external load increment vector, $\mathbf{R}_N$ is the vector of the unbalanced part of the external load corresponding to the previous load step.

Rubber properties are described by the Mooney-Rivlin model of hyperelastic material. The Mooney-Rivlin equation is an adequate model of the behavior of a number of almost incompressible rubber-like materials. Stain energy density function for incompressible material is:

$$\boldsymbol{W}=\boldsymbol{a}_{10}(\boldsymbol{I}^*_1-3)+\boldsymbol{a}_{01}(\boldsymbol{I}^*_2-3),$$

where: $\boldsymbol{I}^*_i$ is the strain tensor invariants; $\boldsymbol{a}_{10}$, $\boldsymbol{a}_{01}$ are the Mooney-Rivlin material constants.

True stress σ for sample uniaxial tension is calculated as follows:

$$\sigma=2\cdot\boldsymbol{a}_{10}\left(\lambda^2-\frac{1}{\lambda}\right)+2\cdot\boldsymbol{a}_{01}\left(\lambda-\frac{1}{\lambda^2}\right), \quad (1)$$

where λ is the elongation in the direction of the load application.

IOP Conf. Series: Materials Science and Engineering **534** (2019) 012023 doi:10.1088/1757-899X/534/1/012023

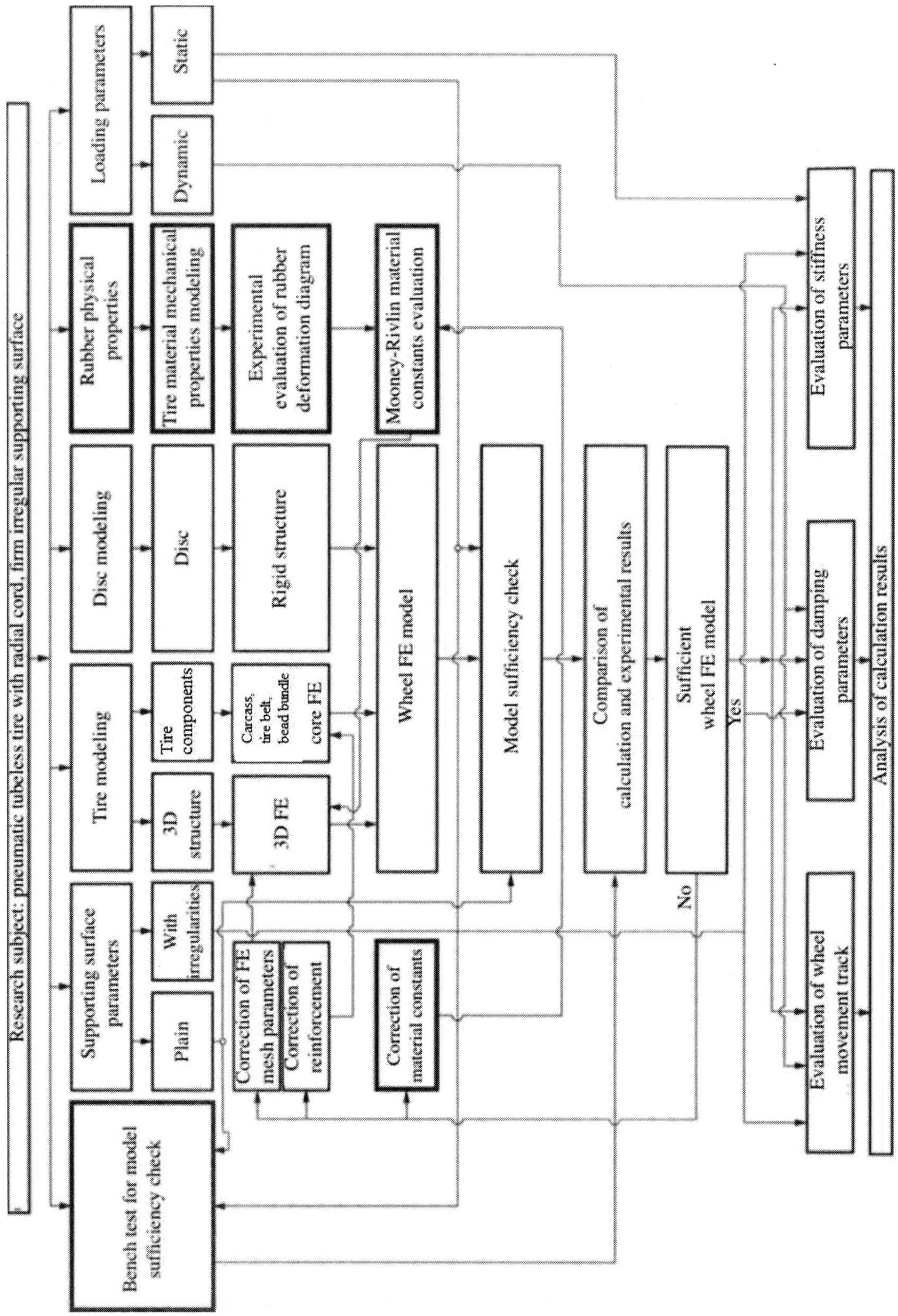

Figure 1. Block diagram of the computational and experimental method for defining stiffness and damping characteristic of the tire.

Mooney-Rivlin material constants were obtained by conducting mechanical tests of rubber samples cut from the object under study.

The contact problem of interaction of the tire and uneven surface is solved by using penalty function method. Friction coefficient as a function of the relative slip velocity v_{rel} is defined as follows:

$$\mu_c = \mu_d + (\mu_s - \mu_d)e^{-D_c|v_{\mathrm{rel}}|},$$

where: μ_c is the friction coefficient; μ_s is the static friction coefficient; μ_d is the dynamic friction coefficient; D_c – exponential damping ratio; v_{rel} is the relative slip velocity.

In order to consider the losses due to hysteresis in the rubber, a model of proportional damping is used:

$$\mathbf{C} = \alpha\mathbf{M} + \beta(\mathbf{K}_T + \mathbf{K}_G),$$

where α and β are the constants, which are defined from equivalent damping.

Figure 1 shows the main stages of the developed computational and experimental method for calculation of the stiffness and damping characteristics of the car pneumatic tire and evaluating its enveloping properties.

Tire finite-element simulation has been carried out (Figure 2). Stiffness parameters of the tire equivalent structural elements have been chosen. Solutions for a number of test quasistatic tasks of tire interaction with a flat surface at different values of the tire air pressure and normal load have been found (Figure 3). The authors have calculated tire deflections, which were compared with the results of the experiment at the "Ground path" test rig, and obtained the mechanical characteristics of the rubber (on the testing machine IP 5081-20).

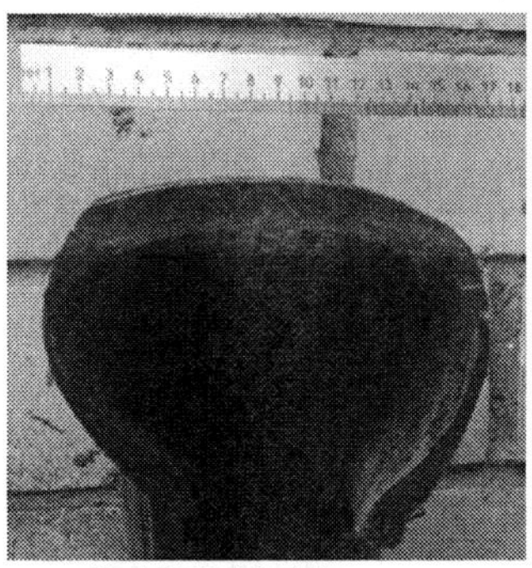

Figure 2. VLI-10 Tire

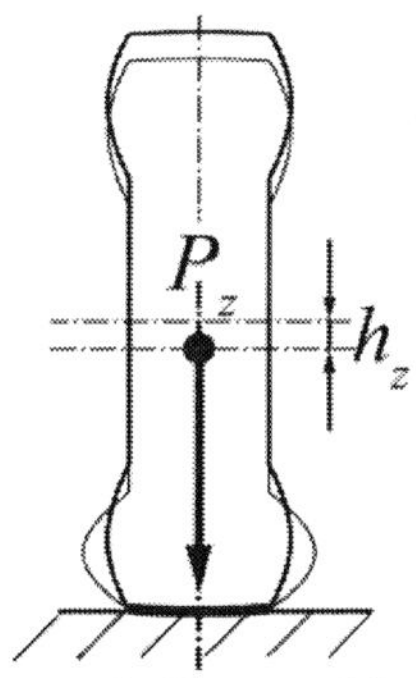

Figure 3. Test problem model.

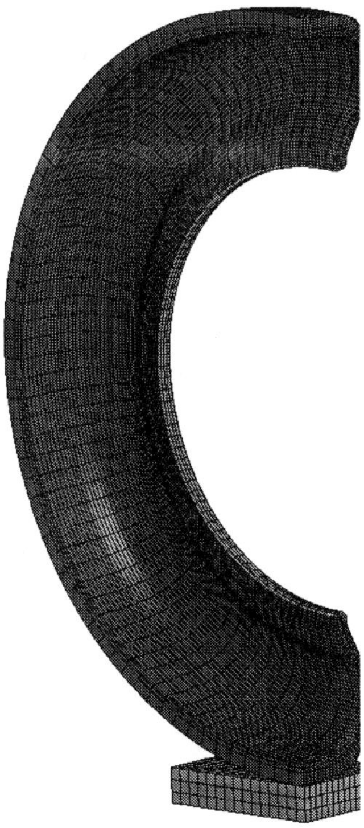
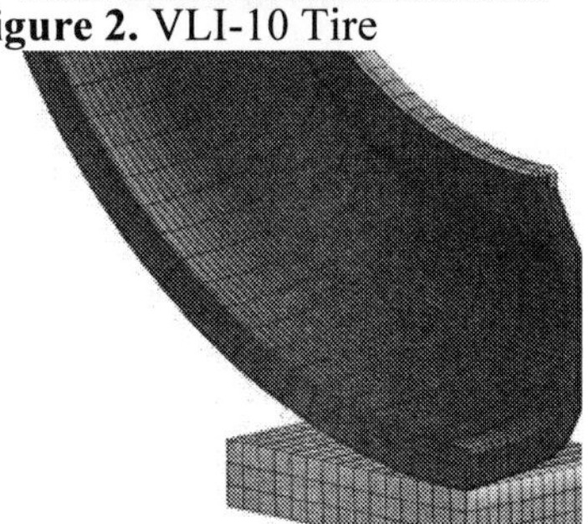
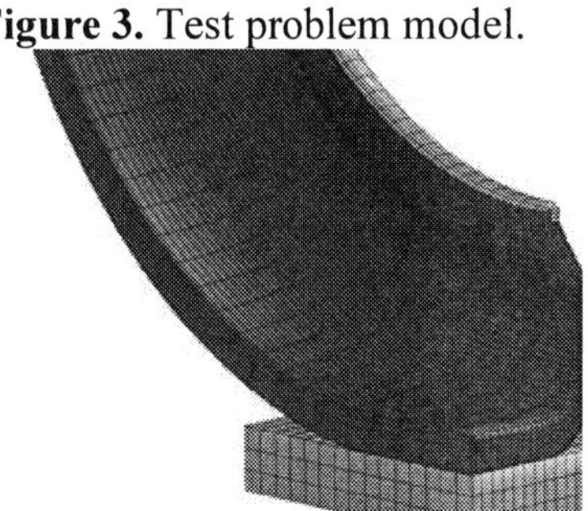

a) b) c)

Figure 4. Tire finite-element model: a) FE model; b) contact area before load application; c) contact patch after application of load.

Figure 4 shows finite-element model of the tire (Figure 4, a) and the relative position of the tire and the flat surface before application of the vertical load (Figure 4, b) and after application of the vertical load (Figure 4, c). Forming of the contact patch between the tire and the surface and significant change of the tire shape are detectable.

The tire deflection h_z dependencies on the normal wheel load P_z at different values of the tire inflation pressure are presented in Figure 5.

The discrepancy between the values of the tire deflections obtained by calculation and experimental methods does not exceed 10%. This confirms the adequacy of the chosen analytic models and the developed principles of modeling, which can be used to study more complex variants of the interaction of the tire with the surface.

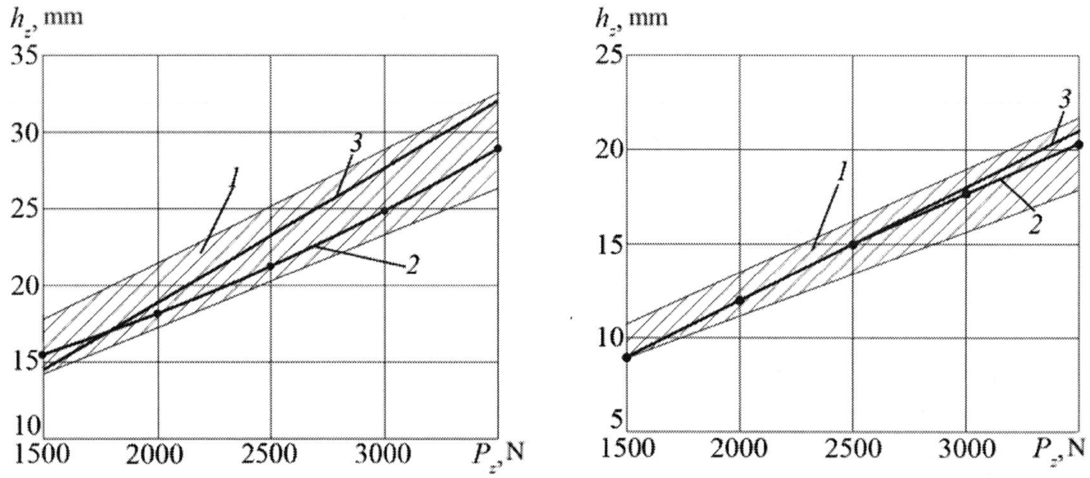

a) б)

Figure 5. The dependence of the deflection of the tire h_z on the normal load P_z on the wheel for different values of the tire inflation pressure: a) 0.1 MPa; b) 0.2 MPa; *1* – experiment*;* *2* – finite element analysis (FEA)*;* *3* – theoretical.

3. Shape of irregularities and tire deformation

Rectangular, cylindrical, triangle and rectangular cleats have been studied (Figure 6). All the irregularities had height q = 20 mm, which is equivalent to the height of the irregularities of the cobblestone pavement. Length s was varied from 40 mm to 120 mm for rectangular cleats and to 160 mm for cylindrical cleats. The cleat tests were performed for different vertical loads P_z and tire inflation air pressures.

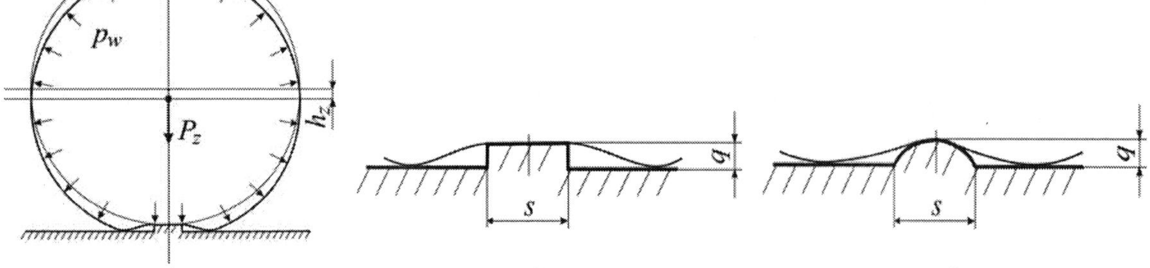

a) b) c)

Figure 6. Diagram of the wheel loading with the vertical force P_z, internal pressure p_w and shape of the cleats: a) test model; b) rectangular cleat; c) cylindrical cleat.

Estimated tire deflection h_z at various load levels for rectangular and cylindrical cleats are presented in Figure 7 and Figure 8. Inflation tire pressure is constant (p_w=0.1MPa).

From the comparison of the dependences presented in Figure 7 it follows that at the same wheel load the tire deflection h_z on the cylindrical cleats is greater than on the rectangular cleats by 25 – 27 % under the same load within the interval of maximum loads from the considered range.

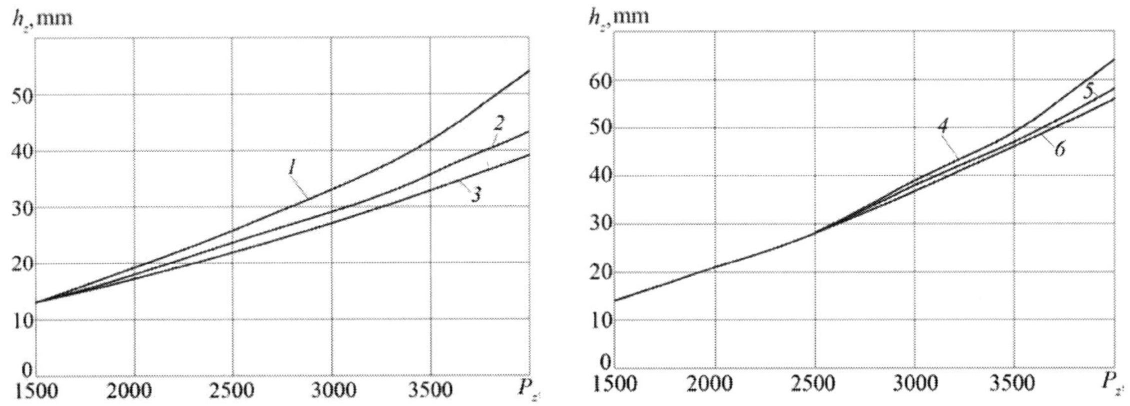

a) b)

Figure 7. The dependence of the deflection of the tire h_z on the vertical load P_z of the wheel interacting with the cleats of length s and of different shapes: a) rectangular; b) cylindrical; 1 – s=40 mm; 2 – s=80 mm; 3 – s=120 mm; 4 – s=40 mm; 5 – s=80 mm; 6 – s=160 mm.

The dependences of the change in radial stiffness for the two considered cleats depending on their length have been obtained (Figure 8). The tire deformation state is shown in Figure 9. Increase in the load leads to decrease in the tire radial stiffness. Increase in the length s of the cleats leads to the increase in the tire radial stiffness.

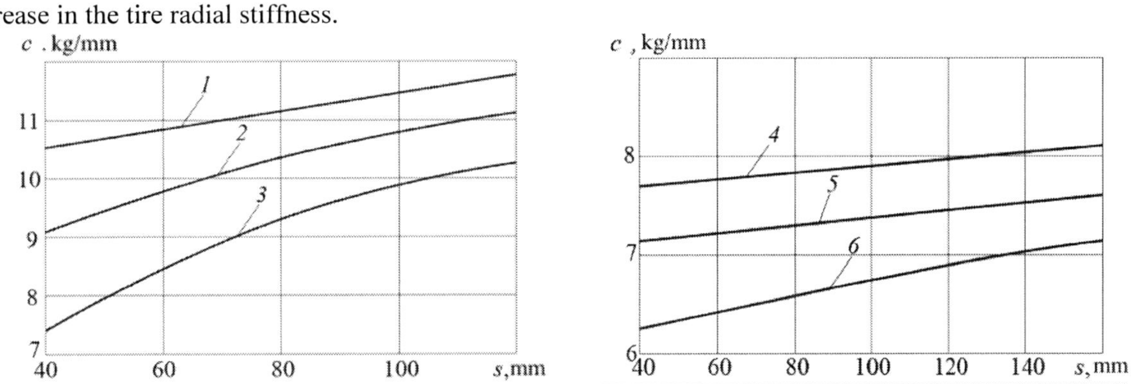

a) b)

Figure 8. The dependence of the normal tire stiffness c when interacting with the cleats with lengths s under different loads: a) rectangular; b) cylindrical; 1 – P_z = 2000 N; 2 – P_z = 3000 N; 3 – P_z = 4000 N; 4 – P_z = 3000 N; 5 – P_z = 3500 N; 6 – P_z = 4000 N.

Analysis of the data, obtained by calculation, demonstrates that the shape of the cleat has significant influence on the tire deflection principles.

The problem of the wheel movement along a firm irregular surface has been studied. The following parameters of irregularities were chosen: s was assumed to be equal to 80 mm, half height of the irregularities q_0= 1,1 mm, the overall length of the segment l=650 mm. The following boundary conditions were formulated: p_w=0,2 MPa; P_z=4000 N. Two modes were considered: in the first mode, the wheel axis linear velocity v_a=1,6 m/s was set, in the second mode, angular velocity n=5 cycles/s was set (in this case the linear velocities of the wheel axis center for both the modes were approximately equal). Calculation results are shown in Figures 10 and 11.

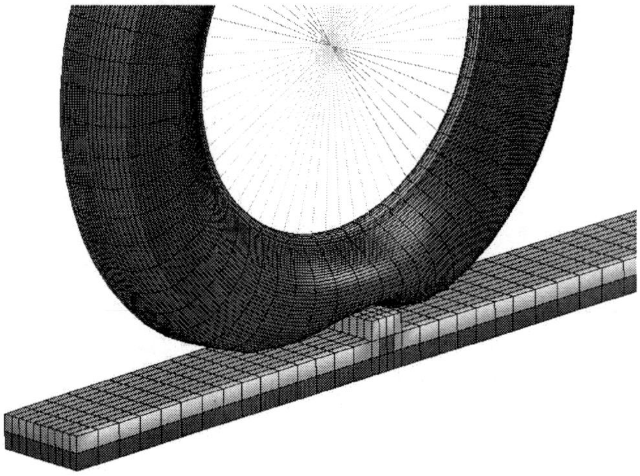

Figure 9. An example of the deformed state of the tire.

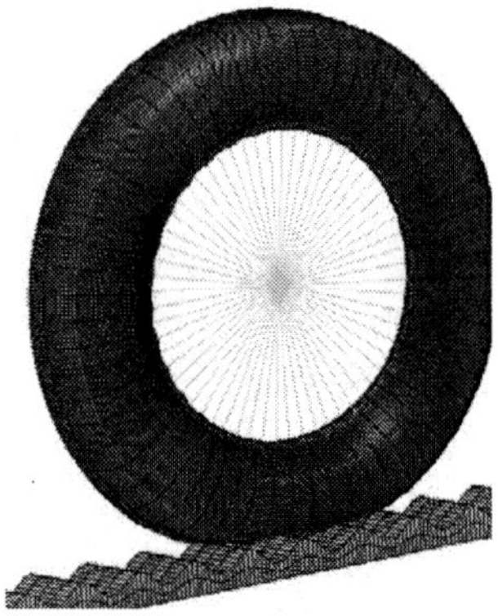

a)

b)

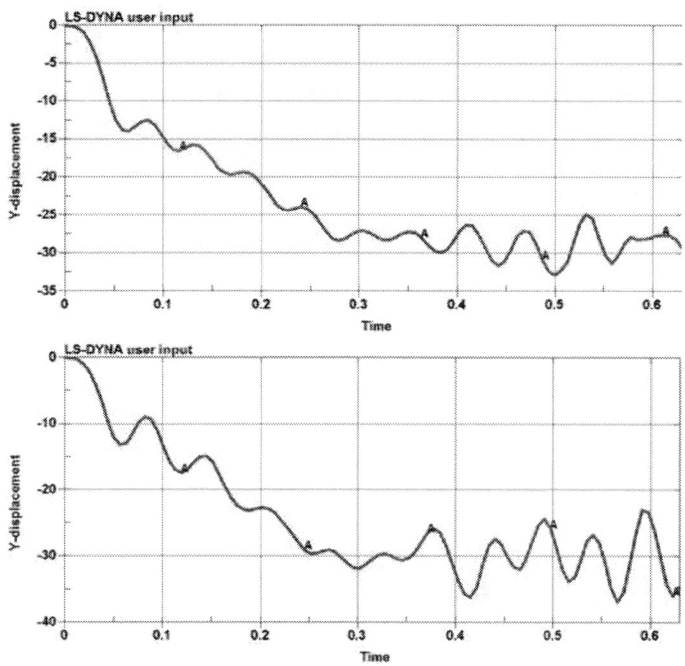

Figure 10. Tire deformation state (sinusoidal irregularities).

Figure 11. Change of the vertical displacement of the wheel axis in time: a) with application of linear velocity; b) with application of angular velocity.

Analysis of the influence of the shape of irregularities on the tire damping parameters has been performed. Figure 12 shows calculation results for peaks and pits with rectangular section.

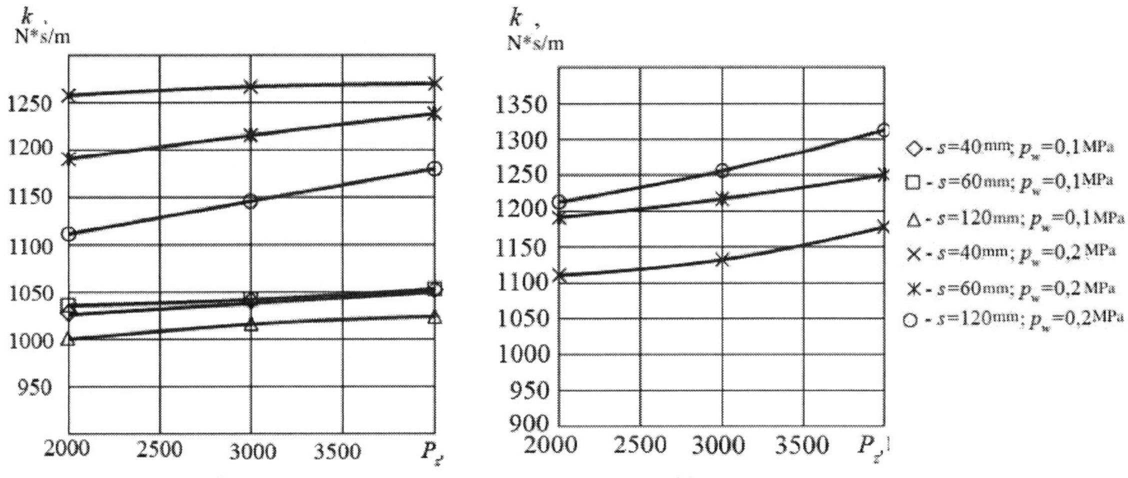

a) b)

Figure 12. Calculation results of tire damping characteristics for different cleats: a) rectangular sectioned peaks; b) rectangular sectioned pits

Based on the results of the research conducted, the following conclusions are made:

1. Shape of irregularities has a significant influence on the tire deflection, its stiffness and damping characteristics.
2. Results obtained from finite-element simulations and approximation results for peaks and pits with rectangular section differ no more than by 5%.
3. A tire analytical model has been developed; the model makes it possible to solve (alongside with static contact tasks) dynamical tasks of the wheel rolling on uneven surfaces.
4. The results have been obtained for the wheel displacement caused by different types of loads (linear and angular loads).
5. The shape of irregularities affects tire damping properties.

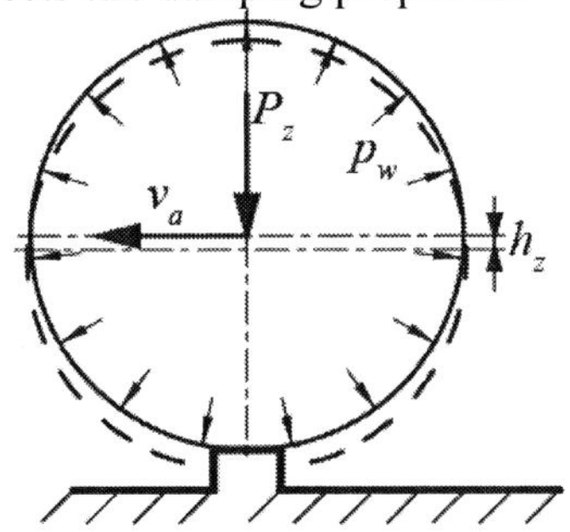

Figure 13. Tire cleat test model.

A series of cleat tests was performed for the case of the tire dynamical loading. The cleat test model is shown in Figure 13. Single cleats of the following shapes were used: square, triangular and rectangular. Models of the cleats are shown in Figure 14.

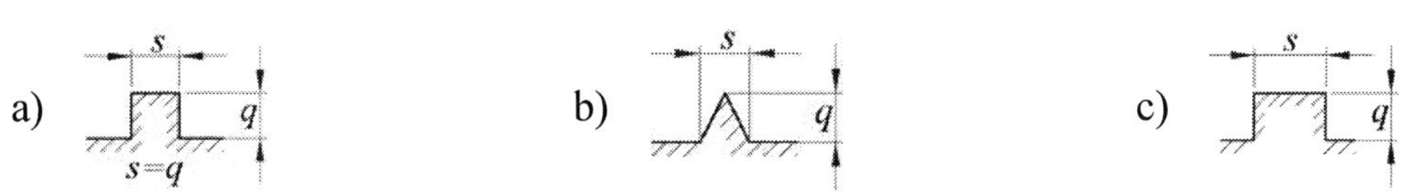

Figure 14. Cross section shapes of the cleats: a) square; b) triangular; c) rectangular.

To verify the adequacy of the developed modeling methods, a computational tire model was developed reflecting the conditions of the experiment. Comparison of the results of the experiment and the calculation confirmed the adequacy of the developed simulation model of the automobile pneumatic tire and proved that it is possible to use the developed methods for the research of the interaction of the automobile wheel with a rigid uneven surface.

Furthermore, testing technique for uniaxial tension of a rubber specimen cut from a tire has been developed in order to obtain Mooney-Rivlin material constants. Flat specimens were used. As a result of the tests, the load – displacement diagram was obtained. The Mooney-Rivlin material constants a_{10} and a_{01} for equation (1) were obtained with the use of the least-square method.

4. Conclusions

1. The created physically and geometrically nonlinear spatial model of the car pneumatic tire considers design features (carcass, tire belt, bead bundle), hyperelastic and incompressible rubber properties and makes it possible to research tire contact interaction with road irregularities.

2. With the use of the developed method for experimental determination of the properties of incompressible material (rubber) constants a_{10}= 1.285 MPa, a_{01}= –0.0004 MPa. of the Mooney-Rivlin material have been obtained. The constants can be used to model the tire interaction with the surface.

3. The research has shown that the tire damping significantly depends on the shape of the road irregularities and loading conditions (vertical load, tire inflation pressure). Difference in the tire deformation when interacting with the cleats of cylindrical and rectangular shapes reaches 20 % at the tire inflation pressure 0.1 MPa (50 % of nominal pressure) and wheel load 3500 N (68 % of nominal load). Study of the influence of the shape of irregularities on the tire radial stiffness showed that the greatest vertical tire deflection occurs in the contact with the triangular cleats with a length of about 20% of the contact patch length. This vertical tire deflection may reach a value of 42 % of the tire deflection on the even surface.

4. Calculation results show that damping ratio is significantly influenced by the tire inflation pressure and shape of the irregularities. For example, decrease in the air pressure by half on a pit results in the decrease of the damping ratio by 18 %. For a peak (of the same length) the damping ratio decreases by 14 %.

5. The developed method for pneumatic tire enveloping properties estimation enables to obtain new values of the tire stiffness and damping properties that can be used in the dynamic analysis of the *road – tire – car – driver* mechanical system.

References

[1] Gorelov V A and Komissarov A I 2016 Mathematical model of the straight-line rolling tire - Rigid ter-rain irregularities interaction *Procedia Engineering* **150** pp 1322-28 (DOI: 10.1016/j.proeng.2016.07.309)

[2] Keller A V, Gorelov V A, Vdovin D S, et al. 2015 Mathematical model of all-terrain truck *Proceedings of the ECCOMAS Thematic Conference on Multibody Dynamics* pp 1285–1296

[3] Dubrovskiy A, Aliukov S, Keller A, et al. 2016 Adaptive Suspension of Vehicles with Wide Range of Control *SAE Technical Paper* 2016-01-8032

[4] Keller A, Aliukov S 2015 Effectiveness of Methods of Power Distribution in Transmissions of All-Wheel-Drive Trucks *SAE Technical Paper* 2015-01-2732Kotiev G O, Padalkin B V, Kartashov A B and Diakov A S 2017 Designs and development of Russian scientific schools in the field of cross-country terrain vehicles building *ARPN Journal of Engineering and Applied Sciences* **12 (4)** pp 1064-71

[5] Keller A, Aliukov S, Anchukov V, et al. 2016 Investigations of Power Distribution in Transmissions of Heavy Trucks *SAE Technical Paper* 2016-01-1100

[6] Gorelov V A, Komissarov A I and Miroshnichenko A V 2015 8×8 wheeled vehicle modeling in a multibody dynamics simulation software *Procedia Engineering* **129** pp 300-07 (DOI: 10.1016/j.proeng.2015.12.066)

[7] Keller A V, Gorelov V A, Anchukov V V 2015 Modeling truck driveline dynamic loads at differ-ential locking unit engagement *Procedia Engineering* **129** pp 280-87 (DOI: 10.1016/j.proeng.2015.12.063)
[8] Volskaya N S, Zhileykin M M and Zakharov A Y 2018 Mathematical model of rolling an elastic wheel over deformable support base *IOP Conf. Series: Materials Science and Engineering* **315** 012028 (doi: 10.1088/1757-899X/315/1/012028)
[9] Keller A and Aliukov S 2015 Methodology of System Analysis of Power Distribution among Drive Wheels of an All-wheel-drive Truck *SAE Technical Paper* 2015-01-2788
[10] Biderman V L, Guslizer R L, Zakharov S P 1963 Car tires: design, calculation, testing, operation (Moscow: Goskhimizdat) p 383
[11] Bukhin B L 1988 Introduction to the mechanics of pneumatic tires (Moscow: Khimiya) p 223
[12] Yatsenko N N 1978 Absorbing and enveloping ability of tires (Moscow: Machinostroenie)p 132
[13] Bohm F 1987 Nichtlineare Schwingungenbeim Rollkontakt von Giirtelreifen (Mitteilung des Curt-Risch-Inst. der Universitat Hannover) pp 23-47
[14] Schmeitz A 2004 A semi-empirical three-dimensional model of the pneumatic tyre rolling over arbitrarily uneven road surfaces *Technische Universiteit Delft* p 322
[15] Yang X Finite element analysis and experimental Investigation of tyre characteristics for developing strain-based intelligent tyre system (Birmingham, Ph. D. thesis) p 242
[16] Vol'skaya N.S., Levenkov Ya.Yu., Rusanov O.A. 2013 Simulation of an automotive pneumatic tire interacting with a solid rough supporting surface *Science and Education: scientific edition of the Bauman MSTU* **5** p 107

IOP Conf. Series: Materials Science and Engineering **534** (2019) 012024 doi:10.1088/1757-899X/534/1/012024

Automation of wheeled vehicles load bearing frames finite-element models loading procedure by using inertia relief method and vehicle multi-body dynamics model

D S Vdovin, I V Chichekin, Y Y Levenkov and M L Shabolin

Wheeled Vehicles Department, Bauman Moscow State Technical University, 105005 Baumanskaya 2th st., 5, Moscow, Russian Federation

E-mail: vdovin@bmstu.ru

Abstract. The paper presents a technique for strain-stress calculation using finite element analysis for wheeled vehicle's load bearing frames with the use of loads, obtained from full vehicle multi body dynamics models (MBD models). In this technique, boundary conditions need not to be defined, as the inertia relief method used. Frame and suspension parts are imposed in MBD model as flexible bodies to increase accuracy of loads calculations. The presented method is capable to automatize loading procedure of finite-element models, thus decreasing computational costs and results processing costs while analyzing numerous load cases. As an example, strength analysis of 6x6 articulated vehicle frame is presented. Proposed method compared to "classic" finite-element frame modelling technique, which use conventional loads formulation and fixed boundary conditions. In load case "standing on ground", calculation results with the use of two methods demonstrate high convergence in zones, located far from suspension and frame mounting joints. For load case "hanging of second axis" results are significantly different, due to ability of MBD-model to capture behavior of suspension links, and calculate true vehicle frame movement in space in particular load case. Loading procedure of frame finite element model is automated, using script, which transfer loads from MDB-model.

1. Introduction

At the moment, methods for vehicle load bearing frames stress calculation, providing quantitative and qualitative convergence with experimental results, are developed and broadly described in literature [1], [3], [4], [6], [7], [6], [7], [6] . Most of these works use finite-element analysis (FEA) [9]. To obtain results, relevant to real strain-stress state of frame under load, it is necessary to create their complex finite-element models, that also takes into account suspension and other components, that contribute to combined system stiffness. In order to simplify the model, simplified rod-and-beam finite elements models of suspension and other vehicle systems usually used. This method has a number of disadvantages, which are, for example, difficulty in modeling of suspension kinematics; leaf springs deformed shape, stiffness properties of metal-rubber mountings, etc.

At the early stage of vehicle design concept development [9] in order to estimate properties of future product, one usually develop a multi-body dynamics model of entire vehicle. While virtual prototyping of different vehicle systems (suspension, transmission, etc.), this models are used as "virtual test benches", examples of these test benches for suspension, steering and the entire vehicle are presented in [10], [11], [12], [14], [15], [16], [16] , [16]. Using virtual prototyping, one can get general and complex properties of the future vehicle (or its system). One of this particular results – load factors, acting in every joints of dynamic model.

Loads, obtained in this way, are applicable for finite-element analysis of load bearing frame with the use of inertia relief method. In inertia relief method, it is assumed, that body or system of bodies are in quasi-static balance state. To achieve quasi-static balance state of the model, equilibrant accelerations are evaluated to eliminate imbalance of loads and moments. Resulting system of linear and angular accelerations makes the sum of all the forces, acting on model, to be equal to zero. Here assignment of boundary conditions in the form of fixed constraints is not required. This makes it possible to transfer loads from MBD model to finite-element analysis of the frame with the use of automated procedures (scripts). Thus, it is possible to analyze a numerous of loading cases, comparing to the manual definition of boundary conditions and loads, that usually used in "classic" methods.

Thereby, the study of the paper is inertia relief method and full vehicle dynamic multi-body model use in automated loading procedure of vehicle frame finite-element model.

To illustrate method, paper describes developed vehicle MBD model with frame, equalizing beam suspension system with flexible parts and non-linear stiffness of metal-rubber mountings. Developed finite-element model of articulated 6x6 vehicle frame also presented. To compare "classic" and proposed method strain-stress state calculation techniques, two analysis with both methods are performed.

2. Algorithm for automatization of frame finite element model loading procedure

In order to perform stress calculation of vehicle frame with the use of the inertia relief method, it is necessary to follow block diagram on figure 1.

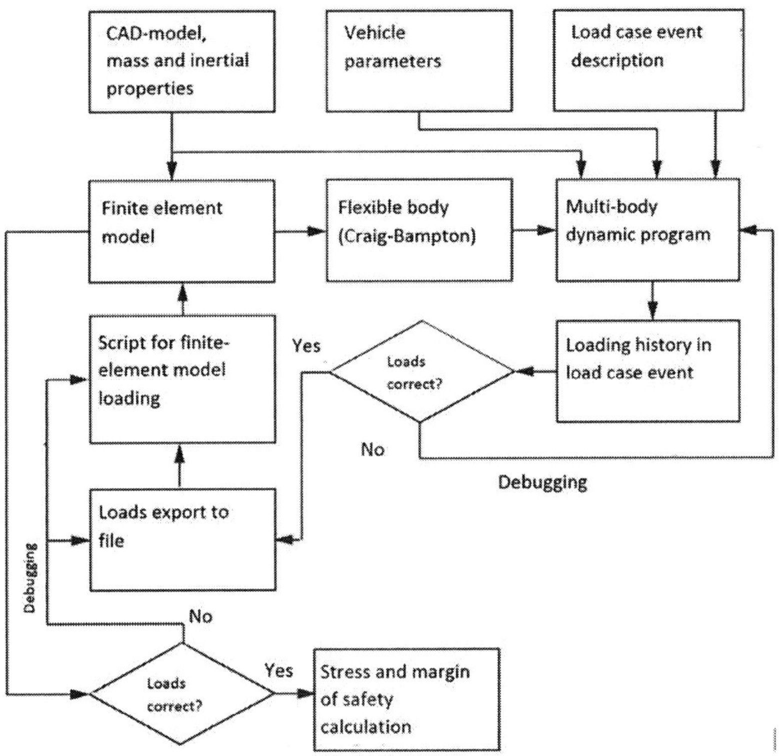

Figure 1. Algorithm for automatization of frame finite element model loading procedure.

Detailed analysis of methods for load adequacy checking and script generation are not presented, because these tasks are strictly specific to chosen software.

Static loads in vehicle frame and front/rear suspensions mounting joints are evaluated with the use of MBD model. Figure 2 shows general view of articulated vehicle MBD model. MBD model consists of: front frame, front leaf suspension, connected to it, rear frame with equalizing beam suspension system, articulation joint assembly, connecting two frames and providing one degree of freedom: rotation of front and rear frames about vertical axis. Model includes steering hydraulic cylinders, modelled as rigid rods.

In the model, the following assumptions are made:

– parts of suspension system, except equalizing beam, are modelled as rigid bodies;

– joints are free of friction;

– wheels deformations are considered in the model of wheel interaction with supporting surface, wheel forces are applied in the center of the wheel.

3. Description of articulated vehicle multi body dynamic model

The MBD model of the articulated vehicle is shown on figure 2.

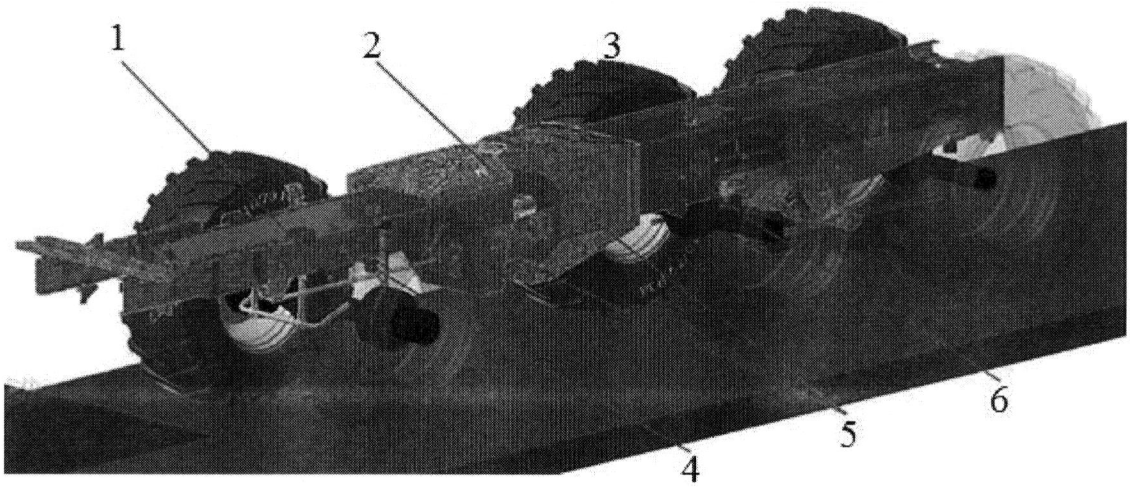

Figure 2. General view of articulated vehicle MBD model: 1 – front leaf suspension; 2 – steering cylinders; 3 – rear suspension with equalizing beam; 4 – front frame; 5 – articulation joint assembly; 5 – rear frame.

General view of front spring suspension model with two leaf springs and stabilizer is presented on figure 3, locations and types of joints, providing connection of parts, are presented on figure 4.

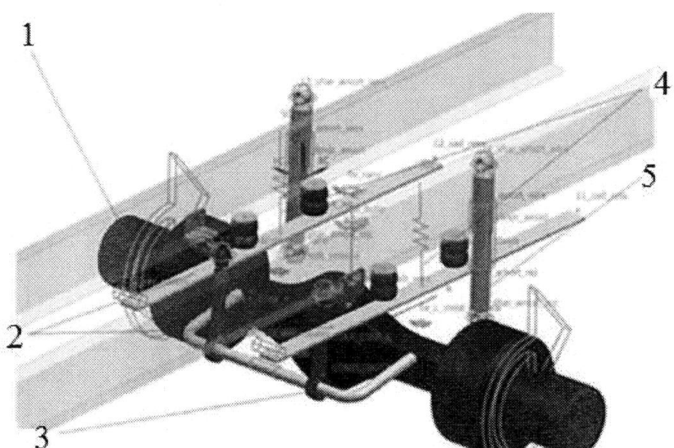

Figure 3. General view of front leaf spring suspension model: 1 – axle; 2 – leaf spring; 3 – stabilizer; 4 – buffer; 5 – shock absorber.

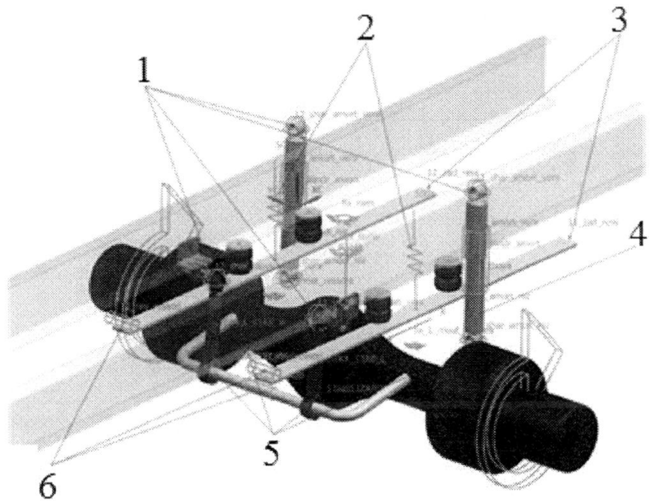

Figure 4. Joints location in the model of front leaf spring suspension: 1 – spherical joint; 2 – buffer spirngs ; 3 – bushing; 4 – damper (shock absorber); 5 – bushing; 6 – cylindrical joint.

Leaf spring model description is presented in work [10].

Suspension stop buffers, modelled as a spring element with nonlinear stiffness, are mounted on the vehicle frame above leaf springs. Stabilizer consists of two rigid units, connected to each other by cylindrical joint with angular spring, whose stiffness corresponds to torsion stiffness of stabilizer.

Figure 5 shows general view of rear suspension with equalizing beam and two axles. Model includes six longitudinal reactive rods (two lower and one upper per each axle) and two lateral reactive rods, one per axis (figure 6).

Figure 5. General view of rear suspension with equalizing beam, two axles, longitudinal and lateral reactive rods.

IOP Conf. Series: Materials Science and Engineering **534** (2019) 012024 doi:10.1088/1757-899X/534/1/012024

All reactive rods connects axles with frame by rubber-metal bushings, which are modelled by standard bushing elements. Equalizing beams are connected to the frame with revolute joint and can rotate about lateral axis. Endpoints of equalizing beams are connected to axles with rubber buffers.

The feature of the developed model is the use of flexible equalizing beam in rear suspension model. Flexible equalizing beam is defined by Craig-Bampton method, based on specially developed finite-element model of equalizing beam (figure 7). Suspension stop buffers fixed on the frame model.

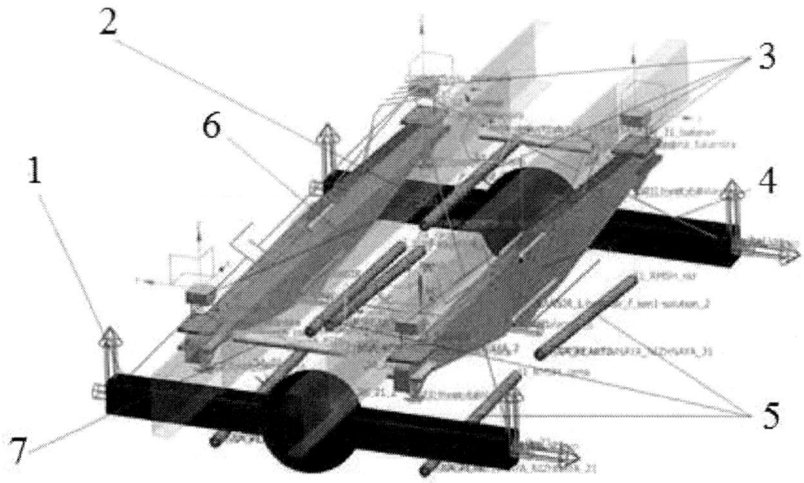

Figure 6. Rear suspension system model: 1 – front axle; 2 – rear axle; 3 – stop buffers; 4 – left equalizing beam; 5 – longitudinal reactive rods; 6 – right equalizing beam; 7 – lateral reactive rods.

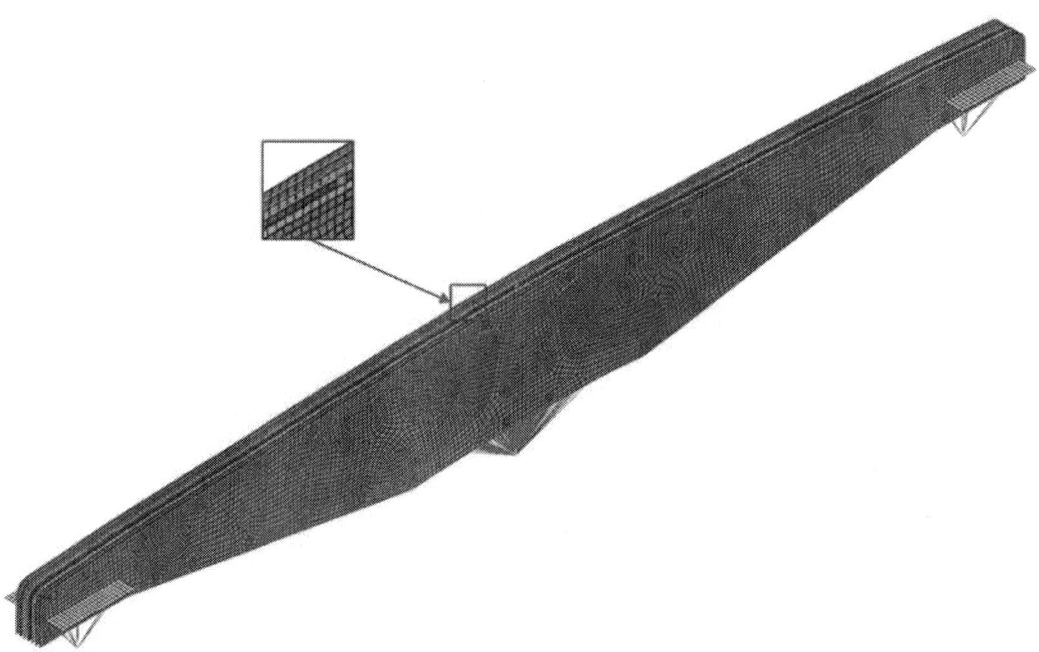

Figure 7. General view of rear suspension equalizing beam finite element model for use as a flexible body in multi-body dynamics model.

Vertical, longitudinal and lateral reactions in tire contact patch with ground are defined with specific model of wheel-ground interaction, and not described in this paper.

4. Load bearing frame finite element model description

For strength analysis of wheeled vehicle frame finite-element model is developed. The model is based on shell-type finite elements. The average size of shell finite element is 30 mm, total number of nodes of front and rear frames is 895973, number of elements is 564100. Bolts and rivets are modelled with short beam elements.

To compare two methods of frame strength analysis, two calulations are performed: 1) "classic" method with the use of simplified rod and beam suspension model and 2) inertia relief method with loads, calculated in multi body dynamics model.

Finite-element model of the frame is presented on figure 8. Full mass of the vehicle is defined with pointed masses, located in gravity centres of massive vehicle aggregates (figure 9).

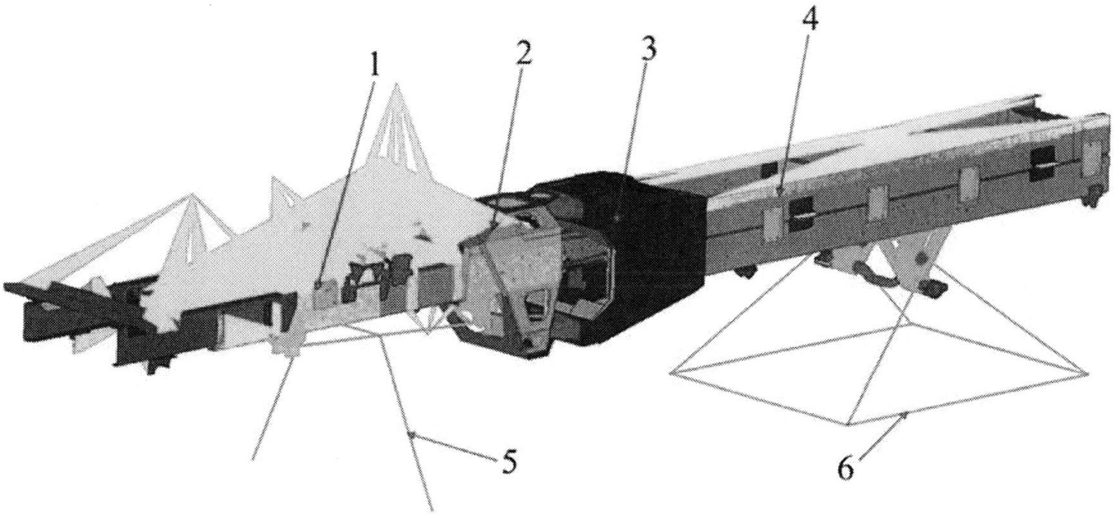

Figure 8. Finite-element model of load bearing frame: 1 – front frame, 2 – front part of articulation joint assembly, 3 – rear part of articulation joint assembly, 4 – rear frame, 5 – simplified model of front suspension, 6 – simplified model of rear suspension.

IOP Conf. Series: Materials Science and Engineering **534** (2019) 012024 doi:10.1088/1757-899X/534/1/012024

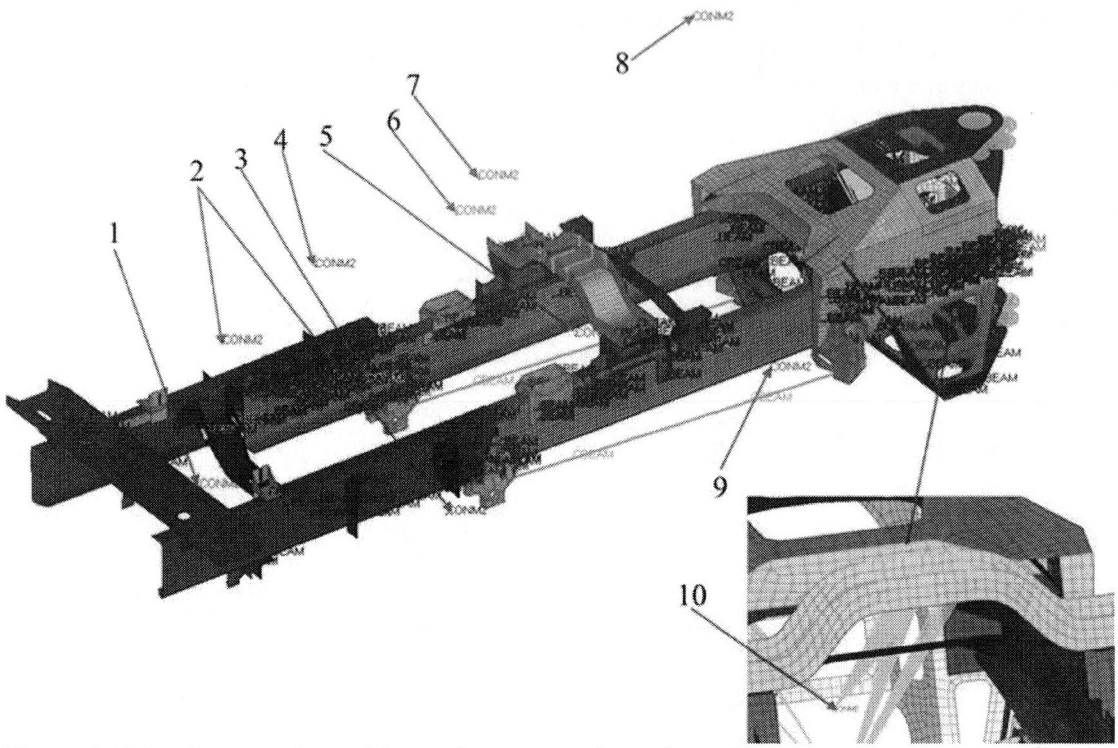

Figure 9. Pointed masses, located in gravity centres of massive vehicle aggregates:
1 – cooling system; 2 – battery and receivers; 3 – engine; 4 – cabin; 5 – gearbox; 6 –pipes, fluids, electrical wires, etc.; 7 – exhaust system; 8 – mass of spare wheel and crane; 9 – cardan shaft; 10 – transfer box.

Joints of front and rear frames are modelled by rod-type elements with assisting RBE2-type rigid elements. Rigid element "RBE2*" allow linear relative translation along vertical axis (figure 10).

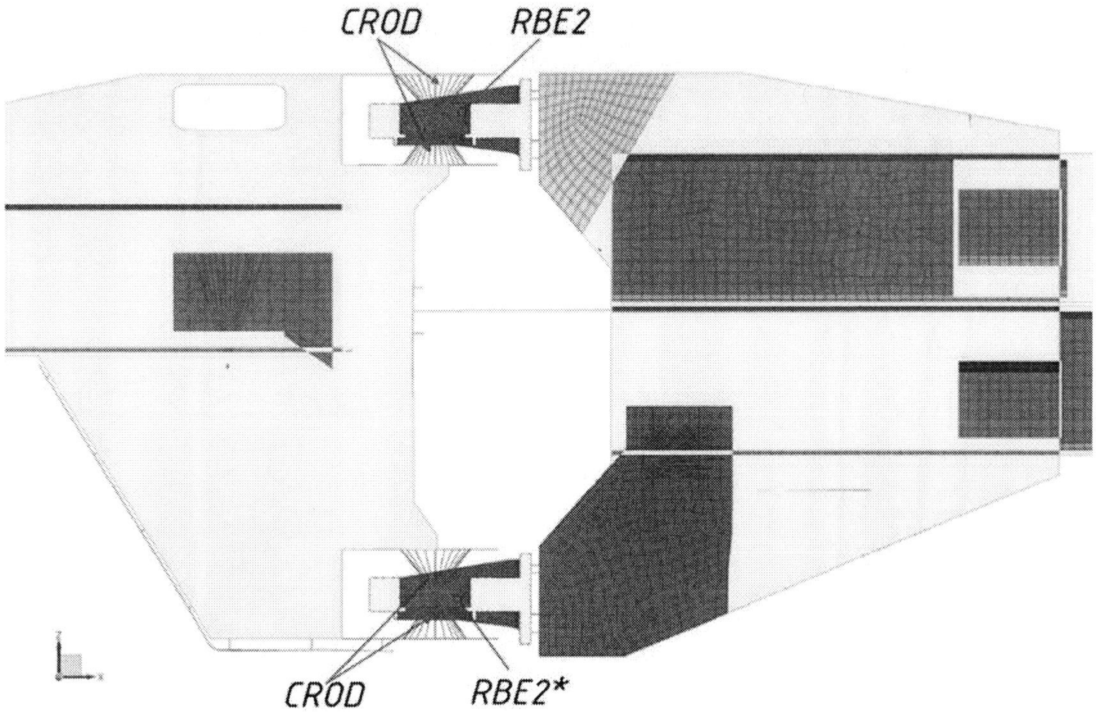

Figure 10. Articulation joint assembly modelling.

Simplified rod and beam models of front and rear suspensions are presented on figures 11, 12. Joints of front and rear suspensions with frame are modelled by RBE2** elements, that permits rotation around Y-axis degree of freedom. RBE3** element in front suspension model have free linear degree of freedom along X-axis. This allows longitudinal motion of leaf spring rear end relative to the frame.

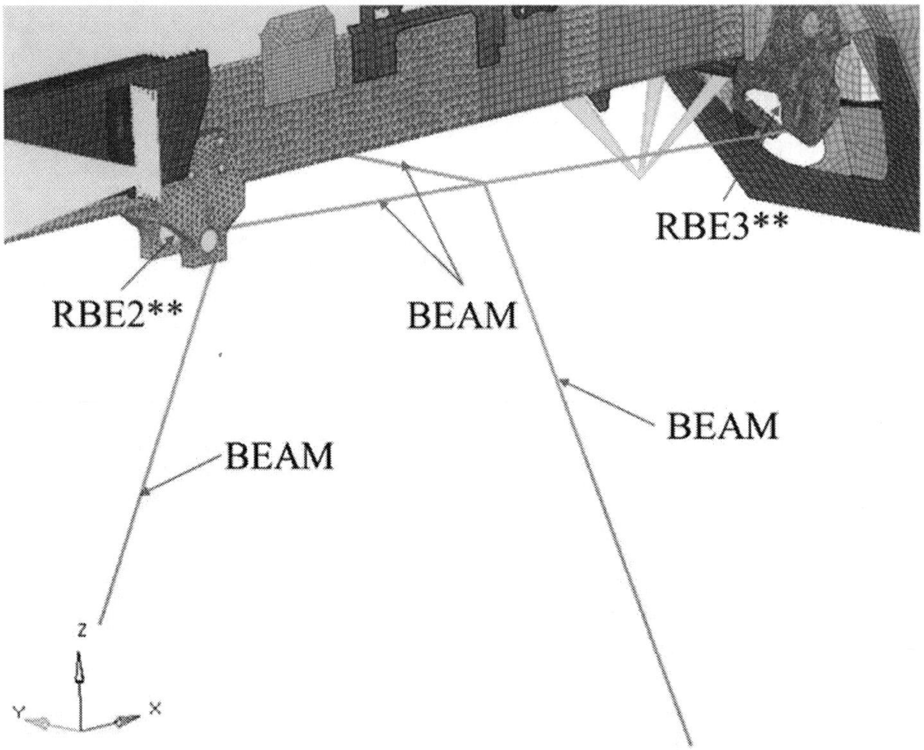

Figure 11. Simplified finite-element model of front suspension.

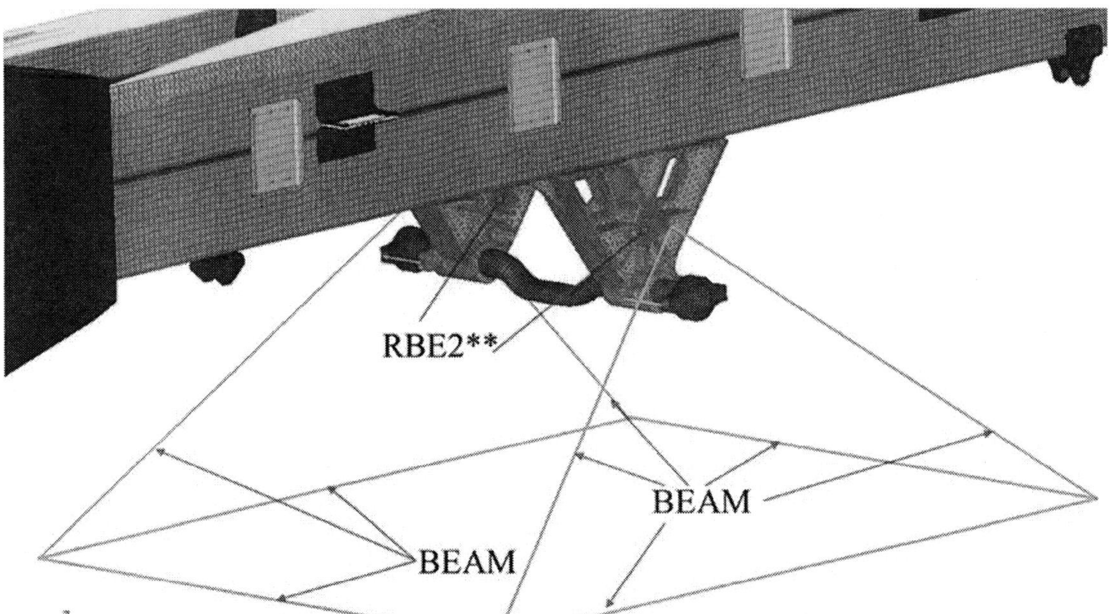

Figure 12. Simplified finite-element model of rear suspension.

Finite-element model of rear frame is presented on figure 13. Payload is modelled by single point mass, located in the center of gravity of rear frame and distributed uniformly on upper longerons with the use of RBE3 element.

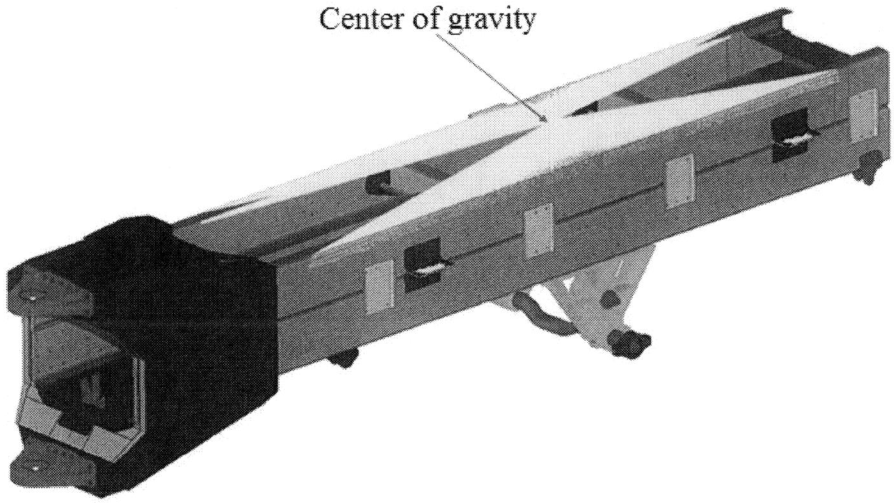

Figure 13. Finite-element model of rear frame with payload, distributed along upper longerons.

Steering hydraulic cylinders in articulation joint assembly and their mountings to front and rear frames are presented on figure 14. Hydraulic cylinders are modelled by rod finite elements, connected to brackets of front and rear frames, via RBE2 elements and intermediate BEAM elements. Material properties of frame are presented in table 1.

Table 1. Material properties of frame parts.

№	Part name	Material	Elastic modulus, E, MPa	Poisson ratio	Yield stress σ_T, MPa	Stress limit σ_B, MPa	Breaking elongation, d5, %
1.	Longerons	20GUT	200000	0,3	480	600	24
2.	Articulation joint assembly parts	10HSND	200000	0,3	390	530	19

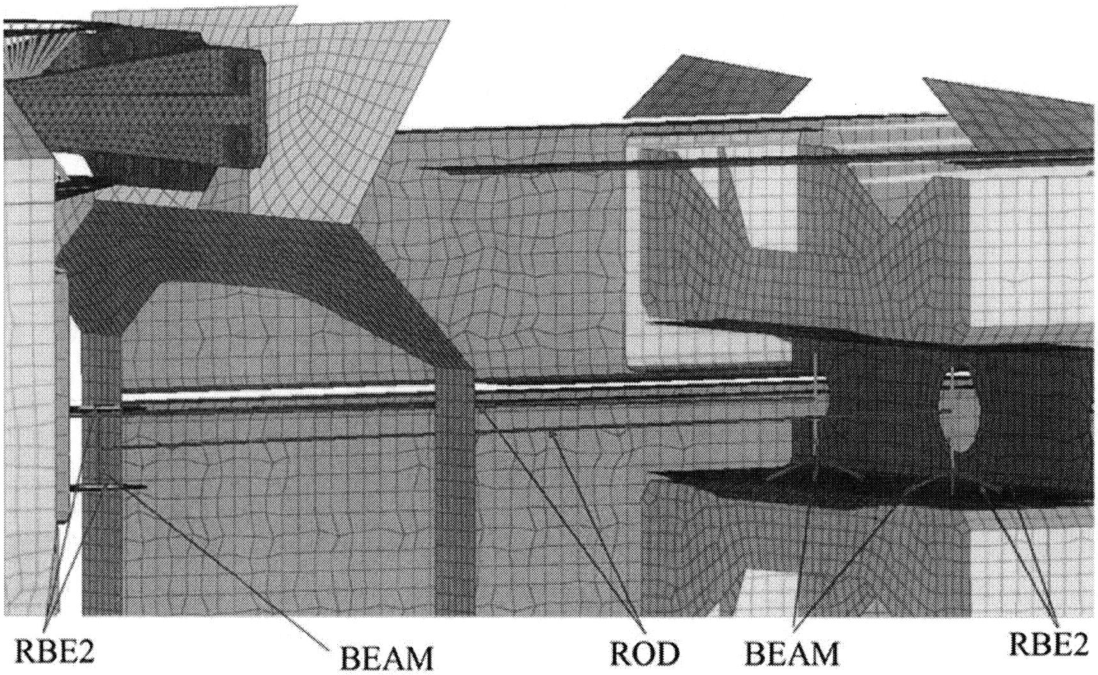

Figure 14. Finite-element model of steering hydraulic cylinders.

Two described method of strength calculation compared on two load cases, typical for 6×6 vehicle: 1) vehicle static standing on plain surface and 2) hanging of second axis. Many other load cases are analyzed as well, but they are not presented in this paper due to its limited volume of article. Also comparison of the stress-strain state is presented for the rear frame only. However, all conclusions made from this two load cases are true for all analyzed loadcases and other frame parts.

In first method, "classic" approach, boundary conditions and loads are presented in figure 15, 16 (where g = 9,81 m/s^2 – gravity acceleration).

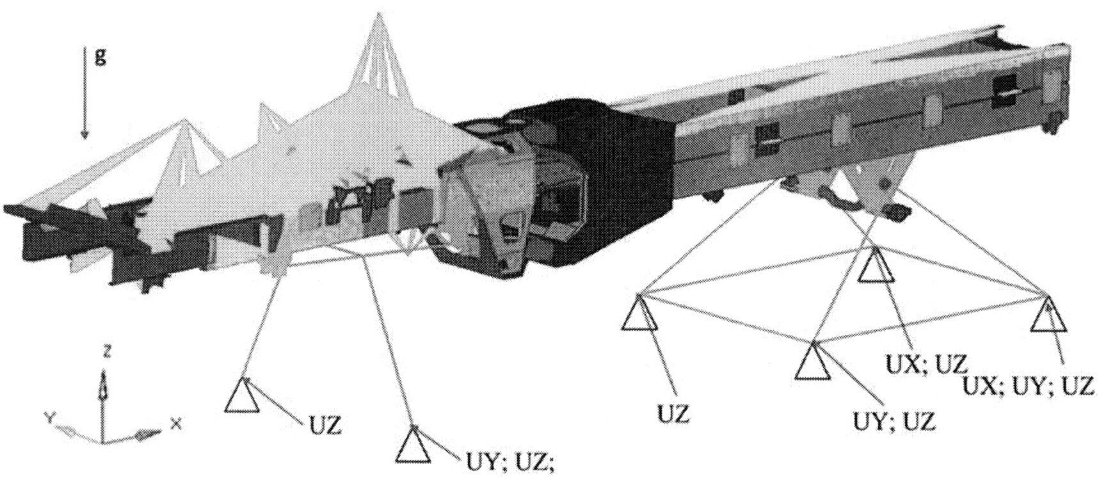

Figure 15. Boundary conditions and loads for load case "standing on ground". "Classic" approach to definition of loads and boundary conditions via simplified rod and beam suspension model.

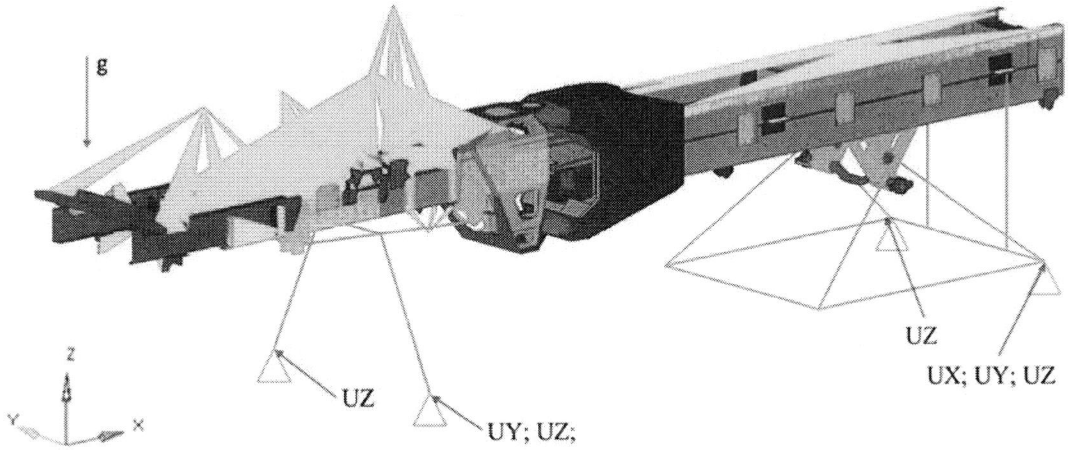

Figure 16. Boundary conditions and loads for load case "hanging of second axis". "Classic" approach to definition of loads and boundary conditions via simplified rod and beam suspension model.

While second axle hanging, rear suspension fits into stop buffer of third axle. Therefore additional beam element and RBE3*** element defined for this load case, which carry Z-axis load from third axle (figure 17).

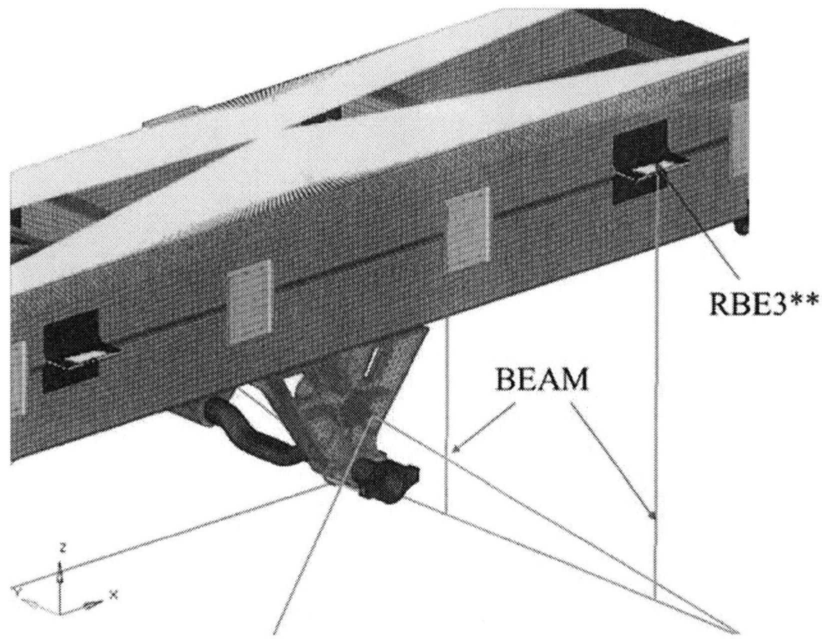

Figure 17. Simulation of equalizing beam suspension, fitted into stop buffer of third axle, in "hanging of second axle" load case.

In second method loads and reactions from suspension joints, calculated using multi body dynamics model, are applied in corresponding points of rear frame (figure 18). Automatization of loading finite element model achieved by specially developed script. This script transfers loads and reactions in key points of frame (joints, rod holders, mountings of hydraulic cylinders, etc.) from multi body dynamic model into finite-element model of the frame.

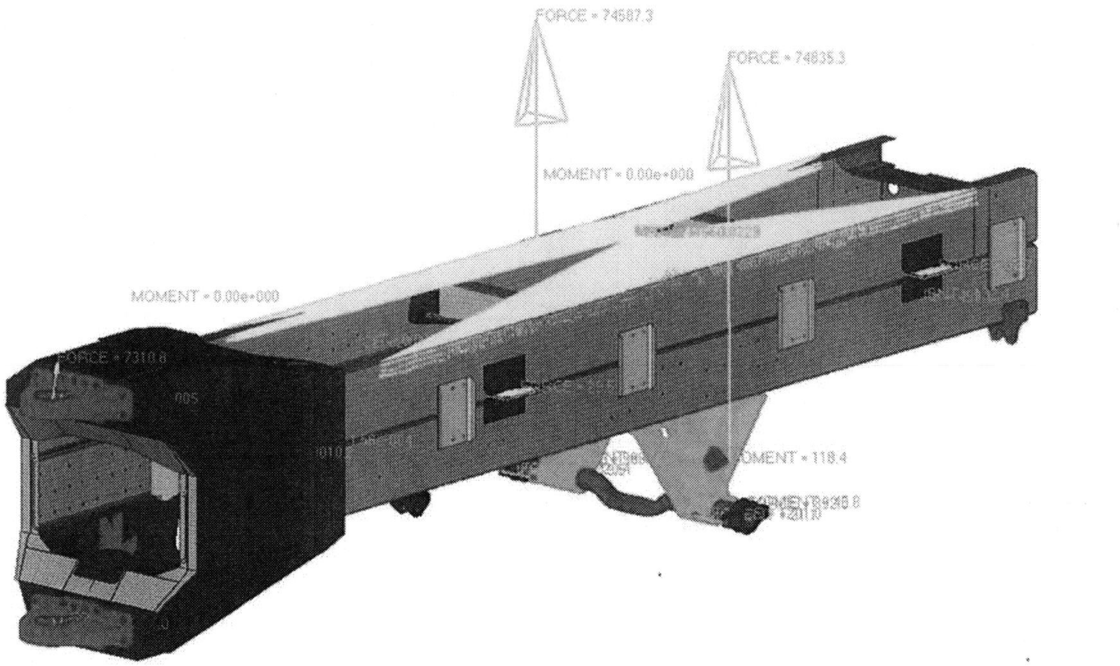

Figure 18. Loads and reaction forces from suspension, acting on rear frame, for load case "standing on ground" obtained by inertia relief method.

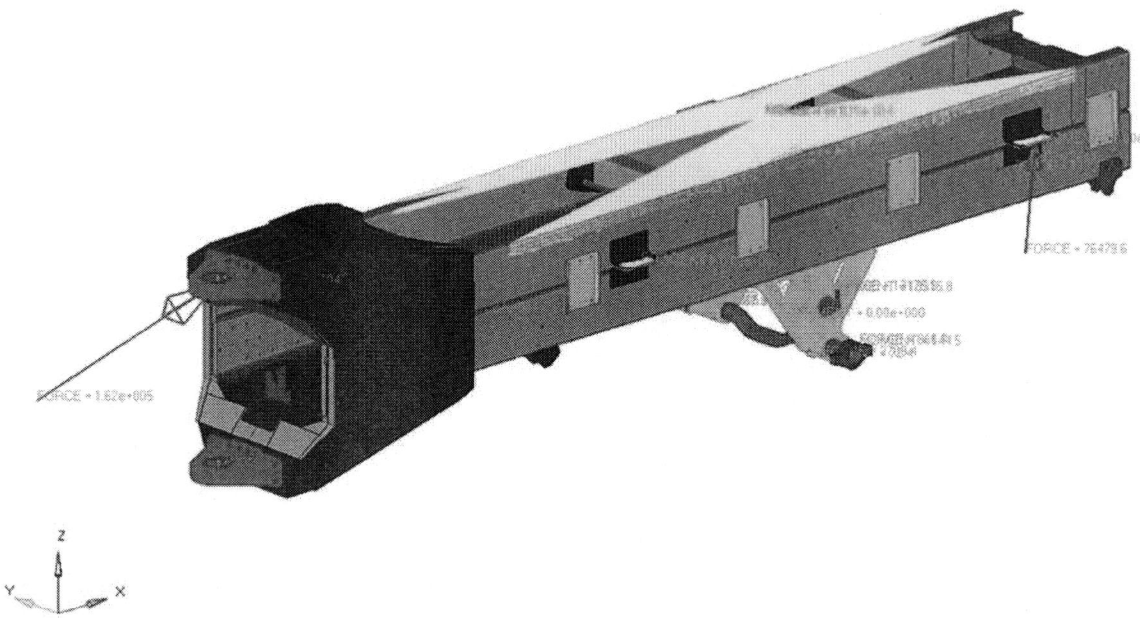

Figure 19. Loads and reaction forces from suspension, acting on rear frame, for load case "hanging of second axle" obtained by inertia relief method.

5. Strength calculation results

Calculated stress state of rear frame, obtained for two computational approaches, are shown on figures 20, 21 – load case "standing on ground"; figures 22, 23 – load case "hanging of second axle". Results demonstrate, that both methods are qualitative convergent. Insignificant difference in results for load

IOP Conf. Series: Materials Science and Engineering **534** (2019) 012024 doi:10.1088/1757-899X/534/1/012024

case "standing on ground" is related to the influence of reactive rod forces, stiffness of rubber-metal joints and loads re-distribution in rear frame from suspension. For load case "hanging of second axle" calculation results have significant quantitative differences, caused by change of spatial orientation of vehicle and its frame (during hanging of second axle, vehicle tilts, resulting in redistributing of internal loads in rear suspension, articulated joint assembly). Figures 22, 23 shows stress distribution of the rear frame.

Negative feature of "classic" method is the necessity to modify suspension rod and beam model in every different load case: load case "hanging of second axis" demands additional elements, for stop buffers modelling at third axle. In other load cases, for example, "hanging of third axle" it is also necessary to modify finite-element model respectively. Modification of finite-element model for every single load case makes analysis of results more complicated: for example, the plotting of envelope stress distribution over number of different finite element models become very labor consuming. Inertia relief method is free of this disadvantage.

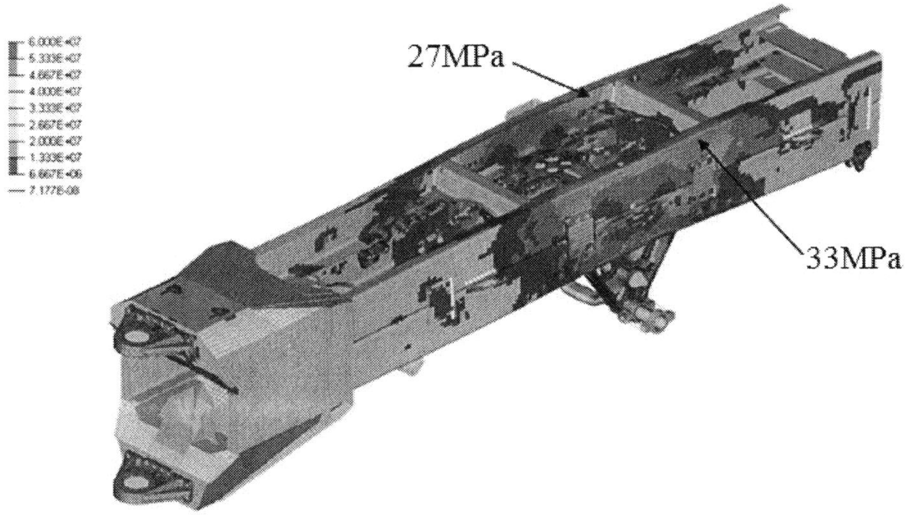

Figure 20. Equivalent von Mises stress, Pa, distribution on rear frame for "classic" calculation method, load case "standing on ground".

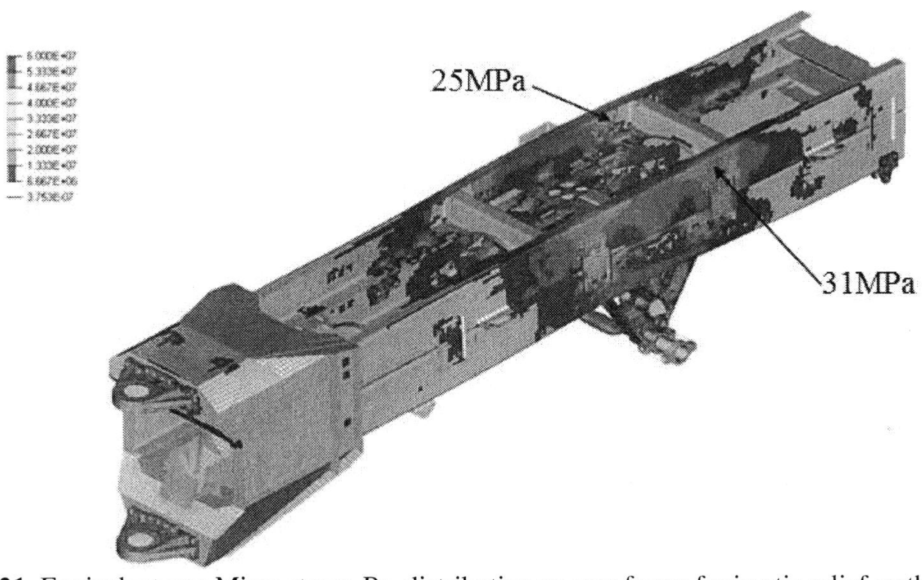

Figure 21. Equivalent von Mises stress, Pa, distribution on rear frame for inertia relief method, load case "standing on ground".

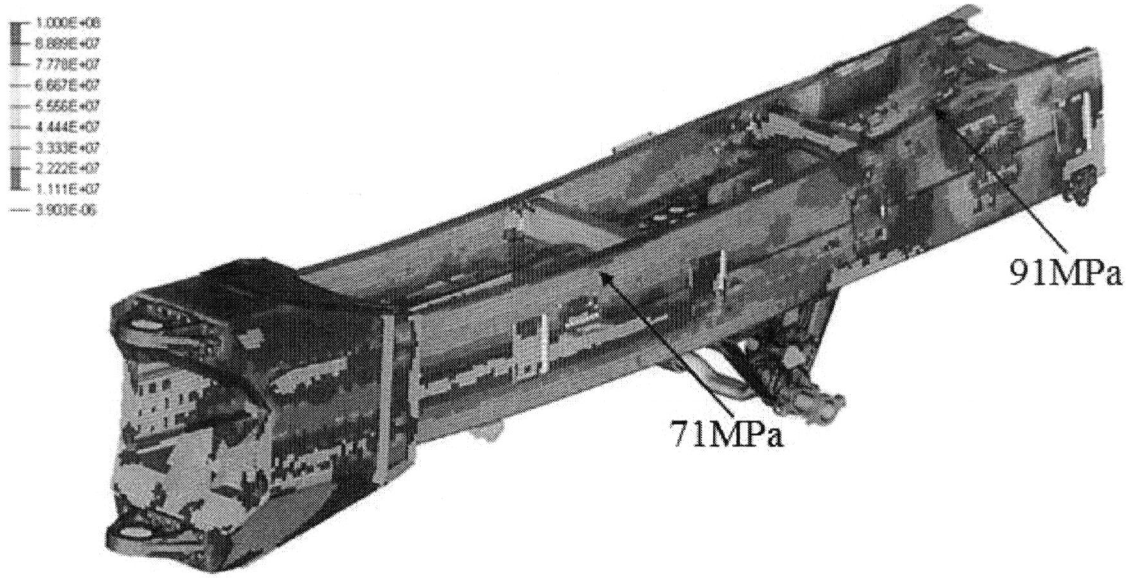

Figure 22. Equivalent von Mises stress, Pa, distribution on rear frame for "classic" calculation method, load case "hanging second axle".

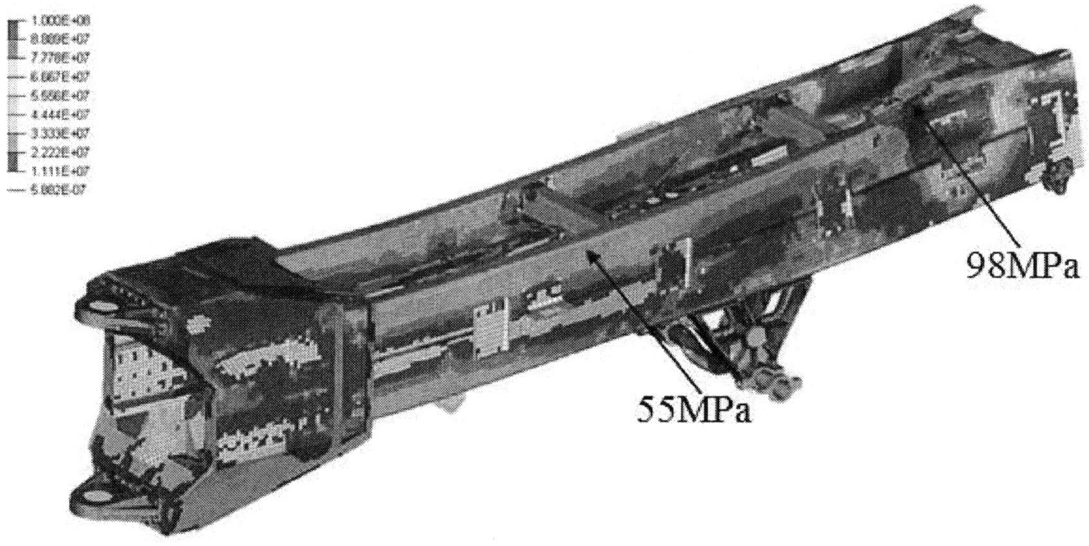

Figure 23. Equivalent von Mises stress, Pa, distribution on rear frame for inertia relief method, load case "hanging second axle".

6. Conclusions

1. Strength calculation results of rear frame have qualitative and quantitative agreement for both methods in load case "standing on plain surface". In load case "hanging of second axle" the results have qualitative agreement, but they have significant quantitative differences, because the multi body dynamic model capture the change of spatial orientation of the vehicle and its frame, general frame and suspension deformations.

2."Classic" approach with simplified rod and beam suspension models does not provide accurate simulation of real suspension kinematics and real force distribution. Thus, it has significant influence on values and directions of loads in local areas of frame and suspension joints.

3. Automated loads transfer from multi body dynamic vehicle model into finite-element model of frame and inertia relief method makes it possible to analyze numerous load cases, which makes it possible to perform optimization tasks of frame more precise, which is crucial, for instance, for topology optimization tasks.

4. Inertia relief method for automated loading of finite element model makes it possible to use single finite element to analysis of full range of load cases. Thus, it is possible to build an envelope stress distribution over all loadcases. This allowing significantly reduce difficulty of numerous results processing.

5. Inertia relief method shown in this paper allows to analyze front and rear frame of articulated vehicle, as well as other systems of vehicle, – separately from each other. This reduce computational time costs and results processing time costs significantly.

6. The proposed method makes it possible to divide front and rear frame analysis tasks between different developers to speed up analysis process. It is also decrease possibility of human error in calculation process.

7. Using "classic" method of loads and boundary conditions formulated via simplified suspension rod and beam models it is necessary to modify finite element models in every load case, also it is necessary to analyze entire frame model, with mass model and suspension models, which increase complexity of calculation significantly.

References

[1] Keller A V, Gorelov V A and Anchukov V V 2015 Modeling Truck Driveline Dynamic Loads at Differential Locking Unit Engagement *Procedia Engineering* (Elsevier BV) **129** pp 280-87

[2] Keller A V, Gorelov V A, Vdovin D S, et al. 2015 Mathematical model of all-terrain truck *Proceedings of the ECCOMAS Thematic Conference on Multibody Dynamics* pp 1285–1296

[3] Kotiev G O, Padalkin B V, Kartashov A B and Dyakov A S 2017 Designs and development of russian *scientific* schools in the field of cross-country ground vehicles building *ARPN Journal of Engineering and Applied Sciences* **12** p 1064

[4] Keller A, Aliukov S. 2015 Efficient power distribution in an all-wheel drive truck *Lecture Notes in Engineering and Computer Science* **2218** pp 1201-1206

[5] Volskaya N S, Zhileykin M M and Zakharov 2018 *IOP Conf. Series: Materials Science and Engineering* **315** (Bristol: IOPScinece) 012028 (doi: 10.1088/1757-899X/315/1/012028)

[6] Gorelov V A and Komissarov A I 2016 *2nd International Conference on Industrial Engineering, ICIE 2016* (Amsterdam: Elsevier Ltd) **150** p 1322 (DOI: 10.1016/j.proeng.2016.07.309)

[7] Keller A and Aliukov S 2015 Effectiveness of Methods of Power Distribution in Transmissions of All-Wheel-Drive Trucks *SAE Technical Paper* 2015-01-2732

[8] Kushvid R P and Chichekin I V 2014 *Car chassis. Design and elements of calculation* (Moscow: MSIU)

[9] Vdovin D S and Chichekin I V 2016 *2nd International Conference on Industrial Engineering, ICIE 2016* (Amsterdam: Elsevier Ltd) **150** p 1276

[10] Levenkov Y Y and Chichekin I V 2016 Determination of the parameters of the spring model for the analysis of loads and evaluation of the strength of suspension elements in the system for calculating the dynamics of solids *Ingenerniy vestnik* **12** 4

[11] Vdovin D S, Chichekin I V and Pozdnyakov T D Virtual test bench for determining loads affecting an automotive steering system *Engjornal* **8** 3

[12] Kong Y S, Abdullah S, Omar M Z and Haris S M 2016 Side force analysis of suspension strut under various load cases *Jurnal Teknologi (Sciences and Engineering)* **78** pp 85-90

[13] Farid M L 2006 *Fundamentals of multibody dynamics : theory and applications* (Basel: Birkhauser)

[14] Lin Liao 2011 *AIAA/ASME/ASCE/AHS/ASC Structures, Structural Dynamics and Materials Conference* (Denver, Colorado) 1

[15] Levenkov Y Y and Vol'skaya N S 2015 The smoothing ability of an automobile wheel pneumatic

tyre in its interaction with hard and rugged bearing surface *Technology of wheeled and tracked machines* **1** 20-26
[16] Vol'skaya N S, Levenkov Y Y and Rusanov O A 2013 Simulation of an automotive pneumatic tire interacting with a solid rough supporting surface *Science and Education: scientific edition of the Bauman MSTU* **5** 107

Logic-based artificial intelligence in systems for monitoring the enforcing traffic regulations

D V Aladin[1,2], O O Varlamov[1,2], D A Chuvikov[2,3], V M Chernenkiy[1], E A Smelkova[1] and A V Baldin[1]

[1] Bauman Moscow State Technical University, 2-ya Baumanskaya Street 5/1, Moscow 105005, Russian Federation
[2] MIVAR, 72 Oktyabrskaya str., Moscow 127521, Russian Federation
[3] Department of Automated control systems, Moscow Automobile and Road Construction State Technical University (MADI), Leningradskiy Pr. 64, Moscow 125319, Russian Federation

E-mail: aladin@dmitry.icu

Abstract. The paper touches upon the issues of monitoring and assessment of driver's actions during driving. Its key idea is to prove that currently there are all necessary conditions to develop control systems of traffic regulations based on artificial intelligence with logical reasoning. The article gives a detailed analysis of advanced driver-assistance systems from the world's leading companies and pays attention to the advantages and disadvantages of such solutions. As an engineering approach to solving the identified problems, it is proposed to use Mivar expert system. The purpose of this decision is to monitor the enforcing traffic regulations and inform the driver about violations. The article proposes the concept of practical implementation of Mivar system included in the driver assistance system. We conclude that Mivar approach provides excellent opportunities for creating the control system for monitoring the enforcing traffic regulations. The proposed solution can be used for reducing the number of accidents caused by drivers.

1. Introduction

The development of science and technology, the continuous improvement of technology and its distribution in industry and in society offer humanity a wide range of opportunities. However, despite the positive aspects, technological progress can have a negative impact on human security. For example, everyone knows the advantages of modern cars, but at the same time, according to The World Health Organization (WHO), car accidents are among ten most common causes of death in the world. In order to improve road safety indicators, various measures are applied, including the tightening of laws regulating the behavior of participants of traffic and the updating transport infrastructure. In Russia by the end of 2016, the number of road traffic accidents (RTA) was over 170 thousand. 90% of these cases were due to violation of traffic regulations by people (drivers and pedestrians). The total number of victims was about 220 thousand people. Improving traffic safety is a critical component to saving lives on the roads.

It is important to note that one of the most radical proposals to prevent accidents is the complete exclusion of people from the process of driving, i.e. the creation of a fully autonomous unmanned transport. Of course, people will design such cars and develop software for them, but they will not directly participate in the process of driving, except, perhaps, some exceptional cases and situations. Moreover, in some promising models of unmanned vehicles there is no steering wheel or pedals for a person.

In general, all specialists understand the way of creating unmanned vehicles, which will include various stages from full human control, through the joint interaction of a person with an automatic control system of the car and to a fully autonomous unmanned vehicle driving on highways (analogue of the "railway"), and secondary roads, and off-road in different weather conditions. And it is generally accepted that the system (complex of systems) of controlling a car should include, at least, an automatic system of trajectory control (holding in the lane, keeping the distance in the stream, etc.), a system of technical vision (collecting information about the road situation in real time), a navigation system (routing) and a Decision Support System to enforce traffic regulations. Of course, there are other important components and systems, for example, interaction of the car with the "intelligent road environment", which will inform about traffic signs, difficulties, weather, etc. via radio channels to an unmanned vehicle, and it will greatly facilitate the collection of information about the road situation and even could eliminate the need to create a complex system of technical vision. For example, in cities and major highways, instead of the usual traffic signs and traffic lights, radio transmitters which report on restrictions and possible routes, distances to obstacles, will be installed. They also will get data from stationary cameras about the road situation and emerging obstacles in real time. In any case, for an unmanned vehicle, it is necessary to create a Decision Support System to enforce traffic regulations. It is generally recognized that for some time the technology of unmanned vehicles will be used in a man-machine mode. For legal permission to use fully unmanned vehicles it is necessary to carry out testing in an automated mode, when a person at any time can interfere in driving. On the other hand, for example, DSS TR can be installed in conventional vehicles and help human driver enforce traffic regulations or monitor the driver's behavior. Many researchers believe that even now the electronic system of the car can "set fines" to their owners-drivers in case of violation of traffic regulations. It is only necessary to regulate the legal aspects, but there are already precedents with stationary automatic cameras, which monitor the enforcing traffic regulations, and "automatic" issuing of fines.

Thus, on the basis of DSS TR a new class of devices can be created– automatic autonomous control systems of enforcing traffic regulations. On the one hand, these control systems will monitor and assist drivers, and on the other hand, such control systems in real traffic conditions will be able to prove the possibility of creating and practical application of DSS TR for fully autonomous vehicles. From a scientific point of view, this is one technical system that will have two purposes:

1) Control system of traffic regulations and driver assistance;
2) DSS TR for fully autonomous cars.

Consequently, the subject of the development of control systems of enforcing traffic regulations based on technologies of logical artificial intelligence is relevant, feasible and practically useful [1].

2. Overview of existing solutions in the field of driver assistance system and AI

Automotive companies (Volkswagen, Tesla, Mercedes-Benz), engineering centers (Google, Siemens, Baidu), universities (Stanford University, Massachusetts Technological Institute, University of Minnesota) and many others are actively engaged in the development of intelligent control technologies for autonomous wheeled vehicles [2–4]. Active work is underway in the field of driver assistance systems, examples of which are intelligent driver assistance systems: Mobileye, Audi Driver assistance systems and Opel SAFETEC. The decision from Mobileye is a hardware and software complex that allows recognizing traffic signs and informing about exceeding the permitted speed. An audio Driver assistance system is a serial product of autopilot in road traffic, which is focused on helping a driver in everyday situations: holding the lane, adaptive cruise control and automatic traffic in traffic jams. Opel SAFETEC is the representative of the integrated security system that combines and coordinates all the systems of active and passive safety, which allows you to recognize traffic signs, monitor the surrounding traffic situation and warn about exceeding the permitted speed. To make driving comfortable, these systems continuously evaluate the road environment with the help of numerous sensors installed inside and around the vehicle. Navigation tools and computer vision are also used for the precise orientation in space.

Now the existing driver assistance systems are quite expensive and therefore they are not installed on all vehicles. But such systems implement mechanisms to maintain a safe distance to the car in front, hold the lane and prevent violations of the speed limit. However, they do not fully solve the problem of preventing traffic violations. First of all, this is due to the fact that these systems do not have modules for assessing the driver's actions for consistency with traffic regulations. Firstly, we are talking about the control of the sequence of actions of the driver and his making difficult decisions, because "simple" traffic violations: journey on prohibited signal of a traffic light or sign; speeding, etc. are easily detected at the "reflex" level of AI systems, for example, with the help of different neural networks or methods of "comparison with decision templates". When making decisions on some sequence of actions (driving through the intersection, overtaking and lane change, interaction with other drivers, decision-making in non-standard situations, etc.), the driver actually builds an algorithm of his actions in real time and in a specific traffic situation. These processes are already being investigated at the "logical" level of AI. It is known that the task of constructing algorithms of actions is complex, important for robots and depends on the computational complexity of the methods of logical inference based on the production rules of the "If – Then" format. In addition, the lack of driver assistance systems of "logical" level is largely due to the fact that the creation of such systems must take into account the following requirements:

- The possibility to adapt the product to the legislation of the countries in which the vehicle is planned to operate;
- The flexibility of updates when changes are made by the regulatory authority in the legislative framework;
- Economic accessibility of the technical platform for the owner of the vehicle.

The developers plan to solve the above-mentioned requirements during the implementation of unmanned vehicles. However, it is now possible to create modules that meet the above requirements and implement them into existing driver assistance platforms. Mivar approach can be the beginning for the creation of a logic-based intelligent system of monitoring the enforcing traffic regulations.

3. Mivar approach to the creation of logical AI

Mivar approach is implemented at the logical level of AI and is based on the idea of combining products and Petri nets. In 2002, it was theoretically shown that the logical inference on products presented in the format of Mivar bipartite oriented networks is performed with linear computational complexity. In 2007, the transition from poorly formalized anthropomorphic terms of artificial intelligence to active reflection theory was substantiated, in which a qualitative scale of "intelligence" of a human and self-organizing software and hardware complexes was proposed. It is important to note sufficient versatility of Mivar technologies that solve problems at a qualitatively new level in the field of AI [5], and a patent on a logical inference in Mivar knowledge base [6] with linear computational complexity was obtained. Autonomous vehicles deal with the problem of creating autonomous intelligent robots for which Mivar technologies were proposed to apply even in 2004 [7], and later the possibility of creating autonomous intelligent robotic systems (IRS) was shown [8].

For further study, it is important to emphasize that in January 2017, in practice, an unmanned autonomous car with a logical AI was demonstrated [2, 9], in which, on the basis of three laptops, it was assembled:

1) a reflex trajectory control system;
2) a simplified system of technical vision;
3) Mivar decision support system with a fairly simple logical model of traffic in the city.

Russian scientists from Moscow Automobile and Road Construction State Technical University take active part in Mivar DSS studies, they developed algorithms for driving an autonomous car as part of intelligent transport systems [10], also, the choice of key technologies for the functioning of the system of inter-object interaction of intelligent vehicles was validated, the problems of determining the mutual position of moving vehicles were investigated, a system of inter-object interaction of intelligent cars was developed, the possibilities of using onboard multiplex networks of motor vehicles by means of road tests and automation of traffic control were shown [11].

Thus, proper intelligent autonomous robots and unmanned vehicles have become a reality through the use of Mivar decision support systems. For example, to calculate the algorithms of service robots [12] and to plan the movement of mobile robots [13] Mivar models, which contain from 250 to 350 production rules, are applied. According to the experts, the model of a journey of a crossroad for an unmanned vehicle will exceed 300 regulations in Mivar network.

Currently, Mivar approach is developing in Bauman Moscow State Technical University as a part of the study of hybrid intelligent information systems [14] with the description of the subject domain in the formalism of the metagraph data model [15] and intelligent systems [16]. Structural development of Mivar approach using the classes and relations was performed, the advantage of Mivar approach over ontologies and cognitive maps was shown [17], Mivar logical inference machine [18] was implemented for the development of various systems using automatic routing of logical inference in Mivar knowledge base [19] on the basis of the logically critical systems (Razumator) [5].

The subject domain of the enforcing traffic regulations is closely related to the analysis of various road accidents, so it is important to note the work of D.A. Chuvikov about the creation of Mivar expert system (ES) "Accident motor vehicle accidents" [20-21], the use of expert modeling in obtaining new knowledge by a person, the practical use of ES "Accident Analysis" for accident analysis and solving the tasks assigned to the forensic center [22-27]. The versatility of Mivar ES shows the possibility of modeling the behavior of an autonomous robot-guide in the environment of V-REP, analytical modeling of a distributed system of data processing method in a background thread and rationale for the selection of software tools for solving problems of mass service [28].

Thus, Mivar technologies allow creating algorithms based on active learning evolutionary bipartite network, controlled by the input data stream. This makes it possible to present the accumulated knowledge in the form of sets of modules, services and procedures, thereby ensuring the properties of scalability, adaptability and flexibility which are necessary for the construction of control system of enforcing traffic regulations. Let us turn to the description of the conceptual model of Mivar system of enforcing traffic regulations.

4. The concept of creation of Mivar control systems for monitoring the enforcing traffic regulations

To solve the problem of informing the driver in real time about the violation of the provisions of the traffic regulations there is an offer to implement Mivar control system (MCS) based on a software product "Razumator" [5]. The MCS is an integral component of the driver assistance system. Mivar system storage is a set of models. Each model describes a relatively simple subject domain: movement in residential areas, passing crossroads, crossings, etc. The structure of the MCS is shown in Figure 1.

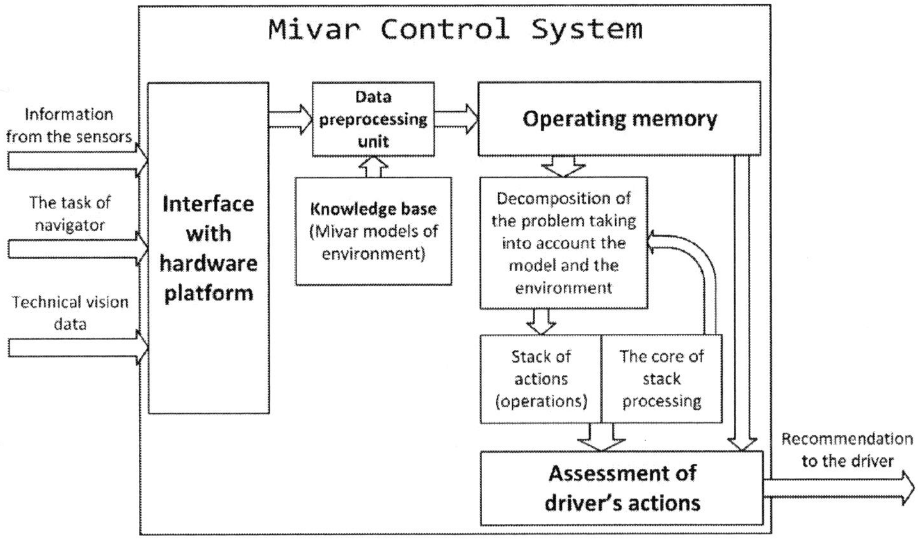

Figure 1. The structure of the MCS.

According to the scheme, the input of the system receives data recognized by vision systems, sensors of the car and the task of the navigator. The data of the external environment is converted into the object format of Mivar model environment. Formalized parameters are stored in the working memory – the memory area, which contains a lot of facts describing the current state of the subject domain, i.e. the traffic situation. According to the obtained factors with the help of Mivar model of the subject domain the algorithm of maneuver is constructed, each step of which is put onto the stack of actions. The resulting set of steps is compared with the actions of the driver and the conclusion is drawn about their consistency with traffic regulations. Let us describe the structure of the MCS in terms of Mivar computing.

M – Mivar multidimensional space. It represents a set of axes, points of space and their values. We select in this space a set of axes $X = \{x_n\}, n = 1..N$, where N is the dimension of M. Let Q_n be the set of elements x_n axis, that:

$$\forall x_n \exists Q_n = \{q_{n_{i_n}}\}, n = 1..N, i_n = 1..I_n, \quad (1)$$

where $q_{n_{i_n}}$ – the element of axis x_n, I_n – the dimension axis x_n.

The set Q_n forms a multidimensional space: $M = Q_1 \times Q_2 \times ... \times Q_n$. In space M, we denote point $m_{i_1,i_2,...,i_N} \in M$ with coordinates $(i_1, i_2, ..., i_N)$. Let $c_{i_1,i_2,...,i_N}$ be point value $m_{i_1,i_2,...,i_N}$. Then $C_M = \{c_{i_1,i_2,...,i_N}\}$ - the set of values of points in the multidimensional space M. Thus, C_M is a representation of a set of facts known about the subject domain in a multidimensional space M.

To transfer from multidimensional space of points in the space of values of points we define the ratio

$$\mu: C_y = \mu(M_y), \quad (2)$$

where $M_y \subseteq M, M_y = Q_{1y} \times Q_{2y} \times ... \times Q_{Ny}$ and $Ny \leq N$.

From the point of view of the multidimensional space M, the traffic situation can be represented as follows. A set of axes $A \subset X$ is the mapping of a subject domain to multidimensional space. For example, A_1 – traffic light, A_2 – prohibiting signs and A_3 – interference on the left. The composition of objects A, which is monitored, is determined on the basis of the amount of data required to make a decision on driving, agreed with traffic regulations. Elements of multidimensional space axes will be a set of Q_A, which consists of variable attributes of objects A. For A_1 axis, the attributes can be: traffic light performance (on/off), signals of the main and additional sections. The C_M set in this case will display the current state of the road situation.

To describe a model of a subject domain in Mivar space, it is necessary to select three axes:

- V – the axis of the elements (objects);

- S – property axis;
- O – the axis of relations.

The elements of the sets V, S and O are independent and form the space VSO. Then space M is represented by a tuple of the form:

$$< V, S, O >. \quad (3)$$

In the multidimensional space VSO in automatic mode Mivar bipartite network in formalism $< P, R >$ (parameters and rules) can be built, in which

- the space objects will correspond to the network settings;
- the relations of space will comply with the rules of networking;
- the properties of the space will correspond to the restrictions of the rules of networking.

The data representation form $< P, R >$ is called Mivar network, in which, on the basis of a logical inference with linear computational complexity, algorithms of "correct" actions of the driver (or DSS of the unmanned vehicle) with enforcing traffic regulations will be automatically built. Then the traffic control system will compare the algorithms with the "performed" actions of the driver. In case of driver errors detection, Mivar control system will give hints to the driver. And in case of a fully unmanned vehicle, Mivar control system will act as a DSS system and drive the car instead of a person.

Having an example of crossing an uncontrolled intersection, we will get acquainted with the work of Mivar network, which evaluates the actions of the driver. Let car A equipped with a control system of enforcing traffic regulations plan to make a left turn (see Figure 2).

When you turn left oncoming cars and cars traveling to the intersection on the right will be obstacles which you need to give way. Also, when crossing the pedestrian crossing, pedestrians must be allowed in first. Under these conditions, the driver of car A is allowed to perform the maneuver.

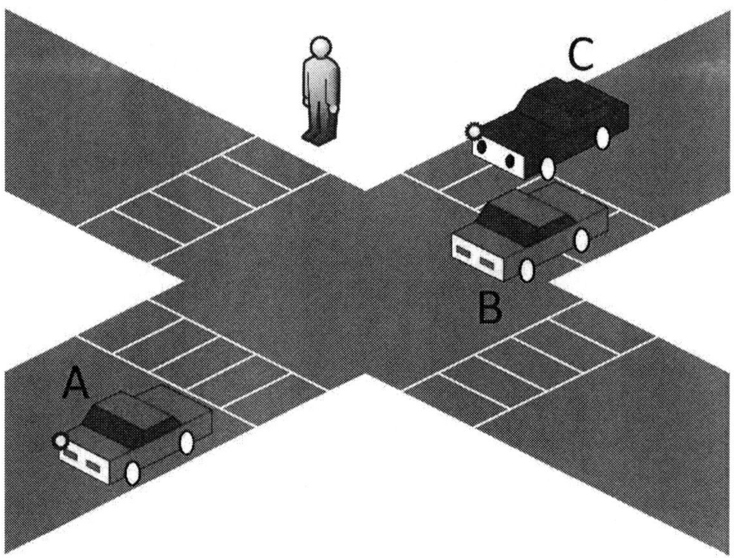

Figure 2. The subject domain "Uncontrolled intersection".

In order for Mivar system to assess the traffic situation, it is necessary to obtain the following data from the sensors of the car:

- P_1 –the task of the navigator (in this case, "turn left");
- P_2 –the status of the roads (current and target road);
- P_3 – obstacles (vehicles at the intersection);
- P_4 –information about pedestrians.

$$P_1, P_2, \ldots, P_4 \in V \quad (4)$$

An example of the parameters describing the set P_4 is shown in Table 1.

Table 1. The parameters describing the set P4.

Name	Description	Valid value
Current road	Pedestrians are at a pedestrian crossing in front of the intersection	Yes/No
Target road	Pedestrians are on the pedestrian crossing after the intersection	Yes/No

To link the collected parameters, the algorithm uses R_1, R_2, and R_3 relationship groups and TP_1 intermediate indicators (a list of objects to observe) and TP_2 (obstacles on the target path).

$$TP_1, TP_2 \in V \text{ and } R_1, R_2, R_3 \in S \quad (5)$$

For the desired parameter in the conditions of a given subject domain, we take the parameter RES (a set of recommended actions) and RES ∈ V.

The resulting Mivar network for assessing the actions of the driver of a vehicle A will look as it is shown in Figure 3.

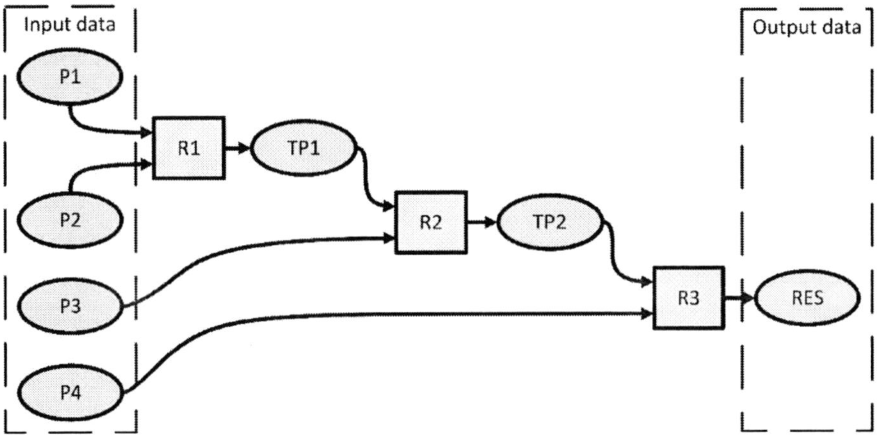

Figure 3. Algorithm graph.

The regulations and parameters of road traffic developed in advance are entered into Mivar knowledge base (another name: Mivar database and rules). According to the acquired knowledge base, recommendations for driving in the current traffic situation are formed depending on the signals transmitted to Mivar traffic control system from the vision system and other data collection systems. It is important to emphasize that for teaching Mivar knowledge bases it is necessary to create formal descriptions of subject domains in the formalism of bipartite oriented Mivar networks, but these issues are investigated in other works. It is important to remind [5] that three software products have been developed for: Expert systems – Wi!Mi Razumator [29]; Understanding the meaning of texts - Tel!Mi; the Decision Support System for autonomous robots – Robo!Razum. WiMi! Technology is used in Tel!Mi to understand the meaning. Product Tel!Mi using texts of instructions allows automatically teaching Wi!Mi Razumator and Robo!Razum. Together, these products can be used for the Internet of Things (IoT) and other cyber-physical systems, as Mivar products Wi!Mi, Tel!Mi, Robo!Razum can work on one common computational module “processor-memory” [5].

Thus, all the prerequisites have been done to create Mivar control systems of enforcing traffic regulations on the basis of Razumator and expert systems that can be used both for Driver Assistance systems and Decision Support systems of unmanned vehicles.

5. Conclusion

One of the most important components of saving lives on the roads is to improve road traffic safety by means of enforcing traffic regulations by drivers. Mivar technologies of the development of expert systems based on logical inference machine Razumator are capable of processing in real time the information received from the system of technical vision about the road situation and build algorithms of the driver's actions that are fully consistent with traffic regulations. Thus, currently, there are all necessary modules to create Mivar control systems of enforcing traffic regulations on the basis of Razumator and expert systems that can be used for driver assistance systems. Such systems will prompt the driver the maneuvers allowed by traffic regulations, and in case of violation of traffic regulations will inform about the violation. In case of legislative authorization Mivar control systems of enforcing traffic regulations will inform state body about the violation and fine a driver similar way to the existing complexes of fixation of violations of traffic regulations. Mivar control system of traffic regulations will be installed in cars to help drivers and evaluate their actions through the whole volume of traffic regulations, in contrast to existing systems which are able to capture only the “simple” violations of the regulations (one rule “If-Then”).

Mivar approach has a high potential for the evolutionary development of software and hardware systems. Therefore, experience and knowledge gained during the development of driver assistance systems (MCS of enforcing traffic regulations) will be further used in the creation of completely autonomous unmanned cars. It is important to note that the powerful tools of Mivar Wi!Mi Razumator, Tel!Mi, Robo!Razum, Mivar scalability models of knowledge and small hardware requirements allow you to get a low cost of MCS TR as well as easily and quickly adapt the product to the legislation of the countries in which you plan to operate the car.

The proposed concept of constructing Mivar control systems of enforcing traffic regulations meets the requirement of economic accessibility of the product for a wide range of buyers. MCS TR can be embedded in new cars in factories or can be installed as additional equipment in the cars which are already in use. Due to the high computational power of Mivar Razumators, there are big opportunities for implementation of high-speed driver assistance systems and decision support systems for unmanned vehicles; this is especially important for real fast-changing road conditions.

6. References

[1] Shadrin S S and IvanovA M 2016 Technical aspects of external devices into vehicles networks integration *International journal of applied engineering research* **11** 7003–7006

[2] Özgüner Ü AcarmanT and Redmill K 2011 Autonomous ground vehicles*Artech house* 289

[3] Nonami K Kartidjo M Yoon K and Budiyono A (Eds.) 2013 Autonomous control systems and vehicles *Intelligent unmanned systems* p 306

[4] Saikin A M Bakhmutov S V Terenchenko A S, Endachev D V KarpukhinK E and Zarubkin V V2014 Tendency of creation of “driverless” vehicles abroad *Biosciences biotechnology research Asia* **11** pp 241–246

[5] Varlamov OO 2017 The Mivar approach as a basis for a qualitative transition to a new level in the field of artificial intelligence *Radio industry* **4** pp13–25

[6] Varlamov O O Hadiev A M ChibirovaM O SergushinG S and Antonov P D 2017 *Automated building route of inference in Mivar's knowledge base* (Russian Federation Patent No.2607995)

[7] Varlamov O O 2004 Systems of processing of information and interaction between groups of mobile robots on the basis of Mivar information space *Artificial intelligence* **4** pp 695–700

[8] VarlamovO O LazarevV M Chuvikov D A and Punam J 2016 On prospects for design of standalone smart robots based on Mivar technologies *Radio industry* **4** pp 96–105

[9] ShadrinS S Varlamov O O and IvanovA M 2017 Experimental autonomous road vehicle with ogical artificial intelligence *Journal of advanced transportation* ID 2492765

[10] Shadrin S S Ivanov A M and Nevzorov D V 2015 Independent wheeled vehicle as part ofintelligent transport systems *Estestv. tekh. nauki* **84** pp 309–311

[11] Shadrin S S Ivanov A M and Karpukhin K E 2016 Applicability of board multiplex networks data of vehicles at driving tests, its development and automation of motion control *Vestnik mashinostroeniya* **7** pp 25–29
[12] Zhdanovich E A ChernyshevP K YufimychevK A Eliseev D V and Chuvikov D A 201 Random algorithm calculation of service robot functioning based on Mivar approach *Radio industry* **3** pp 226–242
[13] Zhdanovich E A PanferovA A YufimychevK A KhadievA M and Eliseev D V2015 Calculation random algorithms of functioning service robot based on Mivar approach *Radio industry* **3** pp 243–254
[14] Chernenkiy V M TerekhovV I and Gapanyuk Yu E 2016 The structure of the hybrid intelligent information system based on metagraphs *Neurocomputers: development, application* **9** pp 3–13
[15] Gapanyuk Yu E Revunkov G I and Fedorenko Yu S 2016 Predicate representation of metagraph data model *Informatsionno-izmeritelniye i upravlyayushchie sistemi [Information-measuring and Control Systems]* **12** pp122–131
[16] Chernenkiy V GapanyukYu RevunkovG Terekhov V and Kaganov Yu 2017 Metagraph approach for hybrid intelligent information systems description *Journal of applied informatics* **12** pp57–79
[17] Chibirova M O 2015 Analysis of existing approaches: ontology, cognitive maps and Mivar nets *Radio industry* **3** pp 55–66
[18] KhadievA M 2015 Development and practical realization of Mivar inference machine *Radio Industry* **3** pp 79–89
[19] Sergushin G S 2015A computer-implemented system for automated route planning inference in Mivar knowledge base *Radio industry* **3** pp 90–99
[20] Chuvikov D A 2017About the expert system "Analysis MVA", based on the concept of the Mivar approach *Problems of artificial intelligence* **2** pp 78–88
[21] Chuvikov D A 2017 Expert system "Analysis MVA" *Russian Federation Patent No.2017660355*
[22] Chuvikov D A 2017Application of expert modeling in new knowledge obtained by man *Radio industry* **2** pp 72–80
[23] Chuvikov D A 2017 The use of expert system "Analysis MVA" and simulation system Virtual Crash 3.0 for solving tasks of criminal expertise centre *Industrial automatic control systems and controllers* **5** pp 23–34
[24] Chuvikov D A 2017 Models of reconstruction and examination of MVA in a formalism of the knowledge base of the bipartite focused Mivar networks *Proceedings of the congress on intelligent systems and information technologies. "IS&IT'17"* **1** pp 308–313
[25] Chuvikov D A Teplov E V Saraev D V Varlamov O O and Punam J 2016 Methods of automation of supervisory control system based on the urban passenger traffic software expertise system *Radio industry* **4** pp 89–95
[26] Panferov A A Zhdanovich EA Yufimychev K A and Chuvikov D A2016 Designing algorithms for service robots on the basis of Mivar approach *International Journal of Advanced Studies* **3** pp 72–86
[27] Varlamov O O 2018 Wi!Mi Expert System Shell – A Novel Tool To Build Knowledge-Based Systems With Linear Computational Complexity *International Review of Automatic Control* **6**

IOP Conf. Series: Materials Science and Engineering **534** (2019) 012026 doi:10.1088/1757-899X/534/1/012026

The experimental system of automatic wheel drift and vehicle rollover prevention

S E Buznikov[1], D S Elkin[2], N S Shabanov[2] and V O Strukov[2]

[1] Simulation department, FSUE "NAMI", st. Avtomotornaya, 2, 125438, Moscow, Russian Federation

[2] Intelligent systems department, FSUE "NAMI", st. Avtomotornaya, 2, 125438, Moscow, Russian Federation

E-mail: sergey.buznikov@nami.ru

Abstract. The research topic relevance is determined by the global significance level of traffic safety improvement. The purpose of the work is to create mathematical model and software for the active vehicle safety experimental intelligent system. To solve this problem, methods of indirect vehicle state variables measurements were used. As a mathematical model, we used a system of equations for the wheels rotation speed on a bend. For estimating the parameters of the mass center and the wheels motion, standard wheel speed sensors and algorithms for processing the measurement data are used. The task of angular drift velocity determining based on wheels rotation frequency measurements is ill-posed and additional conditions are introduced to solve it. To predict the wheel drift and car rollover, extrapolate algorithms are used. These phenomena are prevented by the certain brake formation that limit the object speed at the predicted boundaries level. The research novelty is to create algorithms for an experimental system that provides the solution to the wheel drift and rollover prevention problem in a minimal configuration of technical means.

1. Introduction

The road safety improvement problem is one of the most important problems in modern society and its importance is accepted globally. A systematic analysis of this problem [1] shows that one of the most promising way for its solution is the creation of various vehicle active safety systems.

Full collision avoidance systems claim not only developments in the field of technical facilities but also mathematical models and software. Intellectual properties of modern systems are acquired through the virtual information sensors usage, which allow to increase system information functions in a minimal technical means configuration. Virtual information sensors are algorithmic constructions that replace object state parameters physical sensors. The development of such constructions requires new mathematical models for solving ill-posed problems and algorithms for data processing.

It should be noted that noticeable improvements in the traffic safety should be expected in case of equipping the entire operating fleet, including low price models, with effective active safety systems.

A necessary condition for the unmanned vehicles safe operation is installation of full-featured active safety systems. These circumstances set the trend for the advanced intellectual active safety systems development in the near future.

Among the electronic systems for driving stability improvement historically, an antilock braking system (ABS) stands out, which allows to maintain car controllability and stability during emergency braking [2]. On the basis of ABS an electronic stability control system (ESC) was created and is now included in the configuration of the new cars. This system is mainly aimed at oversteer and understeer suppression of a car [3]. In addition, to increase the driving stability, some manufacturers implement software extension of the basic ABS system, among which are the electronic brake force distribution system and the braking stabilization function in the turn [4]. The traction control system, the extended function of the electronic differential lock, the trailer stabilization function are also referred to the driving stability improvement systems [5].

Despite significant advances in the development of systems for driving stability improvement, each system has shortcomings, caused by both the laid-down operation principles and the physical limitations of the wheel-to-road grip. For example, the ESC system is effective in preventing road accidents of single cars [6], caused by road departure with subsequent overturning [7] and characterized by severe consequences. At the same time, in conditions of dense traffic, ESC systems are often unable to prevent vehicle collisions caused by oversteer or understeer. Among the significant drawbacks of ABS is a significant increase of the braking distance at braking on ice, snow, on roads that do not have a hard coating, on uneven surfaces [8].

This scientific research aims to set and solve the wheel drift and vehicle rollover prevention problem. The developed mathematical model and software is intended for use as part of the experimental intellectual active safety system.

2. Task statement

At the conceptual level, the task of wheel drift and vehicle rollover prevention is reduced to prediction of the dangerous speed limits excess and necessary control actions on the traction, brakes and steering wheel ensuring.

At the content level, the task under consideration is reduced to the dynamic velocity stabilization of the mass center at limited control actions, formed in a hardware/software environment that meets the cost and operating cost constraints.

To assess the solution quality, we use a modified quadratic control quality functional:

$$Q(t_2) = \int_{t_1}^{t_2} C_0 \left[V_m(\tau) - V_{bound}^{B}(\tau)\right]^2 d\tau + \int_{t_1}^{t_2} \Delta C(R,\tau) \quad d\tau + Q(R, t_1) \qquad (1)$$

at $U \in U_{add}$ and $R \in R_{add}$, where

$V_m(\tau)$ – longitudinal velocity of the vehicle mass center;

$V_{bound}^{H}(\tau)$ – high limit of the safe speed, defined as $\min\left[V_{bound\ 0}, V_{bound\ 1}, V_{bound\ 2}\right]$;

$V_{bound\ 0} = \sqrt{0.5abh_m^{-1}|\Psi_c|^{-1}}$;

$V_{bound\ 1} = Re\sqrt{2[m_{12}gb - R_d a_{dT}]k_{sq}|\Psi_c^{-1}|}$;

$V_{bound\ 2} = Re\sqrt{2[m_{34}gb + R_d a_{dT}]k_{sq}|\Psi_c^{-1}|}$;

a and b – wheel track width and wheelbase, respectively;

h_m – height of the mass center;

Ψ_c – resulting angle of the steerable wheels, taking into account the angle of lateral drift;

m_{12} and m_{34}– the masses of the front and rear axle wheels respectively;

g – gravity acceleration;

R_d – dynamic radius of the driving wheels;

a_{dT} – traction-braking acceleration;

k_{sq}– sliding friction coefficient of the wheels in the transverse direction;

$U = (U_1, U_2, U_3, U_4)^T$ – control actions vector on the actuator, including gearbox (U_1), accelerator (U_2), brakes (U_3) and steering wheel (U_4);

IOP Conf. Series: Materials Science and Engineering **534** (2019) 012026 doi:10.1088/1757-899X/534/1/012026

$R = (R_H, R_S)^T$ – technical solutions vector of the control system in the field of hardware (R_H) and software (R_S) means;
$\Delta C(R,\tau)$ – expenses intensity for maintenance of the system technical means;
$Q(R, t_1)$ – system cost at the operation time;
C_0 – penalty coefficient;

$$C_0 = \begin{cases} 0, if\ V_m \le V^H_{bound}; \\ C^*_0 \gg 0, if\ V_l > V^H_{bound}. \end{cases}$$

The best solution to the problem (1) in the above statement is the control actions U (t) sequence realized in the hardware and software environment R, which ensures the stabilization of $V_m\ (t)$ at the level $V^H_{bound}(t)$ in a system with minimal operating costs and initial cost.

3. Research results

To solve this problem indirect measurements are used based on the mathematical models of the object and algorithms for solving ill-posed problems. The resulting virtual information sensors, replacing physical ones can significantly reduce the cost of operation and the resulting system cost.

As an indirect measurements mathematical model of the vehicle wheel drift angular velocity a system of equations for the linear speeds of wheels rotation $V_i, 1 \le i \le 4$ on the bend is used:

$$\begin{cases} V_1 = V_m + 0.5ab^{-1}V_m\Psi_c + \Delta V_{S1} + 0.5a\Delta\omega_m; \\ V_2 = V_m - 0.5ab^{-1}V_m\Psi_c + \Delta V_{S2} - 0.5a\Delta\omega_m; \\ V_3 = V_m + 0.5ab^{-1}V_m\Psi_c + \Delta V_{S3} + 0.5a\Delta\omega_m; \\ V_4 = V_m - 0.5ab^{-1}V_m\Psi_c + \Delta V_{S4} - 0.5a\Delta\omega_m, \end{cases} \quad (2)$$

where $\Delta\omega_m$ – drift angular velocity of the car wheels;
ΔV_{Si} – longitudinal slip speed of the i-th wheel;
V_i – linear rotation speed of the i-th wheel ($1 \le i \le 4$);
V_m – linear velocity of the mass center longitudinal motion.

Differential equation of the course angle:

$$\dot{\Psi}_m = \omega_m + \Delta\omega_m \quad (3)$$

The solution (3), taking into account Euler's formula $\omega_m = V_m \cdot R_m^{-1}$ and $R_m = b \cdot \Psi_c^{-1}$, in discrete time is reduced to the form:

$$\Psi_m(k) = \Psi_m(k-1) + b^{-1}\int_{t_{k-1}}^{t_k} V_m(\tau)\Psi_c(\tau)d\tau + \int_{t_{k-1}}^{t_k} \Delta\omega_m(\tau)d\tau \quad (4)$$

The increment of the course angle $\Delta\Psi_m(k)$ at the k-th step is:

$$\Delta\Psi_m = \Psi_m(k) - \Psi_m(k-1) = b^{-1}\int_{t_{k-1}}^{t_k} V_m(\tau)\Psi_c(\tau)d\tau + \int_{t_{k-1}}^{t_k} \Delta\omega_m(\tau)d\tau \quad (5)$$

In case of rear wheels oversteer, $\Delta\omega_m$ coincides in sign with Ψ_c at $V_m > 0$, which leads to an increase in the module of the increment of the course angle. In the case of front wheels understeer, $\Delta\omega_m$ has the opposite sign to Ψ_c, which reduces the increment module of the course angle. Additional rotation (drift) with the frequency $\Delta\omega_m$ occurs around the front wheels axis center when they are oversteer and around the rear wheels in case of understeering.

IOP Conf. Series: Materials Science and Engineering **534** (2019) 012026 doi:10.1088/1757-899X/534/1/012026

The reason for the car wheels drift is the excess of the centrifugal force acting on the front and rear axle wheels, the sliding friction forces of the corresponding wheel pairs in the tire contact patch with the coating in the transverse direction.

The estimates $\hat{V}_i(k) = \hat{R}_{ci}(k) \cdot \hat{\omega}_i(k), 1 \leq i \leq 4$ are formed from the rotation frequencies of the wheels $\omega_i(k)$ and the tuning data of the wheels free radii $R_{ci}(k)$.

The equations system of wheels rotation speeds for the car on a steep turn, contains 4 equations with 7 variables, and the task of their definition concerns to ill-posed ones.

To convert the original ill-posed problem to the correct one, we introduce the following additional conditions characterizing the physical properties of the object.

<u>The additional condition number 1 is the longitudinal acceleration and wheel slip longitudinal velocities signs consistency.</u>

So, in particular, object is characterized by positive sliding of the wheels on the horizontal surface during acceleration and negative during decelerations, which corresponds to the consistency of the a_m and ΔV_{Si} $(1 \leq i \leq 4)$ signs:

$$\begin{cases} if\ a_m \geq 0, then\ \Delta V_{Si} \geq 0; \\ if\ a_m < 0, then\ \Delta V_{Si} \leq 0. \end{cases} \tag{6}$$

<u>The additional condition number 2 is the limited longitudinal acceleration.</u>

The longitudinal mass center acceleration a_m is limited by the maximum sliding frictional forces of the wheels, which corresponds to the constraint system:

$$a^L_{bound\ 1} \leq a_m \leq a^H_{bound\ 1}, \text{ where} \tag{7}$$

$a^L_{bound\ 1}$ – low mass center acceleration limit at braking, and $a^H_{bound\ 1}$ – high mass center acceleration limit during acceleration.

<u>The additional condition number 3 is the determined magnitude and sign of the angular drift velocity.</u>

The magnitude and sign of the wheels drift angular frequency $\Delta\omega_m$ depends on the ratio of the centrifugal force and the wheels sliding frictional force in the transverse direction for the wheels of the front and rear axle and reduces to the following condition:

$$\begin{cases} if\ V_m \leq \min[V_{bound\ 1}, V_{bound\ 2}],\ then\ \Delta\omega_m = 0; \\ if\ V_{bound\ 1} > V_{bound\ 2}\ and\ V_m > V_{bound\ 2},\ then\ \mathrm{sgn}(\Delta\omega_m) = \mathrm{sgn}(\hat{\Psi}_c); \\ if\ V_{bound\ 2} > V_{bound\ 1}\ and\ V_m > V_{bound\ 1}, then\ \mathrm{sgn}(\Delta\omega_m) = -\mathrm{sgn}(\hat{\Psi}_c), \end{cases} \tag{8}$$

where the boundary understeer and oversteer wheels velocities $V_{bound\ 1}$ and $V_{bound\ 2}$ are determined from the equality conditions of sliding frictional forces to the corresponding centrifugal forces of the front and rear wheels.

The solution of the original problem with the additional conditions in discrete time is reduced to the form:

$$\hat{V}_m(k) = \begin{cases} V_m(k) + 0.5|\Delta V_{si} + \Delta V_{sj}|,\ if\ a^L_{bound} \leq a_m(k) \leq a^H_{bound}; \\ \hat{V}_m(k-1) + \Delta T a^H_{bound},\ if\ a_m(k) > a^H_{bound}; \\ \hat{V}_m(k-1) + \Delta T a^L_{bound}, if\ a_m(k) < a^L_{bound}, \end{cases} \tag{9}$$

where $|\Delta V_{si} + \Delta V_{sj}| = \min[|\Delta V_{s1} + \Delta V_{s2}|, |\Delta V_{s3} + \Delta V_{s4}|, |\Delta V_{s1} + \Delta V_{s4}|, |\Delta V_{s2} + \Delta V_{s3}|]$;

$$\hat{\Psi}_c(k) = \Psi_c(k) + a^{-1} b V_m^{-1}(\Delta V_{si} - \Delta V_{sj}) + b V_m^{-1} \Delta\omega_m; \tag{10}$$

IOP Conf. Series: Materials Science and Engineering **534** (2019) 012026 doi:10.1088/1757-899X/534/1/012026

$$\Delta\hat{\omega}_m(k) = \begin{cases} 0, \ if\ \hat{V}_m\ (k) \le V_{bound\ 1}\ and\ \hat{V}_m\ (k) \le V_{bound\ 2}; \\ b^{-1}\hat{V}_m\ (k)[\hat{\Psi}_c(k) - |\Psi_{bound\ 2}|\mathrm{sgn}(\hat{\Psi}_c)],\ if\ V_{bound\ 1} > V_{bound\ 2}\ \text{и}\ \hat{V}_m\ > V_{bound\ 2}; \\ -b^{-1}\hat{V}_m\ (k)[\hat{\Psi}_c(k) - |\Psi_{bound\ 1}|\mathrm{sgn}(\hat{\Psi}_c)],\ if\ V_{bound\ 2} > V_{bound\ 1}\ \text{и}\ \hat{V}_m\ > V_{bound\ 1}, \end{cases} \quad (11)$$

where $|\Psi_{bound\ 1}| = 2[m_{12}gb - R_d a_{dT}]k_{sq}V_m^{-2}$; $|\Psi_{bound\ 2}| = 2[m_{34}gb + R_d a_{dT}]k_{sq}V_m^{-2}$.

The traction-braking acceleration a_{dT} is determined from the equation of the longitudinal motion of the vehicle mass center with the identified parameters:

$$a_{dT} = a_m + k_x\ m_0^{-1}V_m^2 + k_{fr}g + \alpha_p\ g, \quad (12)$$

k_x – wind drag coefficient;
m_0 – vehicle weight;
k_{fr} – tire rolling friction coefficient;
α_p – pitch angle.

To determine the wheels sliding friction coefficient in the transverse direction k_{sq}, the circle Kamma property is used:

$$k_{Sdi}^2 + k_{Sqi}^2 = (k_{Si}^*)^2, 1 \le i \le 4 \quad (13)$$

where k_{Sdi} and k_{Sqi} are the sliding friction coefficients of the i-th wheel in the longitudinal and transverse directions, respectively;
k_{Si}^* is the top (maximum) value of the i-th wheel sliding friction coefficient.

The k_{Sqi} value, determined from (13) is equal to:

$$k_{Sqi} = k_{Si}^*\sqrt{1 - k_{Sdi}^2(k_{Si}^*)^{-2}} \quad (14)$$

The k_{Sdi} values are determined in accordance with Newton's third law from the equilibrium equation of traction-braking forces and sliding friction forces: the traction-braking force F_i is balanced by the sliding friction force $F_{Si} = F_{Ni} \cdot k_{Sdi}$, where F_{Ni} is the normal component of the dynamic load on the i-th wheel.

Module $|k_{Sdi}|$ is determined from the forces equilibrium equation: $|k_{Sdi}| = |F_i|F_{Ni}^{-1}$.

When the traction-braking forces modulus are small, according to (14), the sliding friction coefficient of the i-th wheel in the transverse direction is $k_{Sqi} \approx k_{Si}^*$.

Identification of k_{Si}^* is performed using software in the process of vehicle movement [9]. The input data for solving this problem are the measured wheel slip S_i and the current sliding friction coefficients of the wheels in the longitudinal direction k_{Sdi}.

The exceeding of the boundary velocities are predicted by extrapolating the boundary velocities $V_{bound\ 1}$, $V_{bound\ 2}$, and V_m by the time τ_e and verifying the inequality fulfillment:

$$V_{bound\ e} < V_{me}, \quad (15)$$

where $V_{bound\ e} = \min[V_{bound\ 0}^e, V_{bound\ 1}^e, V_{bound\ 2}^e]$;

$V_{bound\ 0}^e = \sqrt{0.5abh_m^{-1}|\Psi_{ce}|^{-1}}$;

$V_{bound\ 1}^e = Re\sqrt{2[m_{12}gb - R_d\ a_{dT}]k_{sq}|\Psi_{ce}|^{-1}}$;

$V_{bound\ 2}^e = Re\sqrt{2[m_{34}gb + R_d\ a_{dT}]k_{sq}|\Psi_{ce}|^{-1}}$;

$\Psi_{ce} = \hat{\Psi}_c(k) + \tau_e\Delta T^{-1}[\hat{\Psi}_c(k) - \hat{\Psi}_c(k-1)]$;

$V_{me} = \hat{V}_m(k) + \tau_e\Delta T^{-1}[\hat{V}_m(k) - \hat{V}_m(k-1)]$.

The braking deceleration $a_{dT} = a_d - a_T$ required to prevent drift and rollover is determined from (12) at $a_m(k) = [V_{bound\ e} - V_{me}]\Delta T^{-1}$.

The technical devices of the experimental wheel drift and car rollover preventing system include:
– standard wheel speed sensors;
– actuator of the brake pedal;
– control board;
– autonomous power source.

Currently, the system under development is being tested at the NAMI testing ground in Dmitrov. The analysis of preliminary test results does not disprove the adequacy of the used model and algorithms.

4. Conclusion

The analysis of the conducted studies results allows us to formulate the following conclusions:

– formulated and solved the task of safe speed dynamic stabilization for bends passing, wheel drift and car rollover prevention;

– developed algorithm of traction-braking acceleration control allows to integrate functions of separate regular active safety systems into one, implemented in a minimal hardware configuration;

– experimental system for wheel drift and car rollover prevention can function with incomplete wheel speed sensors configuration, reduced power consumption and higher operational reliability.

Acknowledgments

This scientific article was prepared based on the results of applied scientific research financially supported by the Education and Science Ministry of Russia, agreement is No. 14.625.21.0042, the unique identifier of the project is RFMEFI62517X0042.

References

[1] Buznikov S E 2007 Current state and prospects for the development of automotive active safety systems *Proceedings of the XV International Conference "Security management problems of complex systems"* pp 207-11

[2] Salmon P, McClurea R and Stantonb N 2012 Road transport in drift? Applying contemporary systems thinking to road safety *Safety Science* **50** pp 1829-38

[3] Chu L, Gao X, Guo J, Liu H, Chao L and Shang M 2012 Coordinated control of electronic stability program and active front steering *Procedia Environmental Sciences* **12 pp** 1379-86

[4] Lua Q, Gentilea P, Totaa A, Sorniottia A, Grubera P, Costamagnaa F and Smetb J 2016 Enhancing vehicle cornering limit through sideslip and yaw rate control *Mechanical systems and signal processing* **75** pp 455-72

[5] Ni L, Gupta A, Falcone P and Johannesson L 2016 Vehicle lateral motion control with performance and safety guarantees *IFAC Proceedings Volumes* **49** pp 285–90

[6] Høye A 2010 The effects of electronic stability control (ESC) on crashes – An update *Accident Analysis & Prevention* **43** pp 1148-59

[7] Her H, Yi K, Suh J and Kim C 2013 Development of integrated control of electronic stability control, continuous damping control and active anti-roll bar for vehicle yaw stability *IFAC Proceedings Volumes* **46** pp 83-8

[8] Kececi F and Tao G 2006 Adaptive vehicle skid control *Mechatronics* **16** pp 291-301

[9] Buznikov S E and Elkin D S 2007 Maximum values identification of the car wheels sliding friction coefficients *Certificate of the computer program official registration No. 2007610818 Rospatent*

Vehicle interior air quality standards development

A M Saikin, A V Kozlov and E A Iakunova

FSUE "NAMI", Moscow, Russian Federation

E-mail: elena.yakunova@nami.ru

Abstract. The paper presents results of experiments carried out to refine the procedure for determining the content of harmful substances in the air of vehicle interior coming from the supply system and exhaust system of the vehicle under test, as specified in GOST 33554-2015. Based on the work results proposals were prepared for the working group "Vehicle interior air quality", operating under the auspices of the World Forum for Harmonization of Vehicle Regulations (WP.29) of the United Nations Economic Commission for Europe. Content determination of the air pollutants in the vehicle compartment was carried out under various engine and ventilation system operating conditions when driving along the routes on the testing ground of FSUE "NAMI" and parking in specially selected places with the complete absence of other vehicles in the test area. Based on the test results, a conclusion was made that it is necessary to use a wider range of speed modes and variations in the operation modes of the ventilation system.

1. Introduction

Numerous published studies show that the doses of harmful substances entering the human body during traveling on vehicles significantly increase the daily dose of harmful substances inhaled by humans. For advanced vehicles, including electric vehicles, the problem of air pollution in the cabin will also be relevant while the vehicles with internal combustion engine and tires will continue to move on the roads. Confirming the relevance and importance of current problem, in 2015 the World Forum for Harmonization of Vehicle Regulations (WP.29) established an informal working group "Vehicle Interior Air Quality" (VIAQ). Since July 2017 the work of the group is conducted under the guidance of the Russian State Research Center FSUE "NAMI".

At first, the informal working group VIAQ developed a methodology for determining the volatile organic compounds (VOCs) concentrations emitted from the vehicle interior materials [1].

At present, a technique is being developed for determining the concentrations of harmful substances entering the vehicle interior from the outside. The main sources of air pollution in the vehicle interior are background air pollution measured by monitoring systems; polluted air above the roadway; the exhaust gases of the considered vehicle; exhaust gases of nearby vehicles; tire, roadway cover, clutches and brake mechanisms wear products.

It was decided to use as a basis the standard adopted in the Russian Federation – GOST 33554-2015 "Road vehicles. Pollutants content in the interior of driver cab and passenger compartment. Technical requirements and test methods" [2].

The document requires the determination of formaldehyde (CH_2O), nitrogen dioxide (NO_2), nitrogen oxide (NO), carbon monoxide (CO), saturated hydrocarbons (C_2H_6 – C_7H_{16}) and methane (CH_4) in the vehicle interior. Concentrations of harmful substances should not exceed the maximum single permissible concentrations (MPC_{ms}) established by hygienic standards GN 2.1.6.3492-17 "Maximum permissible concentration (MPC) of harmful substances in the atmospheric air of urban and rural settlements" [3]. MPC_{ms} values of the listed harmful substances are given in Table 1.

Table 1. Regulated harmful substances according to GOST 33554-2015.

Substance	Chemical formula	Hazard Class*	MPC_{ms}, mg/m^3	Engine type**
Formaldehyde	CH_2O	2	0.05	3, 4, 5
Nitrogen dioxide	NO_2	2	0.2	1, 2, 3, 4, 5
Nitrogen oxide	NO	3	0.4	1, 2, 3, 4, 5
Carbon monoxide	CO	4	5	1,2,3,4,5
Saturated hydrocarbons	$C_2H_6 - C_7H_{16}$	4	50	1,2,3
Methane	CH_4	4	50	3,5

Notes: * 1 – extremely dangerous; 2 – highly dangerous; 3 – moderately dangerous; 4 – low-risk
** 1 – spark ignition engines operating on gasoline; 2 – spark ignition engines operating on liquefied petroleum gas (LPG); 3 – spark ignition engines operating on compressed natural gas (CNG); 4 – engines with compression ignition (diesel engines); 5 – compression ignition engines operating on mixed fuel (diesel + CNG).

In accordance with GOST 33554-2015, tests of vehicles are carried out on paved roads with a longitudinal gradient of up to 6%, with a constant speed 50 km/h and idling. The recirculation system must be switched off. The recommended interior temperature is 20 ± 1 °C. Normal climatic factors are ambient air temperature from 15 °C to 30 °C; relative air humidity from 30% to 90%; wind speed 3.5 ± 1.5 m/s; atmospheric pressure from 630 to 800 mm Hg.

The test procedure, taken as a basis is described in the Appendix to GOST 33554-2015, paragraph B 4.2 and B 4.3. In this paper, we investigated the influence of the speed and the ventilation system operating modes on the concentrations of NO, NO_2, CO and total saturated hydrocarbons in the vehicle interior.

2. Test procedure

The test objects were vehicles of M_1 category with a manual transmission: three cars with a gasoline engine, an environmental class 4/5 and two cars with a diesel engine, an environmental class 4/5. The mileage of cars was no more than 30 000 km. The test cars moved along the paved roads of the NAMI Test Center during the "summer-autumn" season.

The tests were carried out in idle mode, at a constant speed, as well as in the drive-coast mode. In each case the influence of ventilation system modes on the concentration of harmful substance in the vehicle interior was investigated. Fan speeds varied from minimum to maximum with air recirculation and air intake from outside.

In constant speed mode, the cars moved with the speeds 50, 90, 110 and 130 km/h. In drive-coast mode, acceleration was made to the speed of 60 km/h, after 15 minutes the vehicle accelerated to 130 km/h with a fully open throttle and then the vehicle was slowed down by engine brake to the speed of 60 km/h. From six to eight "acceleration-deceleration" cycles were made. In all modes, the engine was started from a cold state (20-30 °C).

Measurements of pollutant concentrations were carried out in the express mode (direct measurement in the interior of the test object) during the driving modes. The measured harmful substances are carbon monoxide (CO), nitrogen oxide (NO), nitrogen dioxide (NO_2), total hydrocarbons (C_mH_n).

Quantitative measurements of substances at idle and during driving with constant speed were carried out in accordance with GOST 33554-2015. At the drive-coast mode, the gas analyzer readings were recorded each time when a vehicle had reached the maximum speed (6 – 8 times).

The gas analytical equipment complex used in tests is "P-310A" with a chemiluminescent detector for measuring nitrogen oxide and nitrogen dioxide (NO, NO_2), "Colion-1B" with a photo-ionization detector for measuring gasoline vapor (in terms of benzene C_6H_6), "Optogaz 500.4.CO" with an electrochemical detector for measurement of carbon monoxide (CO).

The final results were calculated by averaging the obtained data for the engine type, recirculation mode and fan speed of the ventilation system.

3. Test results

At idle, the excess of the standardized concentration value was detected only for nitrogen oxide (NO). This concentration was observed in case of diesel cars with recirculation switched on, at the maximum fan speed. For nitrogen dioxide (NO_2), the maximum concentration was also observed at testing of diesel vehicles with the same ventilation system operating mode. High concentrations of nitrogen oxides in exhaust gases of diesel engines caused by the peculiarities of fuel combustion. The maximum concentration of carbon monoxide (CO) was obtained for gasoline vehicles, but the ventilation mode coincided with the previous one (Table 2).

Table 2. Harmful substance concentrations at idle.

Substance	MPC_{ms}, mg/m^3	MPC_{ad}, mg/m^3	Max. value	Engine type*	Recirculation	Fan speed
NO	0.4	0.06	**0.498**	D	Enabled	max
NO_2	0.2	0.04	0.072	D	Enabled	max
CO	5	3	0.53	P	Enabled	max

Notes: P - petrol engine, D - diesel engine

When moving at a constant speed, the results were ambiguous. For cars with a gasoline engine, the maximum concentrations of harmful substances occurred when driving at speed 90 km/h (for NO) and at speed 130 km/h (for NO_2 and CO). Most maximum concentrations of harmful substances were found in the air recirculation mode and maximum fan speed.

In the case of diesel cars, the maximum concentrations of harmful substances were achieved at three different speeds: 50, 90 and 110 km/h. The operating mode of the ventilation system corresponded to the results obtained for diesel cars – with recirculation switched on and fan operation at maximum speed (Table 3).

Table 3. Concentrations of harmful substance when moving at a constant speed.

Substance	MPC_{ms}, mg/m^3	MPC_{ad}, mg/m^3	Max. value	Speed, km/h	Engine type*	Recirculation	Fan speed
NO	0.4	0.06	0.031	90	P	Enabled	max
NO_2	0.2	0.04	0.013	130	P	Disabled	0
CO	5	3	1.01	130	P	Enabled	max
NO	0.4	0.06	0.021	50	D	Enabled	max
NO_2	0.2	0.04	0.031	110	D	Disabled	max
CO	5	3	0.42	90	D	Enabled	max

Notes: P - petrol engine, D - diesel engine

When driving in the acceleration-deceleration mode, the maximum concentrations of harmful substances are obtained for the recirculation and fan operation at medium and maximum speed (Table 4).

Table 4. Concentration of harmful substances in the acceleration-deceleration mode.

Substance	MPCms, mg/m^3	MPC_{ad}, mg/m^3	Max. value	Engine type*	Recirculation	Fan speed
NO	0.4	0.06	0.081	D	Enabled	med
NO_2	0.2	0.04	0.03	P	Enabled	med
CO	5	3	1.16	P	Enabled	max

Notes: B - petrol engine, D - diesel engine

For total hydrocarbons at all speeds and ventilation modes observed concentrations were 0 mg/m^3.

4. Conclusions

The tests helped to identify the ventilation system modes, which provide the maximum concentrations values of harmful substances. For idling and driving at a constant speed the mode of ventilation system providing maximum harmful substance concentrations is recirculation and fan operating at maximum speed.

A particular speed mode at which the concentrations of all harmful substances were the highest failed to be detected, but it was possible to show that higher concentrations of harmful substances in the vehicle interior are observed when the vehicles moved at higher speeds than required by GOST 33554-2015, specifically, at speeds of 90, 110 and 130 km/h.

It is also worth noting that the concentrations of harmful substances in the vehicle interior air were very low – an excess of MPC_{ms} was obtained only for nitrogen oxide at idle. Such concentrations do not correspond to concentrations detected in vehicle interior while travelling along the carriageway [4]. It indicates that the movement of a single car does not reflect the real situation and it is worthwhile to consider other methods of conducting similar studies that take into account the real conditions of traffic movement in the traffic flow.

References

[1] Information on https://wiki.unece.org/download/attachments/44269597/ECE-TRANS-WP29-GRPE2017-10e.pdf?api=v2
[2] Information on http://docs.cntd.ru/document/1200136720
[3] Information on http://docs.cntd.ru/document/556185926
[4] Yakunova E A, Saykin A M 2016 Air quality in the passenger compartment *Engineering Bulletin* **2**

IOP Conf. Series: Materials Science and Engineering **534** (2019) 012028 doi:10.1088/1757-899X/534/1/012028

Creation concept of objective control systems for traffic safety of vehicles

A M Saikin and S E Buznikov

FSUE "NAMI", Moscow, Russian Federation

E-mail: sergey.buznikov@nami.ru

Abstract. Traffic safety is one of the most important problems for modern society. The relevance of research aimed at creating effective means for monitoring the vehicle technical condition and driver condition is determined by the importance of improving traffic safety. The study aims to propose the scientific substantiation of the creation concept of objective control systems for traffic safety of vehicles. To conduct the study system analysis methods and modern control theory were used. The safe driving task is formulated as collision prevention with moving and stationary obstacles. A structured set of collisions contains all typical collisions and overturns caused by technical malfunctions, pavement condition and driving errors. The system creation concept assumes both the measurement and recording of the state vector, its dynamic boundaries and warning signals demonstration to the driver about approaching dangerous boundaries or exceeding them. Modern foreign and domestic objective control automotive systems were compared. System acquires intellectual properties by using mathematical models and algorithms for indirect measurements performed with a minimal quantity of primary information sensors. The research novelty consists in the scientific substantiation of building automobile systems for objective control based on virtual information sensors providing the solution of the dynamic stabilization problem in its most fully algorithmically resolved form.

1. Introduction

The road safety problem affects the vital interests of almost all members in modern society and it is characterized by a global level of significance.

To solve this problem, which is in the field of legislative and executive authorities view, various state and public organizations are constantly working. Leading foreign automakers are carrying out large-scale research in the field of active and passive safety. Some foreign manufacturers equip produced cars with standard systems of objective control and collision parameters registration during an accident [1]. Tachographs [2], DVRs [3] and other monitoring systems [4] have become widespread mean of checking vehicle operating conditions. For an objective assessment of the accidents conditions the usage of "black boxes" is predicted [5, 6, 7], they will automatically register the accident site [8] and call emergency services. Sets of parameters that should be recorded in order to obtain an adequate incident picture as well as the construction principles of the system are open for discussion.

The purpose of this study is scientific substantiation of construction concept for vehicle traffic safety objective control systems.

To determine the motion parameters vector a structured typical collisions set is used, which makes it possible to decompose the general collision avoidance problem into a group of particular problems and formulate the state variables dynamic stabilization problem [9].

In case of state variables dynamic stabilization problem it is proposed to use indirect measurements based on mathematical models and algorithms for solving ill-posed problems.

A comparative analysis is carried out to compare the intellectual level of information-measuring systems within three generations. The results of sensor properties and intelligent systems analysis show the objective advantages of virtual information sensors usage.

2. Problem statement

Automobile objective control systems (OCS) of vehicle traffic safety are intended as for analysis of vehicle movement parameters and driver actions in cases of an accident and for decision making informational support for the driver in the process of accident-free operation.

The collision avoidance problem is reduced to the problem of vehicle state variables dynamic stabilization in the form:

$$x^L_{i\,\mathrm{b}}(X,U,t) \le x_i(t) \le x^H_{i\,\mathrm{b}}(X,U,t), 1 \le i \le n \text{ at } U \in U_{per}, \text{ where} \quad (1)$$

$x_i(t)$, $1 \le i \le n$ – the i-th component of the state vector X;

$x^H_{i\,b}$ и $x^L_{i\,b}$ – higher and lower boundaries of the state vector X i-th component;

U – the control actions vector, including traction, brakes, steering wheel and gearbox;

U_{per} – the permissible range of control actions.

The components x_i of the state vector X and their boundaries are determined from the conditions for preventing typical collisions. The output x_i out of dynamic boundaries means the occurrence of prerequisites for at least one of the typical collisions in a structured set.

It should be noted that object state vector dynamic stabilization problem is one of the most complex control problem. Its solution complexity is caused both by the incomplete observability of the object and by the uncertainty of some dynamic boundaries. For the practical solution of such problems, operator participation in the control is required, who will complete the unobserved components of the state vector and estimate their dynamic boundaries.

The results of this problem solution are largely determined by the individual qualities of operators, the level of their training, accumulated experience, intuition and other hard-to-formalize properties.

3. Research results

Various schemes can be used to solve the problem of monitoring object state variables. The first generation of monitoring systems (Figure 1) arose in the in the 30s-40s of the XX century. They assumed the measurement of the state coordinates X_1 ... X_4 with the help of specialized information sensors S_1 ... S_n, converting the measured physical quantities into the electrical signals parameters $y_i = f_i(x_i)$ with the subsequent inverse transform $\hat{x}_i = f_i^{-1}(\hat{y}_i)$ in the information display devices (IDD).

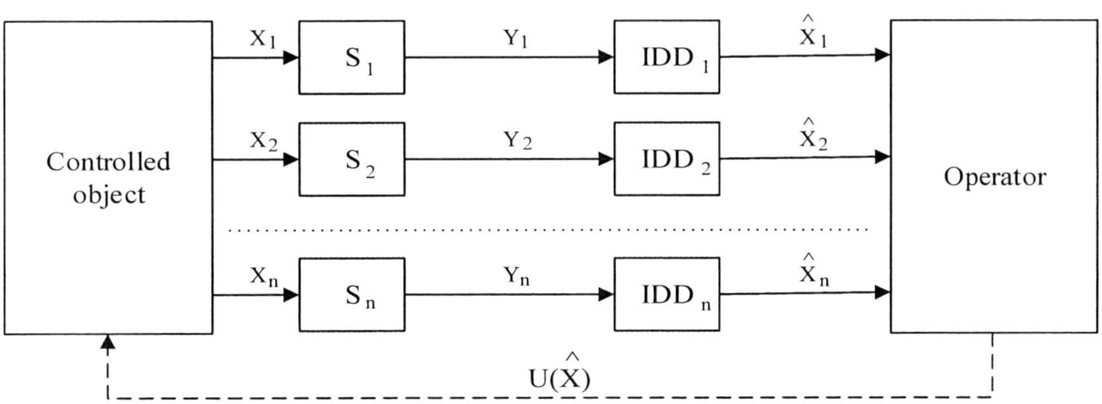

Figure 1. Scheme of the 1st generation system for object state monitoring.

First generation of monitoring systems had several shortcomings in particular the rigid requirements to the stability of direct and inverse transformations and the boundaries stationarity fixed on the IDD scales.

The second generation of monitoring systems (Figure 2) appeared in the 60s of the XX century. They implied the usage of a microcontroller unit (MCU) with software reverse transformation and the possibility of using dynamic boundaries. The disadvantage of such monitoring systems is a significant number of primary information sensors and communication cables, which worsen the majority of system indicators.

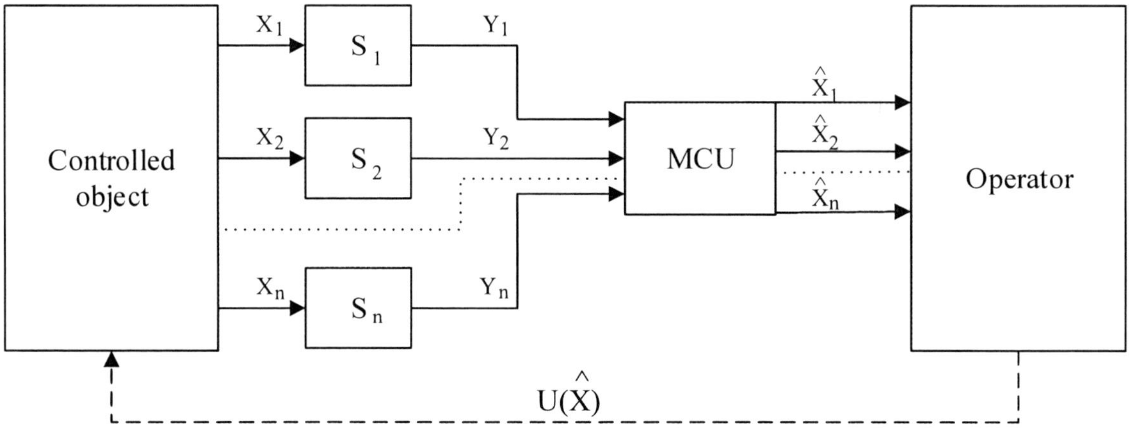

Figure 2. Scheme of the 2nd generation system for object state monitoring.

The third generation of monitoring systems (Figure 3) appeared at the beginning of the XXI century, it assumes the replacement of physical information sensors by virtual ones. Minimum configuration of technical means is used for their work, which is sufficient for indirect measurements of the state coordinates and their dynamic boundaries.

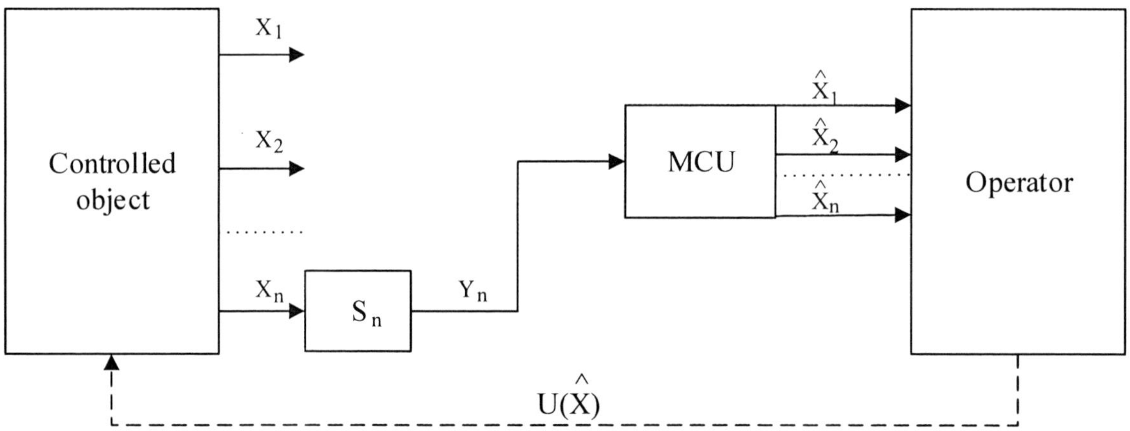

Figure 3. Scheme of the 3rd generation system for object state monitoring.

The intellectual level of the monitoring systems estimated by the ratio of the functions number to the number of required technical means. We found that for the first generation of monitoring systems the assessment of the intellectual level did not exceed 0.5, for the second generation it is around 1.0, and for the third generation it is $\gg 1$.

The conceptual core of the 1st and 2nd generation systems are physical sensors of information, which allows considering these systems as sensor systems. The conceptual core of the third generation systems is shifted to the field of mathematical support, which makes it possible to consider that they are intellectual systems.

Comparative characteristics of sensor and intelligent systems of objective control are given in Table 1.

Table 1. Characteristics of sensor and intelligent objective control systems.

№	Objective control system characteristics	Tachograph	Videorecorder	EDR	ERA "GLONASS"	Developed objective control systems
1	Registration of traffic rules violation	No	No	No	No	Provided
2	Warning about dangerous vehicle states	No	Sound warning about lane departure	No	No	Provided
3	Determination of vehicle location	No	Limited by the conditions of radio visibility	Limited by the conditions of radio visibility	Limited by the conditions of radio visibility	Provided for various conditions
4	Driver condition monitoring	No	No	No	No	Provided
5	Car control state monitoring	No	No	Provided	No	Provided
6	Wheels, suspension and steering condition monitoring	No	No	No	No	Provided
7	Monitoring of the vehicle position on the lane	No	Provided	No	No	Provided
8	Monitoring of the dangerous approach to the obstacle on the lane	No	Provided	No	No	Provided
9	Road pavement condition monitoring	No	No	No	No	Provided

Comparison of properties and characteristics of well-known foreign, domestic and developed systems shows that the main advantages inherent in intelligent systems are manifested in the consumer indicators and characterized by the stability of properties.

The predicted characteristics of the objective control system being developed are superior to those of known systems, some functions of this system are not available in known systems.

The video cameras usage together with the software image processing is necessary to determine distances to the nearest obstacle on the lane in the front and back hemispheres [10].

By processing images of video cameras, it is supposed to determine the speed of obstacles movement and the degree of controlled vehicle deviation from the center of the lane.

In this case, the problem of accident causes recording as well as vehicle state variables dynamic stabilization to which the collision avoidance problem reduces, solved both in real time and based on the results of an objective analysis of the driver's actions in the accident, acquires the property of solvability and determinism of state coordinates and their dynamic boundaries.

It should also be noted that the current traffic rules and penal system of their violations also reduces to an approximate solution to the simplified dynamic stabilization problem for the observed variable state (speed, position on the lane, technical malfunctions, crossing of road marking, turns, stops and parking, etc.).

The simplification of the dynamic stabilization problem in accordance with the algorithm determined by the traffic rules is the blurring of recommendations on the choice of safe speed, distances, decelerations, etc. Penalties for the detected violations are episodic, different from the time integration of penalty functions, characterizing the time during which the vehicle stays in dangerous state, when prerequisites for certain typical collisions remain.

Conceptual difficulties arises when the legislator tries to determine the term "dangerous driving".

The states variables exceed their dynamic boundaries in the time interval mean the existence of prerequisites for collisions and, accordingly, for accidents.

The identification of such intervals and certain variables corresponding to the type of possible collision is carried out by the OCS automatically and allows to evaluate the quality of control based on the results of the driver actions.

At present, none of the known designs of the OCS allows solving the "dangerous driving" identification problem.

In addition, the developed OCS basic design with the basic composition of technical means demonstrates state variables and their dynamic boundaries, as well as indicates their excess in real time for the driver. In comparison with foreign counterparts, it provides information support for decision-making by the driver to ensure driving safety.

4. Conclusion

The results analysis of the conducted studies allows us to formulate the following conclusions.

The trivial solution of the OCS creation problem is reduced to equipping the object with numerous sensors of physical variables that form the conceptual core of traditional sensor monitoring systems.

This direction, developed by leading foreign companies, inevitably leads to the deterioration of almost all system indicators, including the total cost, the cost of operation, installation, maintenance, consumed energy, influence on adjacent systems, fault tolerance, etc. and requires highly qualified service, possible only in the branded centers.

A non-trivial solution of the OCS creation problem with the usage of new technologies is consist in the utilizing of indirect measurements based on knowledge of the object in the form of formalized properties descriptions that form the conceptual core of intelligent systems.

It should be noted that many of the indirect measurements problems are initially incorrect and for the solution they need to be redefined by the properties of the object.

Acknowledgments

This paper bases on the results of applied scientific research, which is carried out with the financial support of the Education and Science Ministry of Russia, Agreement No. 14.624.21.0048. The unique identifier of the project is RFMEFI62417X0048.

References

[1] *Federal Motor Vehicle Safety Standards № 405* Event Data Recorders 2012 U.S. Department of Transportation. National Highway Traffic Safety Administration 49 CFR Part 571

[2] *Operation of tachographs equipped with a block of cryptographic information protection: Operation of tachographs* OOO "Tachograph". URL: www.tahocard.ru/info/ekspluatatsiya-takhografov-s-skzi/

[3] Lelyavin P A, Ponamarenko M Y 2012 Automobile Recorder *Transport of the Russian Federation* **9** pp 68-69

[4] Kurilova A S 2012 Modern transport tracking and monitoring systems used in motor transport enterprises *KANT* **2**(5) pp 80-83

[5] Chernov E I 2013 Automobile "Black Box" *News of SFU. Engineering Sciences* p 26

[6] Han I 2016 Car speed estimation based on cross-ratio using video data of car-mounted camera (black box) *Forensic Science International* **269** pp 89-96

[7] Chung Y 2018 Injury severity analysis in taxi-pedestrian crashes: An application of reconstructed crash data using a vehicle black box *Accident Analysis & Prevention* **111** pp 345-353

[8] Stepchenko A V 2017 ERA-GLONASS and "Black boxes" in cars. A far future or today's reality

Science without borders **4**(9) pp 20-23
[9] Buznikov S E 2018 System analysis of vehicle active safety problem *IOP Conf. Series: Materials Science and Engineering* **315** p
[10] Saikin A M, Buznikov S E and Karpukhin K E 2016 The analysis of technical vision problems typical for driverless vehicles *Research Journal of Pharmaceutical, Biological and Chemical Sciences* **7**(4) pp 2053-2059

IOP Conf. Series: Materials Science and Engineering **534** (2019) 012029 doi:10.1088/1757-899X/534/1/012029

Charging balance management technology for low-voltage battery in the car control unit with combined power system

V V Debelov[1], D V Endachev[1,2], D M Yakunov[3], O I Deev[3]

[1] FSUE "NAMI", st. Avtomotornaya, 2, 125438, Moscow, Russian Federation
[3] Moscow Polytechnic University, 107023, st. B. Semenovskaya, 38, Moscow, Russian Federation

E-mail: [1] vladimir.debelov@nami.ru, [2] denis.endachev@nami.ru

Abstract. The article presents the results of mathematical models and algorithms development complex, which is aimed at ensuring a positive charge balance of an automobile low-voltage rechargeable battery in car control systems with a hybrid power system. The aim of the work is to create and implement a set of software and mathematical models on one of the existing passenger car model, using the default components and systems of the car, to achieve a positive energy balance of the automotive low-voltage power network, and also to solve the problem of battery discharge by normalizing the turn-on time of powerful consumers leading to fast battery discharge. As criteria for the optimal performance of the proposed models and algorithms, the SOC parameters are considered - state of charge (%), battery charge (A * h), charge and discharge current of a battery (A), voltage on a battery during its charge when a charging current is absent or in a quiescent state (V), ambient temperature (°C). Authors defined: composition of a control system, the necessary logic and a set of mathematical models implemented by electronic systems to control the charging balance of the automotive low-voltage power supply network, connections and functions of systems and components. As a result of a full-scale experiment, charge and discharge graphs, experimental data, current characteristics and dynamics of their changes during operation, experimental time-current dependencies during the operation of each component of a low-voltage power grid were obtained. The results of the work support to improve energy balance of an electrical power network, increase the efficiency and service life cycle of batteries, formulate system and technical requirements for control systems and their subsequent implementation. Versatility solutions are provided for improving management of charge balance car battery.

1. Introduction

A modern car is full of electronic systems that allow solving a number of different tasks: to control the power unit, adjust the gears, ensure stability and safety on the road, make a comfortable climate in the vehicle, control interior and exterior lighting, play media files, control various drives and mechanisms. All this is made possible by equipping the car with electronic systems that provide power to the on-board network and maintain the positive energy balance of the electrical network to which consumers are connected.

On a modern automobile, it is customary to locate several sources of electrical energy [7], which provide the on-board network with consumers and ensure the charge of car batteries [8]. Thus, one of the devices produces electrical energy by converting the mechanical rotation of electric machines or generators shaft to electrical energy, other devices convert it to the required power level of the on-board network, and the third perform the function of accumulating electrical energy for subsequent start and powering the on-board network [3, 4].

Despite the fact that it seemed that all existing automobile systems for generating, converting and storing energy are capable of providing the necessary current and voltage to power consumers, in real-life conditions this does not always become possible. To provide the required level of output current for all systems, we would have to increase the characteristics, and as a result, the dimensions of electric machines and generators This would result as a higher torque requirement for the powertrain, and as a result, as higher power and fuel consumption, which would also complicate the implementation of current standards for toxicity and harmful emissions. Increasing batteries capacity of the same type would increase their size and weight, which would require large fuel resources when the vehicle is moving.

To solve the problem of improving the efficiency of vehicle operation and reducing fuel costs, reducing harmful emissions and other tasks aimed at ensuring the overall energy efficiency [1] of a vehicle, it is necessary to create intelligent electronic power management systems. These systems must ensure the optimal energy balance of the car when using energy sources with lower mass and dimensions, and often contribute to a decrease in the energy capacity of the accumulator storage cells while maintaining the functionality of all connected control systems. This is achieved, first of all, due to the efficient distribution of energy between sources and consumers, monitoring the state of the battery and its timely charging, reducing the work duration of powerful consumers, distributed control of energy converters.

2. The task of developing methods and algorithms for ensuring the charging balance of an automotive low-voltage rechargeable battery

Annual tightening of the requirements for harmful emissions cannot be achieve without reducing the operating time of the power unit and proper distribution of electrical energy between sources and consumers. In this connection, cars with combined power system have been widely developed [12]. Cars are equipped with an internal combustion engine that connected with an electric machine. It operates in both engine and generator mode. In the generator mode, the electric machine provide the high-voltage on-board network and contributes to the charging of the high-voltage battery [5]. The low-voltage network is powered by a high-voltage network through DCDC converters. To the low-voltage on-board network, consumers of electrical energy and a rechargeable battery are connected in parallel scheme, which serves as a storage buffer and dampens a unusual changes in the network.

Damping function becomes impossible without the use of batteries with improved charging and time-current characteristics. AGM or helium batteries are often used. These types of batteries can accumulate charge faster due to faster chemical processes. But at the same time, as with lead-acid batteries, they have a feature, that at higher temperatures, above +20 degrees Celsius, it is necessary to reduce the charging current and voltage to prevent boiling of the electrolyte. At low temperatures, it is required to increase the output voltage and charging current to help the battery warm up and increase its charge rate.

Due to the fact that the battery during the operation of the vehicle assumes the functions of a damper when electric power generator or DCDC converter fails to provide the necessary current to power the consumers, it is constantly required to monitor and charge the battery, disabling powerful consumers to prevent its discharge and subsequent degradation. There is a range of charge for batteries where the most efficient use is from 65% to 95% for most modern helium batteries and AGM battery.

During the charging of the battery, the operating voltage is no less important, since at values of charging voltage less than 12.5-13.5 volts, the battery stops charging and becomes a source, transferring the accumulated energy to the on-board network. In order to ensure the required voltage level of the on-board network at which the battery is charged, it is necessary to regulate the output voltage of the generator or DCDC converter which provide the on-board network.

When solving the problem of ensuring the charging balance of an automobile low-voltage battery, it is necessary to solve several key problems connected with its operation:

- It is necessary to monitor and diagnose the state of the battery in order to monitor and detect deviations of its operating parameters from normal values;
- It is necessary to create an optimal control method for the generator set or the power converter in order to adjust the charging current and battery voltage for its effective charge in different temperature conditions and at different values of the load current;
- It is necessary to develop a method that allows disconnecting high-power consumers in case of a negative charge balance of a low-voltage rechargeable battery in order to equalise the currents produced and given to consumers;
- It is necessary to develop a model [11] that monitors the state of the battery at rest in order to detect timely leakage currents and disconnect devices that lead to the consumption of electrical energy when the ignition is off, for example, keyless vehicle access systems.

Solution of the tasks should be a set of mathematical functions, combined into a single model that ensures the implementation of the tasks and implemented as a program code in the automotive electronic unit that acts as a power controller and a car battery charge.

As part of the tasks solution, it is necessary to create a block diagram of the modern power vehicle system structure, including the main sources and converters of electrical energy, as well as sensors that allow measuring the operating parameters of a low-voltage rechargeable battery during its operation. It is necessary to present the allowable intervals for changing the operating parameters of the battery during its operation.

It is necessary to present a generalized concept of mathematical model implementation for controlling the charge balance of a car battery and describe the main connections of its individual elements.

It is necessary to assess the possibility of applying the technology in harsh climatic conditions and to assess the boundary modes of application.

It is required to test the method as part of the existing vehicle sample, to investigate the effectiveness of the proposed charge method and to obtain experimental dependencies of the car battery charge processes when regulating the currents and voltages in the vehicle's on-board network during its operation. When high-powered consumers are turned on, it is necessary to estimate the speed of the battery charge system's response to an instantaneous change in load. To assess the possibility of applying the proposed method on other vehicles is required.

3. Modern power system structure of a passenger car equipped with a combined power system

Modern power system of cars with combined plants is often presented in the following form.

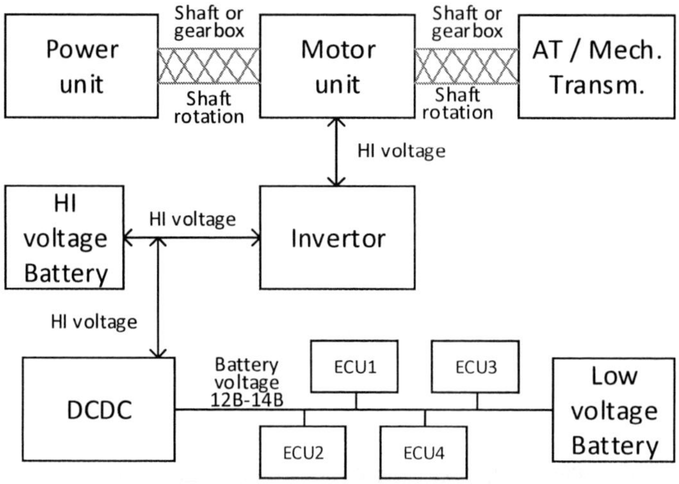

Figure 1. Modern energy system of cars with HES.

The power unit has a direct rigid connection with the electric machine through a shaft or gearbox. Then the electric machine with its shaft is connected to a manual gearbox or an automatic gearbox (automatic transmission).The electric machine is connected to the inverter through a high-voltage cable that powers the electric machine at the time of start-up, and when the power unit is up and running, it supplies the power inverter, which, using the BMS charge control system, charges the high-voltage traction battery (HI voltage battery) to the required level [9] . The DCDC DC-DC converter is powered by a high-voltage network, performing the functions of converting electrical energy from a higher voltage to a lower voltage and vice versa (reversible energy converters). DCDC converter performs two functions. The first function is power supply of the on-board low-voltage network 12V-14V which charges the low-voltage rechargeable battery, all consumers are connected to it. The second function is to help the high-voltage battery to start at low temperatures, when the converter operates in a reverse mode [10]. When the ignition is on, but the power unit isn't working, the power supply of the on-board electronics is combined. Part of the energy is converted by DCDC from a high-voltage battery to low voltage, and another part comes from a low-voltage battery. When the ignition is off, while DCDC is inactive, the entire on-board electronics are powered by a low-voltage rechargeable battery. The battery is used to power the BMS system [6] and control units, including the electronic control unit (ECU) of the traction inverter, until the DCDC converter turns on or the power unit is started up.

4. Block diagram of the main elements involved in managing the charge balance of a low-voltage battery

In order to ensure the most stable balance of electrical energy within the framework of the existing vehicle architecture, it is necessary to choose the concept of ensuring the charge balance of an automotive low-voltage battery. To do this, initially it is necessary to develop a structural diagram (Figure 2) of the main elements involved in managing the charge balance of the battery.

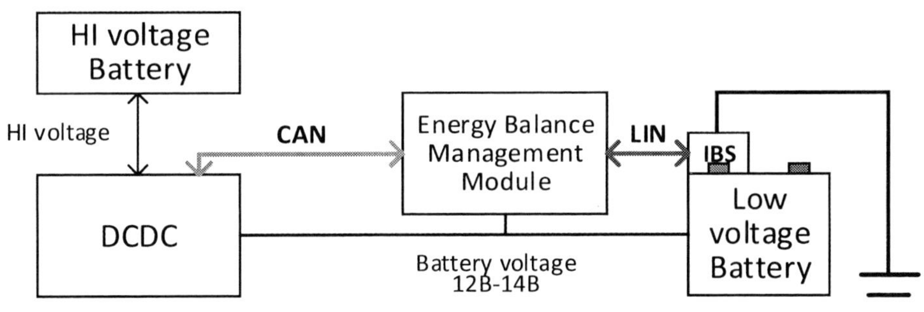

Figure 2. Structural diagram of the main elements involved in managing the charging balance of the low voltage battery.

In the presented scheme, the high-voltage battery is connected to the DCDC converter, which is controlled by the charge balance control unit by generating current and voltage requests at the low-voltage output of the converter. The low-voltage battery is connected with the DCDC converter in parallel to the on-board power supply through an intelligent battery sensor (IBS). The battery sensor allows you to monitor the degree of battery charge (SOC,%), battery life level (SOH,%), battery voltage (U, V), battery pole temperature (T, °C), as well as inflowing and flowing current (I, A). A complete list of parameters, ranges of change and accuracy of measured values for the most common types of sensors for rechargeable batteries with wake-up function are presented in Table 1.

According to the known parameters of the battery, the electronic control unit calculates the values of the setpoint of currents and voltages for the DCDC converter, and also determines operation duration of powerful consumers and, if it's necessary it increases the current in the high-voltage network during start-up by charging the low-voltage battery.

Table 1. The most common measured parameters for an intelligent current sensor of battery.

Parameter name	Measurement range
Duration of sleep mode until the next wake up	30 .. 7620 s
Discharge current when it wake up	-250 .. 0 mA
Charge current when it wake up	0 .. 2500 mA
Charge when the sensor wake up	0 .. 125 Ah
Battery current range	-2000000 .. 2000000 mA
Battery voltage	0 .. 50 V
Battery model temperature	-40 .. 125 °C
Error of current	0 .. 1 (logic)
Error of voltage	0 .. 1 (logic)
Error of temperature	0 .. 1 (logic)
Wake up by current	0 .. 1 (logic)
Wake up by charging	0 .. 1 (logic)
Wake up by charge level	0 .. 1 (logic)
Wake up by timeout	0 .. 1 (logic)
Wake up by voltage	0 .. 1 (logic)
Sensor Sleep Time	0 .. 65534 min
Min Battery Quiescent Current	-65000 .. 0 mA
Battery charge	0 .. 100%
Battery helth	0 .. 100%
Internal resistance of battery	0 .. 200 mOhm
Available capacity	0 .. 249 Ah
Effective capacity	0 .. 249 Ah
Nominal capacity	0 .. 249 Ah
Charge current	0 .. 65535 Ah
Discharge current	0 .. 65535 Ah
Voltage open circuit	0 .. 65.535 V
Charge at open circuit	0 .. 127,5 %

5. The structure of the mathematical model of the charging balance

The structure of developed mathematical model is created and the main connections between the developed program modules are described (Figure 3). The structure of the model consists of the following main elements:

- Charge Balance Manager (MS)
- Condition diagnostics model of the low-voltage power supply network (MD)
- Model of energy control converters DCDC (MC)
- Power Disconnect Management Model (MR)
- Model of the residual battery charge control (MQ)

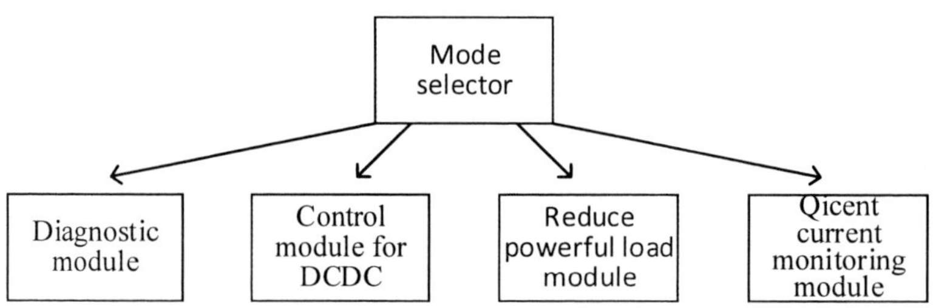

Figure 3. Structure of mathematical model of charging balance.

Mode manager determines which of the models will be active in various modes of vehicle operation. There is following division according to the modes of modules operations is presented in Table 2.

Table 2. Modes in which the battery charging control software modules are activated.

Mode	MS	MD	MC	MR	MQ
Sleep	On	On	Off	Off	On
Preheat	On	On	Off	On	On
Standby	On	On	On	On	On
Radio	On	On	Off	On	On
Climatic	On	On	Off	On	On
Ignition	On	On	On	On	Off
Drive	On	On	On	On	Off
PowerOff	On	On	Off	Off	On
Transportation	On	On	Off	On	Off

The mode manager and the diagnostic model must always be active when the electronic power management unit wakes up by timer, external or internal interruption, then errors that occur during vehicle operation are stored in non-volatile memory (NVRAM) and become available for reading through the tester.

The energy converter control model is active when the high-voltage system is in operation and it is used to select the most optimal converter operation modes according to the current state of the battery for supplying the on-board network and charging the battery with currents that depend on the current charge and battery temperature and also an operating mode.

The model of disconnecting high-powered consumers is turned on at the moment when the DCDC converter is working at the limit of its capabilities, and the calculated current balance in the power grid is shifted to the negative side. Then it is decided to disconnect consumers, which do not affect the safety of the vehicle, but contribute to the unloading of the low-voltage power supply network. In the case when the power supply balance is not reached, then an error flag is set that informs the driver by means of a message on a special display that a problem has occurred and you should immediately contact the service center to fix the problem.

The model of the control of the residual battery charge is used to track the leakage currents in the car after turning off the ignition and powering the electronic components. This is due to the fact that some of the electronic modules are able to turn on after the main ignition is turned off. In the case of an uncontrolled leakage current, there is a risk of discharging the battery to a state when further engine start is not possible or the battery is completely degraded. The model contributes to the formation of a request for de-energizing consumers which wake up in a sleep mode in order to exclude full discharge of battery.

6. Battery monitoring and diagnostics

Monitoring and diagnostics [2] of a battery state are performed in several stages for which the critical parameters of the battery and the state of its charge are determined [XX].

The first stage of diagnostics is that the absence of software error flags is checked by the measured parameters of the intelligent sensor, which are represented by the self-diagnostic functions of the sensor element.

The second stage includes checking the boundary modes of the battery, as well as determining its actual aging and battery life.

The aging and unsuitable battery check is performed by comparing the SOH (state of health) parameter of the battery with the set thresholds.

Old battery is determined based on the following conditions:

IOP Conf. Series: Materials Science and Engineering **534** (2019) 012029 doi:10.1088/1757-899X/534/1/012029

$$MD\_Aged\_Bat = Bat\_Pct\_SOH < Bat\_Pct\_Aged\_Thr, \quad (1)$$

where MD_Aged_Bat is a sign of battery aging, Bat_Pct_SOH is the age or battery life, in percentages, Bat_Pct_Aged_Thr is a threshold at which it is determined that the battery has aged and does not correspond to the stated specifications.

An unsuitable battery is determined on the basis of the following conditions:

$$MD\_NotSuffBatt\_Bat = Bat\_Pct\_SOH < Bat\_Pct\_Aged\_Thr, \quad (2)$$

where MD_ NotSuffBatt_Bat is a sign of battery unsuitability, Bat_Pct_SOH is the age or battery life expressed in percentages, Bat_Pct_ NotSuffBatt_Thr is the threshold at which it is determined that the battery is not suitable for further use as a part of the car.

The battery voltage check is performed by estimating the voltage value in the measured time interval.

Low voltage on the battery is determined from the conditions:

$$MD\_UnderVoltage\ (10s) = (U\_IBS\_Voltage < Bat\_U\_UnderVolt\_Thr), \quad (3)$$

where MD_UnderVoltage is the logical function of the output signal versus time when the logical condition is met, U_IBS_Voltage is a battery voltage, Bat_U_UnderVolt_Thr is the voltage threshold at which the error is determined in a given time interval.

High voltage on the battery is determined from the conditions:

$$MD\_OverVoltage\ (10s) = (U\_IBS\_Voltage < Bat\_U\_OverVolt\_Thr), \quad (4)$$

where MD_OverVoltage is the logical function of the output signal versus time when the logic condition is met, U_IBS_Voltage is a battery voltage, Bat_U_OverVolt_Thr is the voltage threshold at which an error is determined in a given time interval.

The voltage of a discharged battery is determined from the conditions:

$$MD\_DeepDschrg\ (60s) = (U\_IBS\_Voltage < Bat\_U\_DeepDschrg\_Thr), \quad (5)$$

where MD_DeepDschrg is the logical function of the output signal versus time when the logic condition is met, U_IBS_Voltage is a battery voltage, Bat_U_ DeepDschrg_Thr is the voltage threshold at which the error is determined at a given time interval.

High and low voltage, which is determined in conditions when all consumers are disconnected from the battery.

$$MD\_DisconectUppThr\ (10s) = (U\_IBS\_Voltage < Bat\_U\_DisconectUppThr\_Thr), \quad (6)$$

$$MD\_DisconectLowThr\ (10s) = (U\_IBS\_Voltage < Bat\_U\_DisconectLowThr\_Thr), \quad (7)$$

where MD_DisconectUppThr and MD_DisconectLowThr are logical functions of the output signal versus time when the logical condition is met, U_IBS_Voltage is a battery voltage, Bat_U_DisconectUppThr_Thr and Bat_U_DisconectLowThr_Thr are the voltage thresholds at which errors are determined in a given time interval.

The determination of a battery charge state is performed in the third stage of checks (Figure 4) and is performed by comparing the current charge level with permissible thresholds. It is customary to single out several basic battery charge modes, for example:

- A fully discharged battery (SOC <MD_Stt1UppLim);
- Fast battery charge (MD_SOC_UppLim2> SOC> MD_SOC_UppLim1);
- Normal battery charge (MD_SOC_UppLim3> SOC> MD_SOC_UppLim2);
- Full battery charge (MD_SOC_UppLim3 <SOC)

An error state is detected in the case of a long state, when SOC = 100% or SOC = 0%.

IOP Conf. Series: Materials Science and Engineering **534** (2019) 012029 doi:10.1088/1757-899X/534/1/012029

A feature of the presented state diagram of the control algorithm is that in the event of a change in any of the modes, an instantaneous response of the control system occurs, allowing to compensate the destabilized state.

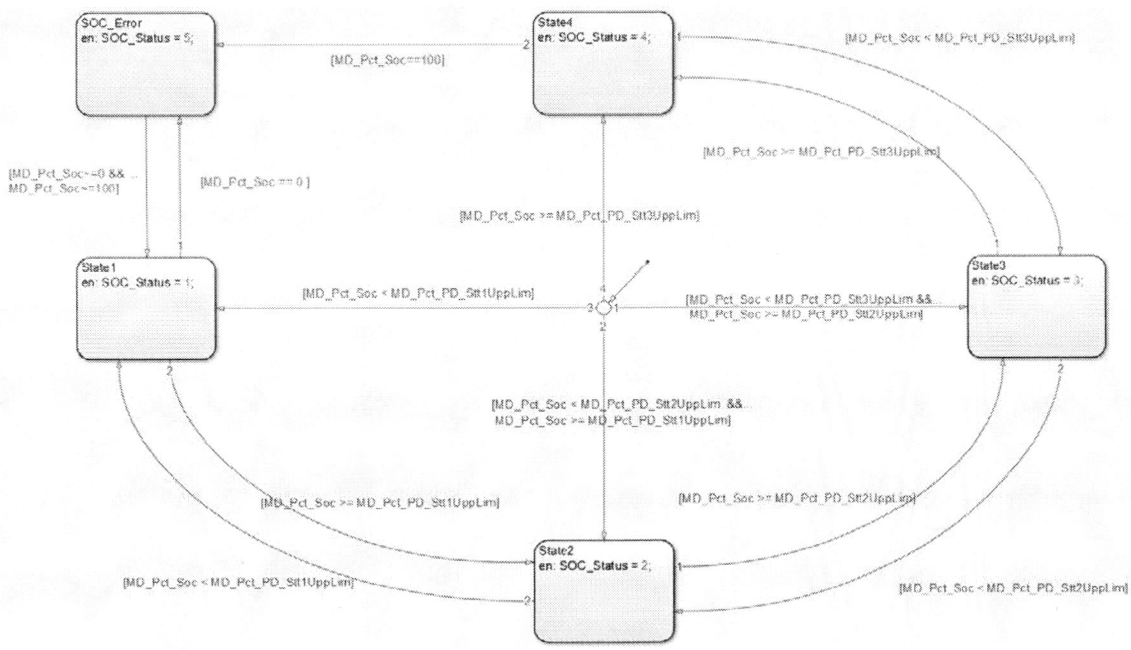

Figure 4. State diagram of determining the state of a battery charge.

7. Control method of the DCDC converter to charge the battery and achieve a positive charge balance

The method consists in calculating the current balance of the power supply, followed by the determination of the required current limit and voltage set point, which vary depending on what changes in temperature and battery charge, how many consumers are turned on and what is the actual load of DCDC converters.

Current calculation is performed based on what current is required from DCDC converters to ensure battery charge, which maximum battery charge current is permissible at current values of charge and temperature, and which current flows in or out of the battery itself. This method of forming the current setting for a system of DCDC converters can be described by the formula:

$$I_{req} = \sum I_{DCDC} + I(t^o, SOC)_{BAT.LIM} - I_{BAT} \tag{8}$$

where I_{BAT} - is the current flowing in or flowing out of the battery, $\sum I_{DCDC}$ - is the total current of the DCDC converters, $I(t^o, SOC)_{BAT.LIM}$ - is the current limit, according to the battery charging characteristic model for which the example of implementation is presented in Figure 5.

Viewing "2-D Lookup Table" block data [T(:,:)]:

Breakpoints	Column	(1)	(2)	(3)	(4)	(5)
Row		**-40**	**-20**	**20**	**60**	**100**
(1)	**0**	70	70	45	30	30
(2)	**25.09803...**	65	65	40	25	25
(3)	**49.80392...**	60	60	40	20	20
(4)	**74.90196...**	55	55	35	15	15
(5)	**100**	50	50	30	10	10

Figure 5. The model implementation example of battery charging characteristics.

The figure shows that as the charge increases, the inverter's current setting decreases. This is due to the fact that as the charge increases, the internal resistance of the battery changes, which leads to a decrease in the input current. Therefore, to charge the battery in this case a large current does not require. With a low charge, the current is artificially limited to prevent battery degradation. At the same time, it should be a little bit higher than the rated charging current in order to dampen the switching on of consumers and not interrupt the process of battery charging. If the battery is damaged or completely discharged, the current has a maximum value and compensates the defect of the battery. As the battery temperature rises, the charging current decreases to prevent boiling of the electrolyte.

The voltage setting for the DCDC converter is selected depending on the mode of the car battery operation. For the following modes, the principle of setting the voltage of the converters is used, which is described further in the text.

The condition in which an error is detected requires a permanently high voltage value of 14.6 V if the error occurred due to a low charge and a constant low voltage of 13.5 V if the error occurred due to a constant high charge.

In the case of low charge, you need to charge the battery quickly. For this purpose, the voltage setpoint is higher than usual, 15.6V for a fully discharged battery.

As the battery starts charging, the voltage setting changes to a lower value of 14.8 V in order to charge the battery and then go back to normal operation.

In the normal operation mode of the car battery, the battery charging model is used, for which the voltage setting depends on temperature and battery charge as shown in Figure 6. As the temperature rises, the voltage requested from the DCDC converters also decreases as the battery charges and also voltage reduces. At the highest possible temperature and with a fully charged battery, the voltage setting is approximately equal to the open circuit voltage of the battery in steady state at 13.2 V at room temperature in order to protect the battery from destruction.

Viewing "2-D Lookup Table" block data [T(:,:)]:

Breakpoints	Column	(1)	(2)	(3)	(4)	(5)	(6)	(7)
Row		**-40**	**-13**	**1**	**5**	**50**	**54**	**77**
(1)	**0**	15.65625	15.34375	15.15625	15.125	14.4375	14.15625	13.53125
(2)	**49.80392...**	15.65625	15.34375	15.15625	15.125	14.4375	14.15625	13.53125
(3)	**60**	15.46875	15.15625	14.96875	14.90625	14.3125	14.0625	13.5
(4)	**74.90196...**	15.21875	14.875	14.71875	14.65625	14.09375	13.9375	13.4375
(5)	**80**	15.125	14.78125	14.625	14.59375	14	13.90625	13.4375
(6)	**87.05882...**	15.09375	14.75	14.5625	14.46875	13.90625	13.8125	13.375
(7)	**94.11764...**	15.0625	14.71875	14.5	14.375	13.84375	13.71875	13.28125
(8)	**100**	15.0625	14.71875	14.5	14.375	13.84375	13.71875	13.28125

Figure 6. Voltage setting for battery charging depending on the electrolyte temperature and low voltage battery charge.

In normal mode, in case of rapid changes in charge, for example, as a result of switching on powerful consumers, when the DCDC converter transmits the maximum current for a given control mode. Further adjustment of the maximum output power is achieved by not significantly adjusting the voltage setting. The total power will be represented by a formula where the output value of the mode is about 2-5% more, which allows to provide additional current in the power supply system and additional power to consumers.

$$P_{total.max} = P_{requested.max} + P_{extra.power} = U_{req} \cdot I_{req} + U_{extra.power}(SOC) \cdot I_{req} \quad (9)$$

The additive voltage, additionally requested in the counting function (Figure 7) of the maximum power, is represented by the function of the dependence of the output voltage on the battery charge. The additive voltage decreases as the battery charges.

IOP Conf. Series: Materials Science and Engineering **534** (2019) 012029 doi:10.1088/1757-899X/534/1/012029

After the selection of the current and voltage setting of the DSDS of the converter is made, it is necessary to calculate the current and voltage setting individually for each converter, if there are several such converters. If the converter is used alone, the current and voltage setpoints are calculated based on the maximum power of the device and its load factor. The setpoint calculation function also takes into account that if the converter is overloaded or there is an error, then the current and voltage setpoint are calculated for the second converter, taking into account its mode of operation.

Viewing "Lookup Table" block data [T(:)]:

Breakpoints Column (1)

Row		--
(1)	***49.80392...***	0.5
(2)	***60***	0.40625
(3)	***70.19607...***	0.3125
(4)	***80***	0.1875
(5)	***89.80392...***	0.09375

Figure 7. Additive voltage setting function for the converter.

For greater clarity, Figure 8 shows the state diagram where the current and voltage settings are calculated for two DCDC converters that supply the on-board network from the high-voltage network. In the first relation, the total current of the converters is checked and if the requested current is bigger than the total, then the current setting is equal to the maximum current of the converter. In the second condition, the request current and the load of the first converter are checked. If the converter current is less than the total, but more than one of the two converters, then the current setting for the first converter is equal to the maximum current based on the load, and at the second converter the setting is the remainder of the current difference between the requested and the first converter.

IOP Conf. Series: Materials Science and Engineering **534** (2019) 012029 doi:10.1088/1757-899X/534/1/012029

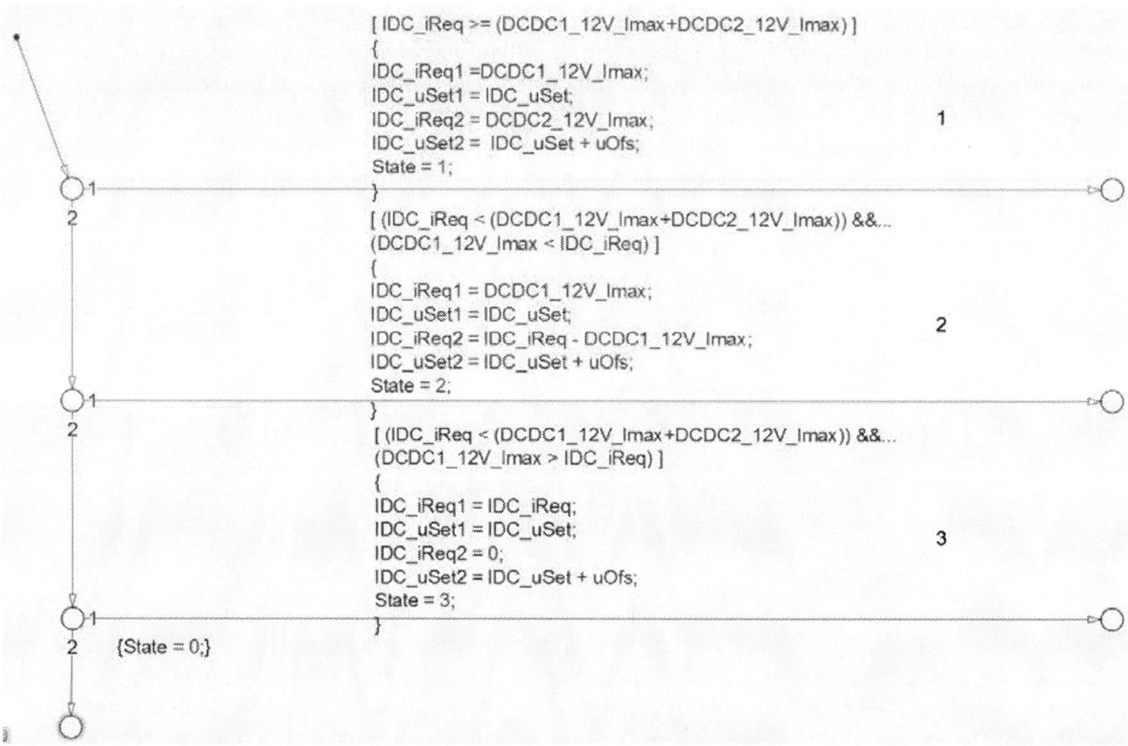

Figure 8. Calculation of the setpoint currents and voltages for two DCDC converters.

The third mode provides for the case when the requested current is less than that any of the converters can give, then the current setting of the second converter takes a zero value, and the first is loaded with a current corresponding to the calculated setpoint. In case of a converter error or its 100 percent load, with a zero current request, which means, overheating, only one converter is operated, which is limited by the maximum output current based on the converter power.

8. Battery status monitoring in waiting mode

In order to monitor the state of battery charge, a model is needed that monitors leakage currents. To do this, it is necessary to wake up the electronic unit containing the model of ensuring the energy balance of the car in order to check the leakage current with the nominal rest mode and then again set the time for the current sensor until a request is made to wake up the electronic control unit. As the transition to the low-power mode occurs, the electronic modules included in the on-board power supply network disconnect powerful loads and later other loads, up to the transition to the hibernation mode with waking up on an event or a timer.

IOP Conf. Series: Materials Science and Engineering **534** (2019) 012029 doi:10.1088/1757-899X/534/1/012029

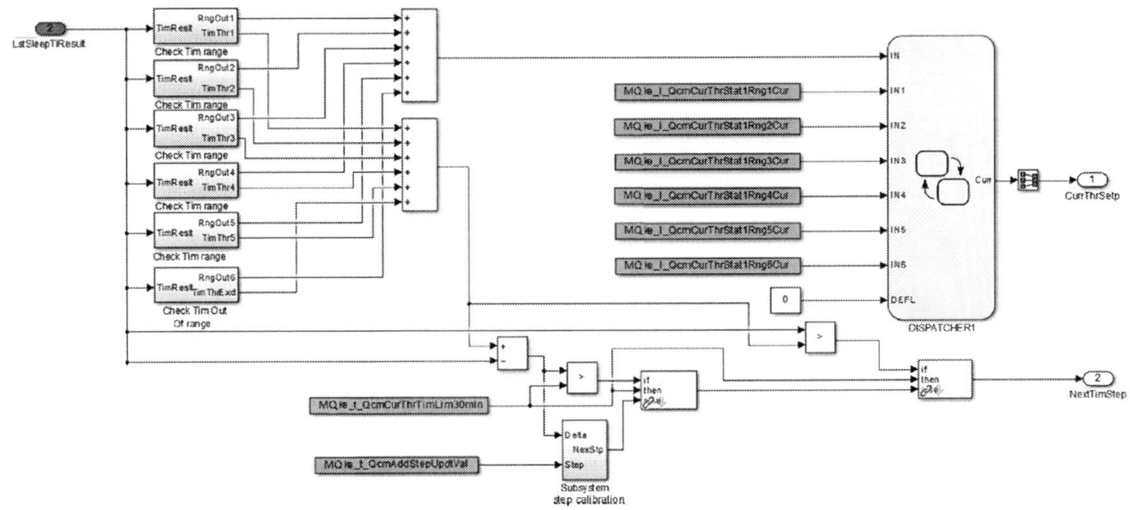

Figure 9. Model for calculating the controller waking time and leakage current threshold.

Figure 9 shows a model that tracks the duration of the time during which the controller was in power saving mode. After that, a current setpoint is formed, which is set as a threshold for the battery sensor comparator, and the time until the next waking up is set. The time until the next waking up is regulated by the step size and is necessary to compare periodically the current value with the allowable one, check the residual charge, and in case of a critical battery discharge, de-energize devices that periodically wake up and consume electrical energy while the rest remain off.

The charge monitoring function is represented by a model for calculating waking time and a set of comparators that turn off other consumers to prevent critical discharge of the battery in power saving mode. In normal mode, when the DCDC converters are active, the function is disabled.

9. A method that allows disconnecting powerful consumers when charge balance of a low-voltage battery is negative

In the course of operation, when the current of all consumers exceeds the sum of maximum converters current and charging battery current, then in order to bring the consumption of the entire system to the optimum mode, high-power consumers are disconnected in series. They are determined based on the priority of their use in the car. So, for example, heating nozzles or hoses of headlight washer has the lowest priority, and the ABS / ESP hydraulic unit or electric power steering has the highest priority and are disconnected in the most recent case when the battery charge is not sufficient to carry out a response to the disturbance and there is a risk of loss board voltage, leading to a sudden shutdown of all on-board electronics.

To avoid sharp disturbances or premature reactions, a moving average type filter is used. Filtration is performed for charging and flowing current through a low-voltage battery. Next, the following parameters are calculated:

1) The sign of the process, where charge is positive, and discharge is negative.
2) it is determined how long it takes to charge the battery to the set value.
3) It is determined how much charge in amperes / hours flows in and out of the battery.
4) The actual battery charge is calculated taking into account how old the battery is.
5) For a known current, the maximum charge time is limited.

Next, using the obtained values, a comparison is made with the permissible limits of the residual charges of the battery for the low-voltage power supply network, and on this basis, a group of consumers is selected that needs to be turned off in order to achieve a positive charge balance of car battery (Figure 10).

IOP Conf. Series: Materials Science and Engineering **534** (2019) 012029 doi:10.1088/1757-899X/534/1/012029

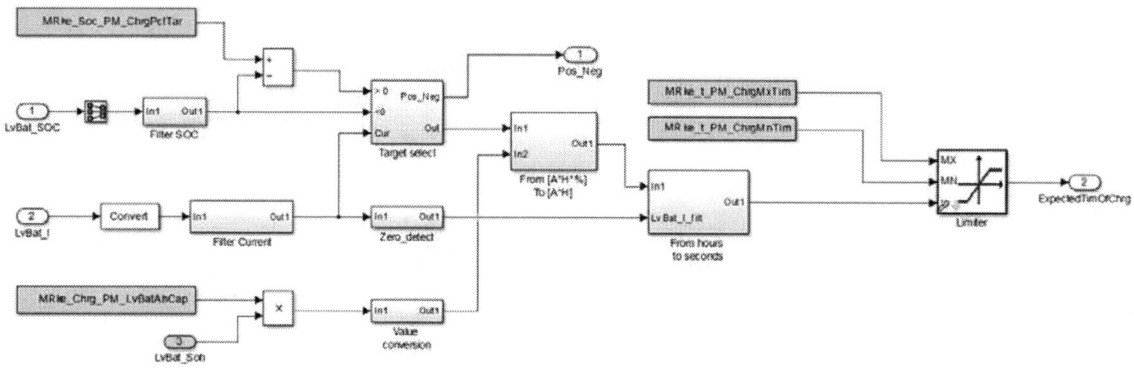

Figure 10. Model for calculating the characteristics used in the analysis of energy balance of power grid and the choice of a group of disconnected consumers.

When selecting a group of disconnected consumers, a flag is formed depending on whether this mode is critical or not, which is used by the system (Figure 11) to inform the driver about an abnormal situation in the power supply.

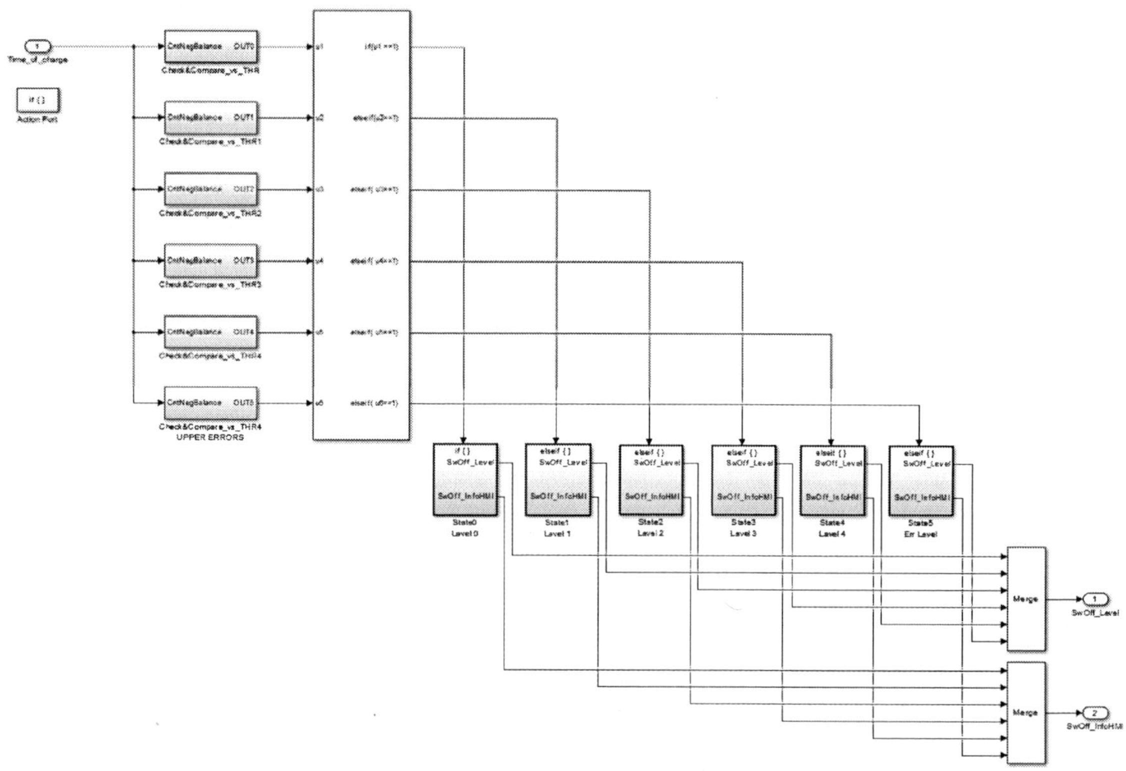

Figure 11. Model of request formation to disconnect a group of consumers with a negative charge balance and low-voltage battery.

10. Experimental studies

Testing the proposed method of controlling the charge battery balance was carried out using a sample of a battery, a DCDC converter, an intelligent current sensor, different power loads, and a control unit into which the proposed control model was integrated. The characteristics of the test battery and DCDC converter are presented in Table 3 and 4.

Table 3. Technical characteristics of the battery EFB 6CT-100.

Maximum current	930A
Rated voltage	12,5V
Open circuit voltage (NRC)	11,7…13,2V
Battery Capacity	100Ач
Operating temperature range	-25…+50 C
Leakage current	50 мА

Table 4. The dependence of the OCV on SOC and density of electrolyte for EFB 6CT-100.

The state of battery charge, %	Discharge ratio Kr	OCV, V (2 hours after charging)	Electrolyte density, reduced to 25 ° C, g / cm³
100	0	12,7 or higher	1,28
75	0,25	12,5	1,25
50	0,5	12,3	1,22
25	0,75	12,1	1,18
0	1	11,8	1,12

Table 5. Technical characteristics of the DCDC converter BSC624.

The range of high rated voltage	220-450 V
Low Voltage	14.0 V
Low Voltage Range	8-16 V
Continuous current at rated for example	200 A
Maximum allowable current	250 A
Long conversion mode	2.8 kW
Maximum possible power conversion	3.5 kW
Converter efficiency	94.4 %
Ambient temperature	-40 to +85 °C

Figure 12 shows a plot of an experiment in which a battery is charged with low currents about 3-5% of the nominal value in accordance with the calculated setting, which limits the value of the current flowing into the battery in this experiment. The graph shows that with an increase in the actual current, the current demand for the DCDC converter changes, which allows for a positive battery charge balance in the process of load control. At the moment of connecting the load, the system destabilizes, which leads to a short-term current increase in the power supply system with subsequent return to a stable value. The response time of the control system, taking into account software delays, is about 100 ms, and the error in controlling the value of the output current and voltage is no more than 5%, which is sufficient to confirm the correctness of the proposed model. The presence of current peaks is due to the fact that the nature of switching on loads isn't linear and repeatedly leads to destabilization of the current in the low-voltage power supply network.

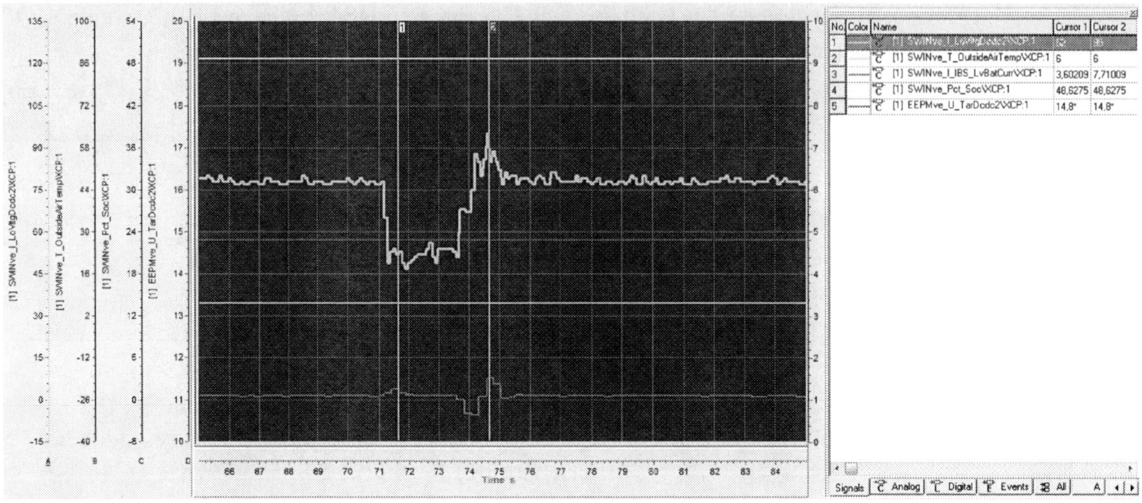

Figure 12. Graph of the experiment with the load change in the power supply system.

11. Results and conclusions

As a result of the work, you can come to the following conclusions:

- Monitoring and diagnostics of the battery state allows you to detect deviations from the nominal modes in time and inform you that intervention of technical services in the analysis of faults causes is necessary.
- The proposed method of controlling a power DCDC converter promotes efficient charging in different temperature conditions and under different loads.
- The method, which allows disconnecting of powerful consumers, contributes to the alignment of current converters and load currents.
- A model that monitors the battery state at rest contributes to the long-term preservation of the battery charge and the prevention of critical discharge below the set value.

The solution of the set tasks was accomplished due to the implementation of mathematical functions combined into a single model and implemented as a program code as part of an automotive electronic control unit.

The presented generalized concept of the implementation of a mathematical model for controlling the charge balance of a car battery can be implemented both for vehicles with ICE and for traditional vehicles with ICE.

It is necessary to assess the possibility of applying the technology in harsh climatic conditions, and to assess the boundary modes of application.

Testing the method as part of the control unit on an existing vehicle sample confirms that a positive battery charge balance is achieved for standard electronic control systems.

12. References

[1] Yakunov D M, Debelov V V and Endachev D V Energy efficiency of electric and hybrid vehicles in conditions of low negative temperatures In the collection: Intellectual transport systems *Collection of works of the International Automotive Scientific Forum IASF-2017* pp 270-78

[2] Kozlovsky V N, Debelov V V, Deev O I, Kolbasov A F, Petrovsky S V and Novikova A P 2017 Perspective diagnostic systems for managing an autonomous transport object *Truck* **6** pp 21-28

[3] Debelov V V, Kozlovsky V N, Stroganov V I and Pyanov M A 2014 The complex of electronic control systems for the movement of a passenger car with a combined power plant Part 1 *Electrotechnical and Informational Complexes and Systems* **1** pp 40–48

[4] Kozlovsky V N, Stroganov V I, Debelov V V and Pyanov M A 2014 The complex of electronic control systems for the movement of a passenger car with a combined power plant Part 2 *Electrical and information systems and systems* **2** pp 19-28
[5] Endachev D V and Skripko L A 2017 Calculation of the degree of charge of lithium-ion batteries used in battery systems of electric vehicles and hybrid cars *Electronics and electrical equipment of transport* **4** pp 8-11
[6] Varlamov D O, Eremenko V G and Yakunov D M 2016 Simulation of a balancing device on boosting DC / DC converters for a LIFEPO4 battery *Practical power electronics* **3** pp 16-20
[7] Barzukov S N, Guskov A V, Zinoviev E V, Shiganov D A, Shorin A A and Eidinov A A 2013 Electric energy storages for motor vehicles with combined power plants *Works NAMI* **252** pp 62-80
[8] Yutt V E, Morozov V V, Sokolov L A, Reznik A M and Ospanbekov B K 2017 Modern current sources and charging stations for electric vehicles Tutorial. Moscow State Automobile and Road Technical University (Moscow)
[9] Bakhmutov S V, Sizov Yu A and Kim ME 2015 Development of a mathematical model and the study of a traction battery of a hybrid power plant *Works NAMI* **262** pp 105-112
[10] Kurmaev R Kh, Struchkov V S and Tsimbalyuk M A 2015 The issue of ensuring the specified temperature regimes of KEU elements when using them on the PBX *Works NAMI* **260** pp 69-80
[11] Stroganov V I, Kozlovsky V N, Shakursky M V and Zayatrov A V 2018 Modeling the basic processes of electric vehicles and cars with a combined power plant *Electronics and electrical equipment of transport* **2** pp 8-13
[12] Kozlovsky V N, Aydarov D V, Vasilyev M M and Debelov V V 2018 Development of electric cars and cars with hybrid energy unit *Gruzovik* **6** pp 18-21

IOP Conf. Series: Materials Science and Engineering **534** (2019) 012030 doi:10.1088/1757-899X/534/1/012030

Development of the unmanned vehicle traffic control system with electric drive

A V Tumasov, D Yu Tyugin, D M Porubov, V I Filatov and A V Pinchin

Nizhny Novgorod State Technical University n.a. R.E. Alekseev, Minina str., 24, Nizhniy Novgorod, 603950, Russian Federation

Abstract. The development of unmanned vehicle traffic control systems is a world trend. The development of the market of unmanned vehicles is stimulated by various contests and competitions. This paper presents the prerequisites and the first steps in the development of unmanned vehicle traffic control systems by the NNSTU n.a. R.E. Alekseev with the support of engineers from the United Engineering Center of the GAZ Group and specialists from PJSC GAZ. This system is developed for operation in difficult road-climatic conditions in the territory of the Russian Federation. As a chassis for installation of the system, an electrical platform was chosen. The component composition of the unmanned vehicle traffic control system is proposed, the location and coverage of the equipment are determined. The software modules of the unmanned traffic control system are implemented. Experimental studies of the system were carried out.

1. Introduction

The market sector of unmanned vehicle traffic control systems is new and rapidly growing both in the world and in the Russian Federation. Large automakers develop systems for unmanned vehicle traffic control based on ADAS (Advanced Driver Assistance Systems). Also, these systems are successfully developed by researches [1 - 5] and IT companies, such as Google (figure 1a), Nvidia (figure 1b), Baidu (figure 1c); and manufacturers of military equipment: DARPA [9]; Oshkosh Defense (figure 1d); Lockheed Martin [11] etc. (Figure 1).

a B c d

Figure 1. (a) Google self-driving car, (b) NVidia self-driving car, (c) Baidu self-driving car, (d) Oshkosh unmanned ground vehicle.

To evaluate the efficiency and increase the level of competition among developers of unmanned traffic control systems the competition of unmanned vehicles has been organized with the support of the US government (DARPA Grand Challenge [9]). These competitions revealed all the shortcomings of unmanned vehicles and significantly influenced the development of the unmanned industry. It is worth noting that most of the participants were represented by relationship of big automaker, who has provided vehicle, and the university engaged in realization of unmanned functions on this vehicle.

In 2014, the United States approved the first national standard in the field of autonomous vehicles - SAE J3016 «Taxonomy and Definitions for Terms Related to On-Road Motor Vehicle Automated Driving Systems [12]». The standard contains definitions, terminology and classification of vehicle automation levels. The document under consideration is the basis for the subsequent development of standards and determines the level of interaction between the community dealing with automated and autonomous vehicles.

According to this standard [12], the highest level of automation of vehicles is reached by Google (Level 5 - Full Automation).

The main difference of Google's development from other developers of unmanned vehicles is that in the motion process the data obtained from LIDAR is verified with 3D high-resolution maps recorded earlier. Due to this, it is possible to detect an obstacle on the path by the comparison. Therefore, at this stage of technology development, the autonomous movement of the car is possible only in those areas where detailed three-dimensional maps were pre-recorded.

In the Russian Federation, the direction of development of unmanned vehicle traffic management systems is less successful, but there is a number of companies which have a big potential in the development of these systems such as Yandex [13], Cognitive technologies [14], FSUE "NAMI" [15], PJSC KamAZ [16], Avrora robotics [17] and others.

Among the Russian developers of unmanned vehicles, one of the most successful companies is Yandex [13]. The purpose of the development is the use of unmanned vehicles in the field of transportation services (taxi, trucking, etc.). To date, Yandex has developed two prototypes of unmanned vehicles based on the Toyota Prius and Kia Soul. The hardware consists of the following types of sensors. Inside the car there are cameras of the front and rear view, from which the recognition of cars, pedestrians, road signs and marking, as well as the boundaries of the carriageway, takes place. On the roof of the car there are LIDARs: Velodyne HDL-32 and Velodyne VLP-16. These sensors with the help of a laser emitter scan the surrounding space. Based on the information received, a three-dimensional map is compiled. With the help of this map, exact distances to certain objects around the vehicle are calculated. The vehicle is also equipped with sensors that determine its location, speed and direction of travel. These are GPS / GLONASS receivers, a block of inertial meters and sensors that measure the odometer data of the vehicle.

To reduce the gap in the technological development of unmanned vehicle traffic control systems, adapted to use in Russian climatic and road conditions, the staff of the NNSTU n.a. R.E. Alekseev together with the Joint Engineering Center of GAZ Group currently work on the development and implementation of unmanned traffic control systems for commercial vehicles of the GAZ Group.

2. Development of the unmanned traffic control system

As the chassis for installation of the unmanned vehicle traffic control system, an electric platform for commercial vehicles was chosen (Figure 2).

Figure 2. Appearance of the electrical platform.

IOP Conf. Series: Materials Science and Engineering **534** (2019) 012030 doi:10.1088/1757-899X/534/1/012030

This choice is based on a number of advantages of electric transport over traditional types of vehicles such as the increase of safety and environmental friendliness through the use of modern technologies that ensure the use of electricity as the main source of energy. Almost all large automakers make progress in this area, and advanced countries, such as the USA, China, Japan, Germany and Great Britain - have long-term programs to support the development of electric transport and consumer incentives. Currently, support is expected from the government of the Russian Federation.

The control of the chassis electrical units does not require the installation of actuators as in the case of mechanical, hydraulic and pneumatic units. It increases the safety and reliability of the vehicle systems, as well as the speed and cost reduction of the various vehicle systems.

This platform is the basis for the line of commercial vehicles of the GAZ Group, however, for the implementation of the unmanned traffic management system, the bus was chosen. The use of route transport as an object on which unmanned operation solutions will be applied is explained by a number of reasons: the main route of the vehicle has been determined in advance, the ability to move along designated lanes for public transport, the possibility of charging batteries at the final points of the route. At the first stage of development the hardware of the system has been determined (Figure 3).

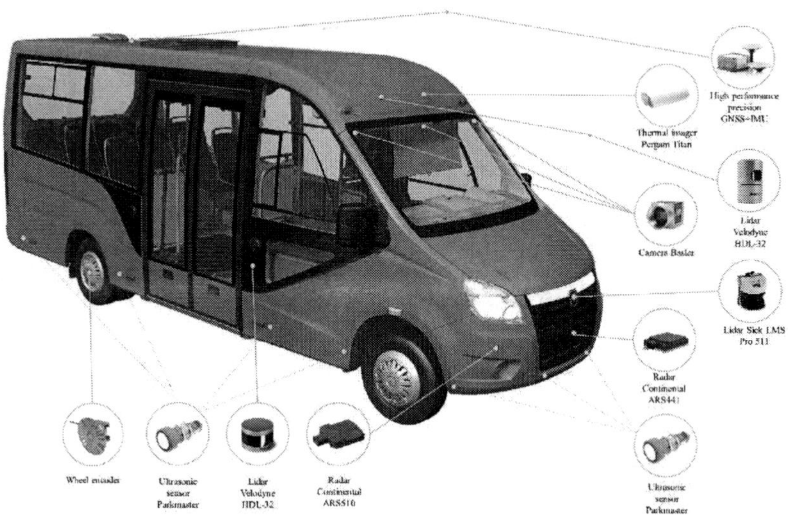

Figure 3. The hardware of the unmanned traffic control system.

In order to collect and analyze information from various types of sensors of the system, an on-board computer with INTEL Xeon Processor E5-2690 V4 and three NVidia GeForce GTX 1080 Ti graphics cards is used.

LIDARs will be used to detect other traffic objects and obstacles along the route, as well as construct a virtual route map (Figure 4). Velodyne HDL-32 will be used to scan the space in front of the vehicle. This sensor has a wide vertical sector of operation (from + 10° to -30°) and a range of up to 200 m. Since the sides and the back side are the least responsible, then for these zones will be applied LIDAR Velodyne VLP-16, having a vertical viewing angle from -15° to + 15° and range up to 100 m. Also the planar LIDAR Sick LMS Pro 511 will be installed at the bumper level of the vehicle. The use of this sensor is due to its high scanning frequency of 75 Hz and a wide temperature range of operation from -30° C to +50° C. All LIDARs are connected to the on-board computer via an Ethernet connection. To implement the system for detecting objects of the road scene, the following software was developed: to detect objects on the video stream neural networks (modified architecture yolo v218) were used, to detect obstacles according to LIDAR data, cluster cloud algorithms were implemented as well as methods of linear algebra. All implemented modules for interaction with each other were integrated into the ROS infrastructure (Robot Operating System [19]).

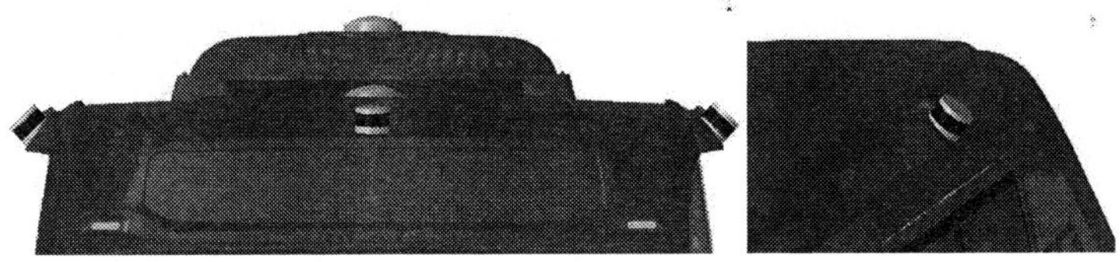

Figure 4. Location of LIDAR Velodyne VLP-16.

An important aspect in the design of unmanned traffic control is the reliability of the entire control system as a whole. Especially this issue is relevant in extreme road and climate conditions. To improve the reliability of the obstacle detection system at the unmanned vehicle, in addition to the LIDAR system, a short-range and long-range radar system will be used. The most suitable by the characteristics are the radars Continental ARS441 and SRR510, operating at a frequency of 77 GHz. The ARS 441 radar has a range of up to 174 m with a horizontal operating angle of ± 10°. The SRR510 radar has a range of up to 70 m with a horizontal operating angle of ± 75°. With the on-board computer, these radars are connected via the Ethernet connector. The application of the radar system will allow to assess the situation in front of the car, to measure the distance to obstacles at long distances. The radar system operation is less affected by weather conditions, color of objects and illumination. Also radars have good resolution in range and speed, instant data processing, which is especially important for high speeds. It should be noted that systems with different types of radars are already being used by large automakers in the development of ADAS systems.

Compliance with traffic rules is a prerequisite for the operation of an unmanned vehicle on public roads. Recognition of the road scene participants, in particular, other cars, pedestrians, road signs, marking, etc. is carried out with the help of the system of technical vision, the hardware part of which is represented by video cameras Basler [20]. In the unmanned vehicle traffic control system, three video cameras are used, completely covering the front view.

To improve the efficiency, reliability and safety of the obstacle detection systems and other road users, the Titan 63221 thermal imager will be used, with a horizontal view of 90°, a vertical view of 65°, a temperature range from -60° C to 50° C. The thermal image, after processing by the computing complex, will improve the recognition of the boundary of the roadway, the estimation of the approaching turns and curvatures on the route in advance. It is worth noting that the thermal imager, unlike video cameras, operates even when the headlights of oncoming cars are illuminated, through smoke, dust, fog and rain.

In order to plan the trajectory of motion along the given reference points (in geo-coordinates) and recalculation of the trajectory parameters into the control commands of the vehicle actuators, a path planning program was implemented. The algorithm is implemented in the context of solving the optimization problem, which consists in finding the optimal control parameters of the actuators while minimizing the trajectory error and reaching the set speed. The program consists of the following blocks: the trajectory calculation unit, the navigation system data acquisition unit, the map data display unit, the route entry unit. The hardware consists of GNSS receiver OS-203-GSM, GNSS antennas AT330 and inertial block Xsens [22]. The system is implemented by a separate scheme for the integration of GNSS data of the receiver and IMU, i.e. they work independently of each other. Periodically, the IMU is updated according to GNSS data. The navigation complex provides high-precision geographic coordinates, speed, altitude, number of satellites and power of the satellite signal. It connects to the on-board computer via the Ethernet interface. The embedded software performs data processing, writes it to a database and associates measurement data with all sensors with geographic coordinates.

The motion is carried out by the kinematic model with prediction (Model predictive control - MPC). The traffic control functions are performed by the top-level logic when starting the route, an obstacle occurs on the route or the end point of the route is reached. At each step of the algorithm (with each new arrival of data from the GPS receiver), a solution is performed to find the optimal control parameters for the actuators when moving from the current waypoint to the direction of the next point. Thus, the model allows to predict the trajectory of the vehicle path taking into account the current position, speed and set of control commands. Solving the problem of finding the optimal control parameters allows to define a set of commands for actuators.

3. Testing of the unmanned traffic control system

The vision system aimed to recognize the objects of the road scene, was tested in real traffic conditions before being installed on the electric bus.

During the tests, the system has performed the detection and recognition of cars, pedestrians, a number of traffic signs and signals of traffic lights. The software system of the vision system is implemented using the developed application OCVStudio, based on the OpenCV [23] library. The initial data for the system is a frame from the optical sensor. It is processed through the neural network.

In the development and training of neural networks, the modified architecture of yolo v2 was applied. Also the prevention of retraining in the extended neural network, addition of filter layers, increase in the depth of convolution and the size of completely connected layers have been performed.

The user interface of the complex displays in real time the information about the presence of cars, pedestrians, a number of road signs, as well as traffic light's signals on the roadway. The system test frames are shown in Figure 5.

Figure 5. Test frames of the vision system.

The criterion for correct operation of the system was the accuracy of recognition of objects in the road scene. The calculated accuracy, which shows the number of correctly recognized road objects in relation to the total number of recognized objects was 78%.

In parallel with the tests of the vision system, experimental studies were carried out in order to determine obstacles on the unmanned vehicle route, to perform the possible localization, and also to plan the route at a specialized polygon of the GAZ Group. The frames of experimental research are presented in Figure 6.

Figure 6. Experimental studies of the unmanned traffic control system.

In order to assess the possibility of identifying obstacles on the way and generating a cloud of points on the road scene for further localization of found obstacles in this cloud of points, tests has been conducted with the LIDARs. Fragments of the tests are shown in Figure 7.

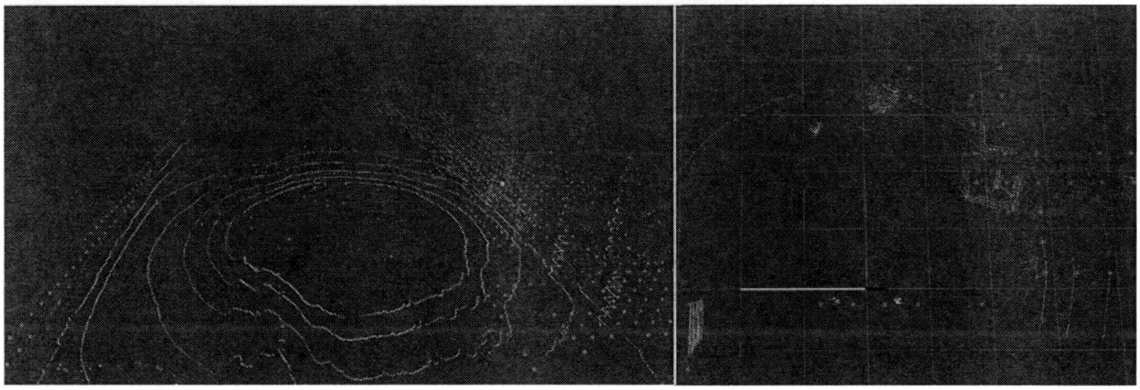

Figure 7. Fragments of tests with LIDARs.

The vehicle stopped when an obstacle in front of the vehicle higher than 15 cm and closer than 8 meters was detected. The motion began when there was no obstacle at a distance of 9 meters. Thus, movement and stopping on the obstacle signal is carried out through the trigger. This eliminates multiple trips to stop and move if the obstacle is located at the boundary of the detection zone.

Based on the results of the LIDAR research, it can be concluded that obstacles on the way are determined correctly and with high accuracy. Also the expediency of using the method of vehicle localization in space on the basis of LIDAR's data was confirmed. In the future, when driving an unmanned vehicle, the pre-compiled map will be compared to the real-time map, which is also formed with the help of LIDAR. Based on the results of this comparison, the localization of unmanned vehicle in the relative coordinate system will take place, where LIDAR is the center.

Planning and control of the motion along the trajectory of an unmanned vehicle was made on the basis of data from the navigation system mentioned above. The trajectory was entered by the user through the interface of the developed software "Tracker" with the display of the route of motion. The motion was carried out at given points with a feedback on the GPS / GLONASS receiver with RTK corrections from the base station (b) and control of the actuators via the MPC software. It was conducted more than 50 races on the route. The chosen route simulated an urban environment consisting of buildings close to each other, metal structures, trusses and pipes. The analysis of the navigation system was made after the test races and consisted of comparing the coordinates from the GNSS receiver, the IMU block and the reference pre-recorded trajectory, as well as monitoring of the signal quality metrics. The continuity and smoothness of the trajectory, the quality of the connection along the entire length of the route, the control of communication between the control units of the executive mechanisms and the on-board computer, the necessity and the quality of the inertial block operation has been determined during the races. A fragment of testing is shown in Figure 8.

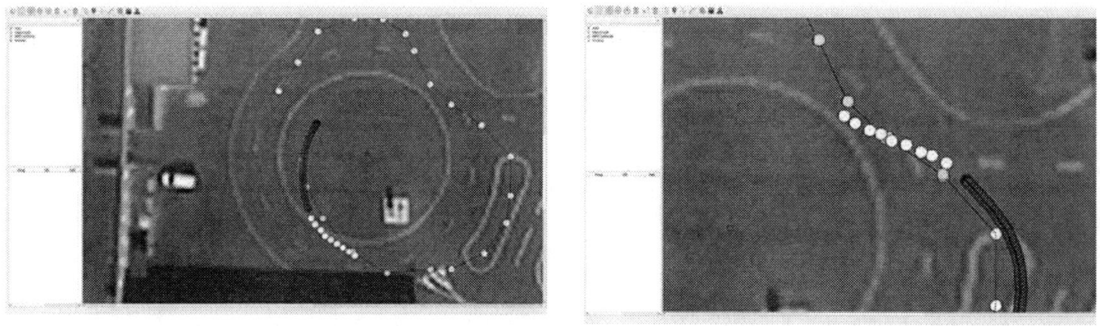

Figure 8. The user interface of the planning and traffic management program.

The quality of tracking of navigation satellites along the entire trajectory corresponded to the norm. The only exception was the intersection with the iron trusses and pipes, which caused a delay in updating the trajectory for 1-3 seconds. The signal quality of the network was satisfactory, however, in some places (especially at the beginning of the route), breakdowns of communication (irregular) were observed. It should be recommended to use Wi-Fi to transmit corrective information from the base station to the vehicle with an additional backup link. The distance between the points for optimal operation was about 2 meters along the entire route.

It should be noted that the navigation trajectory cannot be considered as an absolutely reliable source of information about positioning. Thus, this navigation system can be used in urban conditions, but without the presence of tunnels and other long discontinuities in communication with GNSS satellites. Based on the results of navigation system research, the deviation of the INS trajectory from the GNSS trajectory did not exceed 3 meters. The minimum deviation is 1.334 m, the maximum is 2.647 m. The wide interval between the minimum and maximum values arose due to the loss of communication with GNSS satellites for a long time (more than 10 seconds) due to interference in the traffic area (metal trusses and pipes) and low vehicle speed.

4. Conclusions

The structure of the unmanned vehicle traffic control system was developed, the hardware part was selected. Development of design documentation for various modules of the unmanned traffic control system, software and data transmission network, which unites all subsystems in a single network on the on-board computer, preparation of the electrical platform for the installation of unmanned traffic control systems has been carried out. Experimental studies of the modules of the unmanned traffic control system, namely the vision system for recognition of the road scene objects and the obstacle detection module on the path of the unmanned vehicle have been conducted. Based on the results of these tests, these modules have confirmed their operability.

To assess the positioning of the vehicle in space, the module for planning and controlling the trajectory of motion through satellite navigation has been tested. It should be noted that this navigation system can be used in urban conditions, but without the presence of tunnels and other long-term discontinuities in communications with GNSS satellites. To increase the accuracy of positioning the vehicle in space in the future, it is necessary to ensure joint work with diversification systems to improve the reliability of the system (technical vision, work with laser scanners, etc.). Some work will also be carried out in order to improve the quality of the output trajectory of the inertial system, and repeated testing is planned. In the future it is necessary to improve the quality of the output trajectory of the inertial block, and also to ensure joint work with diversification systems.

The implementation of this unmanned vehicle traffic control system, as well as the skills collected during the development, will contribute to the development of unmanned buses for public transport

IOP Conf. Series: Materials Science and Engineering **534** (2019) 012030 doi:10.1088/1757-899X/534/1/012030

routes. This, in turn, will significantly reduce the cost of transportation of passengers and time resources, increase economic efficiency, and minimize the number of road accidents.

Acknowledgements
This research done with the financial support from Ministry of Education and Science of the Russian Federation (the unique project identifier is RFMEFI57717X0268).

The experimental research conducted with the use of measurement equipment of the NNSTU Centre of collective using «Transport Systems».

References

[1] Bojarski M, Del Testa D, Dworakowski D, Firner B, Flepp B, Goyal P, Jackel L D, Monfort M, Muller U and Zhang J et al 2016 End to end learning for self-driving cars *preprint arXiv* 1604.07316
[2] Chen C, Seff A, Kornhauser A, and Xiao J, 2015 *Deepdriving: Learning affordance for direct perception in autonomous driving. In Proceedings of the IEEE* International Conference on Computer Vision pp 2722–2730
[3] Krizhevsky A, Sutskever I and Hinton G E 2012 Imagenet classification with deep convolutional neural networks. In Advances in neural information processing systems pp. 1097–1105
[4] Hane C, Heng L, Lee G H, Fraundorfer F, Furgale P, Sattler T and Pollefeys M 2017 3d visual perception for self-driving cars using a multi-camera system Calibration, mapping, localization, and obstacle detection. *CoRR, abs*/1708.09839
[5] Zhang J and Sanjiv Singh 2014 *LOAM: Lidar odometry and mapping in real-time* Robotics: Science and Systems Conference (RSS) July
[6] Autonomous car development company Waymo Retrieved May 10 2018 from: https://waymo.com/
[7] The Official NVIDIA Blog. Retrieved May 10, 2018, from https://blogs.nvidia.com/
[8] Baidu Retrieved May 10 2018 from: http://apollo.auto/
[9] DARPA. Retrieved May 10 2018 from: https://www.darpa.mil/
[10] Oshkosh Defense Retrieved May 10 201, from:https://oshkoshdefense.com/
[11] Lockheed Martin Retrieved May 10 2018 from:https://www.lockheedmartin.com/en-us/index.html
[12] SAE International Retrieved May 10 2018 from: https://www.sae.org/
[13] Yandex Taxi Unveils Self-Driving Car Project. Retrieved May 10 2018 from: https://yandex.com/company/blog/yandex-taxi-unveils-self-driving-car-project/
[14] Cognitive Technology Retrieved May 10 2018 from: https://www.cognitive.ru/
[15] Central scientific research automobile and automotive engine institute «NAMI». Retrieved May 10, 2018, from: http://nami.ru/
[16] KAMAZ PTC Retrieved May 10 2018 from: https://kamaz.ru/
[17] Avrora Robotics Retrieved May 10 2018 from: http://avrora-robotics.ru/
[18] YOLO Retrieved May 10 2018 from: https://pjreddie.com/darknet/yolo/
[19] Robot Operating Syste Retrieved May 10 2018, from: http://www.ros.org/
[20] Basler the power of sight Retrieved May 10 2018, from: https://www.baslerweb.com/ru
[21] Pergam Retrieved May 10 2018, from: https://www.pergam.ru/
[22] Xsens Retrieved May 10 2018 from: https://www.xsens.com/
[23] OpenCV Retrieved May 10 2018 from: https://opencv.org/

IOP Conf. Series: Materials Science and Engineering **534** (2019) 012031 doi:10.1088/1757-899X/534/1/012031

Study of the optimal reinforcing structure of the compressor wheel from composition material of the transport turbocharged engine

V M Fomin[1], D V Apelinskiy[2] and A N Netrusov[3]

Moscow Polytechnic University, 107023 Moscow, Bolshaya Semyonovskaya str., 38, Russian Federation

E-mail: [1]mixalichdm@mail.ru, [2]apelinskiy_mami@mail.ru, [3]a.netrusov@mail.ru

Abstract. The purpose of the study was to find the optimal configuration of the main parameters of the reinforcing layer of a compressor wheel made of composite material. The optimization problem for an anisotropic system was solved using the Gauss-Seidel method. According to the results of calculations, it was established that under the condition of preserving the strength of the wheel, there is the possibility of transition to reinforcing layers in which the fibers are oriented only in the circumferential direction, which makes it possible to significantly simplify the technology of its manufacture. The use of reinforcement for compressor wheels made of composite material can significantly reduce its mass and polar moment of inertia, which contributes to the improvement of strength and performance of the turbocharger and the engine as a whole.

1. Introduction

At present, there is a tendency to the almost universal application of pressurization systems of transport engines using turbochargers for this purpose. With all the known advantages of turbocharging systems, there are a number of problems associated with the further improvement of their performance. One of these tasks is to increase the strength and operational reliability of structural elements of the turbocharger, reducing the centrifugal load [1,2].

Another important task that requires its solution is the high inertia of the turbocharger rotor, which manifests itself in the period of changing the load mode of the engine. At the moment of increasing the cycle fuel supply to the engine, there is a discrepancy between the fuel supply and the turbocharger performance, which causes such a negative phenomenon as "turbo lag". Under these conditions, operational fuel consumption and exhaust gas toxicity increase, engine acceleration deteriorates over the whole range of unsteady and transient modes of its operation, which are most characteristic of any engine used for ground transport. It is known that the duration of these modes has a direct dependence on the moment of inertia of the turbocharger rotor.

Thus, it is necessary to search for opportunities to improve the performance of the turbocharger. First of all, it is improving the resource characteristics and reliability, as well as reducing the inertia of the rotor. This is one of the promising directions in the research practice of the transport engine-building industry.

One of the solutions in the direction of achieving the required qualities is the use of composite materials (CM) for the elements of the turbocharger rotor, which have a lower density, and, consequently, low inertia. Recently, a number of studies on finding such a solution have appeared in foreign and domestic research practice, generally showing its potential opportunities [3-6].

However, despite the large number of proposed design solutions for the development of impellers from CM and a wide range of CM [7, 8], at the moment there are no studies to improve the strength properties of such wheels. The key problem is the search for the optimal reinforcement structure of its matrix, which is one of the factors hindering the practical implementation of these solutions.

Thus, the further development of research to find the optimal reinforcement of impellers from CM is one of the actual directions in the field of creating high-performance turbochargers for transport engines.

2. Object and research methodology

According to the results of a previous study [4], the design of a compressor wheel was proposed. The compressor wheel is made of randomly reinforced composite material with a reinforcing layer in the form of several mono layers of continuous fibers. The search for the optimal configuration of the reinforcement of this wheel in order to identify the maximum indicators of its strength properties was the subject of this study. Taking into account the target orientation of the research, a complex optimization problem was posed, including the selection and description of variable parameters and target optimization functions, localization of intervals of allowable values of variable parameters.

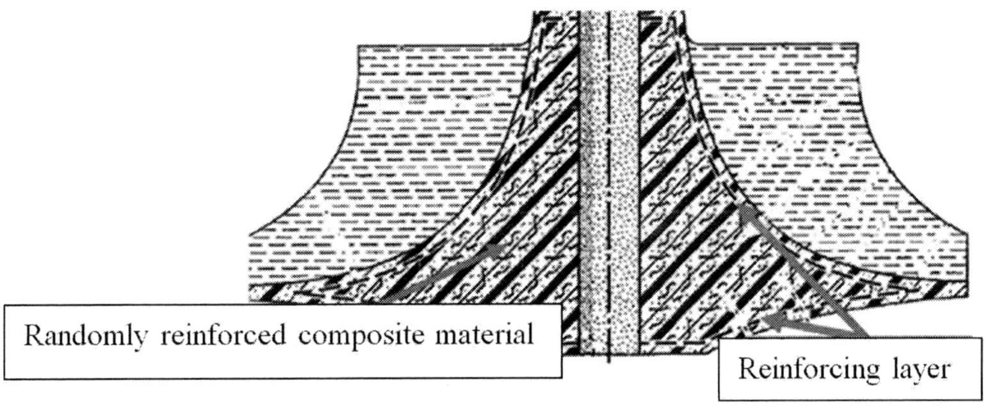

Figure 1. Scheme of the compressor wheel made of composite material.

The reinforcing layer is formed by alternately stacking monolayers in accordance with the chosen orientation scheme relative to each other. Each monolayer is a unidirectional composite material with continuous fibers with different schemes of laying (orientation). For traditional (metal) wheels of standard size with a diameter of 130 mm, the disk wall usually has a minimum thickness of 3.3 mm. Taking this into account, in the process of optimization, the range of variation within the limits was adopted: tmin = 0.2 mm...tmax = 1.4 mm. In this case, the thickness of the reinforcing layer should be comparable with the thickness of the outer layer of the base material, which will ensure the possibility of the formation of a blade crown on the disk. In general, for other sizes of turbochargers this range of variation may vary depending on the geometrical configuration of the wheels.

Polyamidoimide is used as the matrix of the composite (base material) in this study. Polyamidoimide, randomly filled with 30% carbon fiber, was used for the blade crown and the inner part of the disk. High-strength carbon fibers, which have high physical and mechanical properties temperature stability and low density, are adopted as reinforcing elements. The properties of the matrix and fiber materials are given in table 1 and 2. The subsequent calculation of the properties of the reinforced composite material, using the material properties of its components given in the tables, was carried out according to [7].

IOP Conf. Series: Materials Science and Engineering **534** (2019) 012031 doi:10.1088/1757-899X/534/1/012031

Table 1. Properties of polyamidoimide with randomly arranged carbon fiber (30%).

Property name	Value
Density ρ, kg/m3	1480
The modulus of elasticity E, GPA	16,5
Ultimate strength [σb], MPa	221
Poisson's ratio μ	0,39
Coefficient of linear thermal expansion α, 10-6 deg-1	9

Table 2. Properties of high strength carbon fiber.

Property name	Value
Density ρ, kg/m3	1700
The modulus of elasticity E, GPA	240
Ultimate strength [σb], MPa	3250
Poisson's ratio μ	0,2
Coefficient of linear thermal expansion α, 10-6 deg-1	-0,4

The study was conducted by the Gauss-Seidel method (coordinate descent method to find the extremum of the objective function). The problem of linear theory of thermoelasticity of anisotropic medium was solved for each combination of variable parameters. The solution to this problem is implemented by the finite element method in the Femap software package. All calculations are made for the maximum possible speed limit of the turbocharger. The nature of the thermal effect on the structure was taken according to the results of the study [2].

According to the results of optimization the comparative analysis of mass-inertial characteristics of turbocharger rotors with compressor wheels made of composite and traditional (metal) material was carried out.

3. The results of the study and their discussion

The solution of optimization problems is associated with the analysis of a large amount of data and results, which leads to the high complexity of their solution. This implies the need to reduce the number of less significant variable parameters and objective functions, localize the intervals of allowable values of the variable parameters.

Below is a description of the adopted variable parameters. The first of these is the share Vf of the fiber of the reinforcing layer in the composition of the composite material. This value for unidirectional composite materials can be from Vf min = 0.4 to Vf max= 0.8.

The second variable parameter is the fraction m of the thickness of the reinforcing layer, in which the fiber is oriented in the circumferential direction. The interval of variation of this parameter was taken from mmin = 0.5 to mmax = 1.

The third variable parameter is the total thickness t of the reinforcing layer.

In this study, General safety margin of the wheel is adopted as the target function, which for each combination of variable parameters is equal to:

$$n_0 = \min(n, n_w),$$

where $n = [\sigma_b]/\sigma_{eq}$ – the safety margin of the blade crown; [σb] - the ultimate strength of the base material of the matrix (table 1); σ_{eq} – equivalent stresses in the base material, determined in accordance with the Huber-Mises-Hencky energy theory of strength; nw – the safety margin of the reinforcing layer with continuous fibers according to the Mises-Hill criterion.

The Mises-Hill criterion was adopted because of its ease of use. The postprocessor Femap automatically determines the minimum value of the safety margin for all monolayers.

During the study, a preliminary assessment of the dependence of the compressor wheel safety factor on the number of splitting the total thickness of the reinforcing layer into monolayers was carried out. For this purpose, calculations were carried out with the following three wheel reinforcement schemes: 0/900, 0/900/0/900, 0/900/0, where 00 coincides with the circumferential direction in the disk.

It was found that the largest relative difference in safety margins between the second and third reinforcement schemes, determined in accordance with the Mises-Hill criterion, was 3%. On the basis of this, it can be concluded that from the point of view of the accuracy of the calculation there is no significant difference from which number of monolayers the reinforcing layer is formed. In the course of optimization in order to simplify the procedure of computational analysis, we can consider the minimum number of monolayers equal to 2.

Thus, the optimization condition will be as follows:

$$n_0(V_f, t, m) \to \max,$$

$$V_{f\min} < V_f < V_{f\max}, \quad m_{\min} < m < m_{\max}, \quad t_{\min} < t < t_{\max}.$$

Tetrahedral elements and elements of the laminate type were used for the calculation. As the boundary conditions, it was assumed that the compressor wheel was fixed on the surface of the shaft hole in the circumferential direction. In the axial direction - the wheel is stationary along the surfaces of its hub. The finite element model is similar to the model presented in [4].

In the course of the work, separate subprograms were written for performing some operations, such as creating a package of layers, assigning properties of elements, conducting a series of calculations, etc. This significantly reduces the time required for model preparation and calculations.

In parallel with the calculations, the error of their results was estimated. This evaluation was carried out using the Runge rule, based on a comparative analysis of this model, with a finite element model having eight times less nodes. It is established that the stresses arising in the wheel, the error does not exceed 10%.

The analysis of the behavior of the objective function is carried out in the Cartesian coordinate system of variable parameters by alternately "fixing" two of them.

The condition for terminating the optimization process is as follows:

$$\max\left(\frac{|V_{fk+1} - V_{fk}|}{V_{fk}}, \frac{|t_{k+1} - t_k|}{t_k}, \frac{|m_{k+1} - m_k|}{m_k}\right) \le \varepsilon,$$

where ε - the relative error of determining the variable parameters; k - the iteration number.

The permissible error of the values of the variable parameters for neighboring iterations in this study was assumed to be 0.05.

At the third iteration of the solution, the condition for terminating the optimization process was fulfilled. As a result, the vector of optimal values of the variable parameters Vf = 0.5, m = 0.75, t = 0.8 mm was obtained.

During the optimization process, it was determined that the maximum general safety margin of the wheel is 1.67. It is limited by the tensile strength of the material of the blade crown. This is due to the achievement of a reinforcing layer of certain stiffness (more in the circumferential direction), in which the disk material practically ceases to perceive the forces from the centrifugal load. In this case, only the reinforcing layer begins to work. The most stressed zone of the hole is unloaded, and the most vulnerable place is the root part of the blade fig.

The total safety margin of a wheel with a reinforcing layer of continuous fibers exceeds the safety margin of traditional aluminum wheels by 40-50%.

An additional measure aimed at increasing the strength of the wheel may be a decrease in the wall thickness of the disc. Reducing the amount of base material used wheel will reduce the "parasitic" mass and, accordingly, the centrifugal load, which loads the reinforcing layer and the blade crown. This event should be carried out after clarification the minimum required thickness of the outer layer.

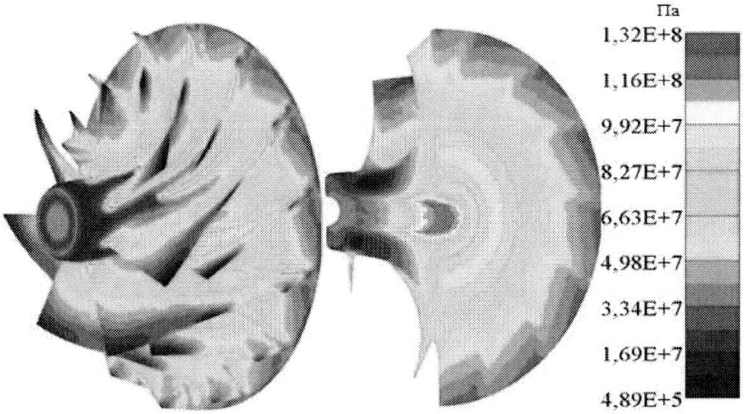

Figure 2. Color graphic diagram of generalized stresses in a compressor wheel made of composite material, with an additional reinforcing layer at optimal variable parameters: Vf = 0,5, m = 0,75, t = 0,8 mm.

Based on the analysis of the results, we can conclude that there is a possibility of increasing the level of manufacturability of the wheel by reinforcing only in the circumferential direction. In this case, the reduction in the overall safety margin will be less than 5%.

The maximum generalized displacement of a compressor wheel with a reinforcing layer is 30–40% lower than that of the aluminum analogue (Fig. 3). This fact allows us to conclude that the destruction of such a composite wheel with increasing operating speed will occur from exhaustion of the strength of the material, and not from contact with the stationary parts of the compressor housing. The values of the generalized displacements on the radius of the shaft hole are 0.038-0.045 mm, which is also less than the aluminum wheels by 30-40%.

It is also worth noting that the maximum generalized displacements of 0.324 mm are localized on the edges of the blades (Fig. 3). It should be noted that the components of the displacement along the z axis in this zone have negative values, which will not lead to contact of the wheel with the compressor housing. Then the determining movements will be radial ones, which do not exceed 0.277 mm on the edges of the blades, which is also less than the permissible limits (0.7 ... 0.8 mm).

IOP Conf. Series: Materials Science and Engineering **534** (2019) 012031 doi:10.1088/1757-899X/534/1/012031

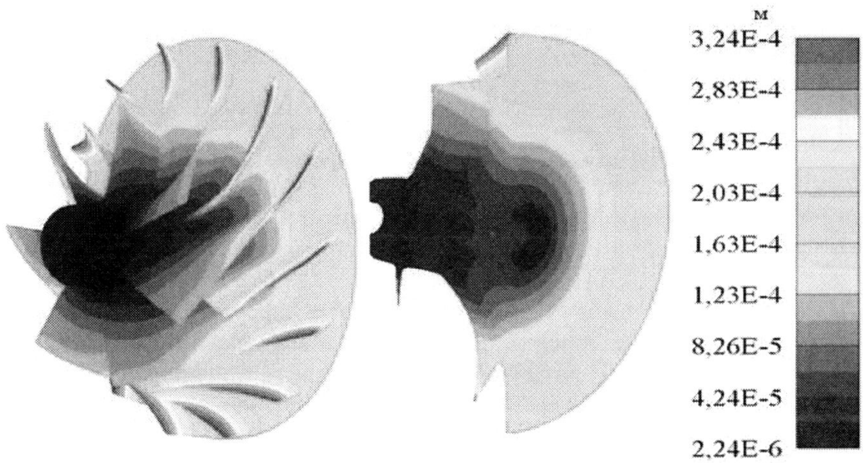

Figure 3. Color graphic diagram of the total displacements of the compressor wheel, made of composite material, with an additional reinforcing layer at optimal values of the variable parameters: Vf = 0.5, m = 0.75, t = 0.8 mm.

At the final stage of the study, a computational analysis was made of the mass-inertia characteristics of the compressor wheel from the composite material with an optimized reinforcing structure. Its mass was 0.266 kg, and the moment of inertia is 360·10-6 kg·m2. For comparison: the mass and the moment of inertia of the wheel of similar design made of aluminum alloy are 0.49 kg and 650·10-6 kg·m2, respectively. Thus, the use of composite material for compressor wheels allows reducing their mass and moment of inertia by almost two times: by 46% and 45%, respectively. The total mass and the moment of inertia of the entire rotor due to the use of such a wheel can be reduced by 10% and 16%.

4. Conclusion

According to the results of the research carried out on the search for the optimal reinforcing structure for the compressor wheel made from a composite material, the following conclusion can be drawn.

1. For the wheel reinforcement structure, the vector of optimal values of variable parameters was revealed: the fiber share of the reinforcing layer Vf = 0.5; the proportion of the thickness of the reinforcing layer m = 0.75; the total thickness of the reinforcing layer is t = 0.8 mm.

2. The total safety margin of the wheel with optimized reinforcing structure was equal to 1.67, which exceeds the safety margin of traditional aluminum wheels by 40-50%.

3. It has been established that while maintaining the strength of the wheel, it is possible to use reinforcing layers in which the fibers are oriented only in the circumferential direction, which can significantly simplify the technology of its manufacture.

4. The use of reinforcement for compressor wheels from composite material allows reducing its mass and polar moment of inertia almost twice, which helps to reduce the moment of inertia of the rotor by 16%. In general, the results of the study determine the obvious prospect of using composite materials with an optimized reinforcement structure for the compressor wheels of turbochargers, associated with the possibility of a significant reduction in the inertia of their rotors while maintaining the acceptable strength properties. The consequence of this is the improvement of the level of operational qualities of the turbocharger (reduction of inertial loads and increase of the service life, improvement of injectivity during transient modes of operation, etc).

IOP Conf. Series: Materials Science and Engineering **534** (2019) 012031 doi:10.1088/1757-899X/534/1/012031

References

[1] Kaminskij V N, Kaminskij R V, Grigorov I N, Teregulov T I and Sibiryakov S V 2016 The calculation of the stress-strain state of the compressor wheel of the turbo-compressor of the transport engine *Naukograd* **4**(10) pp 46-51

[2] Lushchenko V A 2014 *Formation and quality assurance of the bearing assembly of a turbocharger of a highly accelerated automotive diesel during the design and development phase* (Naberezhnye Chelny FGAO VPO) p 121

[3] Netrusov A N and Fomin V.M. 2017 Investigation of the strength properties of impellers made of composite material of diesel engines supercharging units *Traktory i sel'khoztekhnika* **2** pp 21-28

[4] Netrusov A N and Fomin V M 2017 Comparative analysis of the characteristics of the turbocharger rotor from composite and traditional materials *Naukograd* **2**(12) pp 66-73

[5] Netrusov A N and Fomin V M 2017 Composite materials for impellers of automotive and tractor turbo-compressors *Traktory i sel'skohozyajstvennye mashiny* №8 pp 28-36

[6] Hommes D J and Williams C E Composite centrifugal compressor wheel (United States Patent № US 8,794,914)

[7] Karpinos D M 1975 Composite materials Directory (Kiev) *Naukova dumka* p 592

[8] Pyatov I S, Hiboev O V, Buzinov V G, Makarov A R and other 2014 Carbon materials for parts of gas turbine engines and internal combustion engines, problems and prospects *Izvestiya MGTU «MAMI»* №4(22) **1** pp 55-60

IOP Conf. Series: Materials Science and Engineering **534** (2019) 012032 doi:10.1088/1757-899X/534/1/012032

Management strategy of new energy-saving technology in the field of ground transport based on the principle of thermo-chemical regeneration of waste heat energy

V M Fomin[1,3], **V F Kamenev**[2,4] **and D V Apelinskiy**[1,5]

[1] Moscow Polytechnic University, 107023 Moscow, Bolshaya Semyonovskaya str., 38, Russian Federation
[2] FSUE "NAMI", 2 Avtomotornaya St., 125438, Moscow, Russian Federation

E-mail: [3]mixalichdm@mail.ru, [4]kamovf@mail.ru, [5]apelinskiy_mami@mail.ru

Abstract. The article describes the strategy of energy saving management in the field of operation of ground transport, developed on a fundamentally new basis for the effective transformation of the energy potential of waste heat energy. In the course of the work, the energy technological scheme is investigated and proposed based on the organizati on of thermochemical conversion of the initial hydrocarbon fuel using waste heat into a new type of synthesized fuel with higher rates of heat of combustion. The mechanism of this effect has not yet found its detailed consideration in the modern theory and practice of power engineering. It is based on the law of energy conservation, which in thermochemistry is interpreted as the law of Hess and its consequences.

1. Introduction

The energy carried away with the exhaust gases of the power plant of the vehicle determines the share of chemical energy of the fuel that is not realized in the engine duty cycle of the power plant. If we take into account that in some cases the heat loss with exhaust gases amounts to 50% of the chemical energy of the consumed fuel, then this aspect of the problem of fuel and energy savings deserves close attention.

It seems advisable to use some of this heat to minimize the thermal losses of the duty cycle of the power plant. One of the promising areas associated with the use of secondary heat of waste gases is the use of recycling tools. Heat-exchange regenerators and recuperators are used in the existing technological schemes for utilization of the energy potential of waste gases from stationary and aviation power plants. There are methods to regenerate heat, used in large stationary installations, based on the use of waste heat boilers or other means [1, 2]. All these technological methods of waste heat utilization and high mass-dimensional indicators of technical means used for its implementation, limit the possibilities of their practical application on ground vehicles.

This determines the feasibility of developing, on a fundamentally new technological basis, more acceptable for the transport system of heat utilization. Work towards finding acceptable utilization tools for power plants of ground vehicles, a priori, should be developed along the path of creating efficient and most compact systems. In general, for the effective conversion of the energy potential of the waste heat of the power plants into useful work, it is desirable to create conditions under which the greatest part of this heat could be reused in the duty cycle of the power plants, extremely minimizing thereby thermal losses of a cycle and, as a result, causing economy of the consumed fuel.

2. The principle of thermochemical recovery of waste heat

Taking into account this target orientation, an energy technology scheme based on the ability of some hydrocarbons to convert into a new type of energy carrier with a higher energy potential relative to the initial product has been investigated and proposed. Such a scheme has the most important advantages over other recycling schemes. During the thermochemical conversion of the initial hydrocarbon product, a mixture of gases is obtained: carbon monoxide and hydrogen. This is the so-called "synthesis gas", with higher calorific value index. The mechanism of this effect has not yet found its detailed reflection in the modern theory and practice of power plant engineering. Therefore, we consider this question in more detail.

From the point of view of energy expediency, it is desirable to organize the conversion at a relatively low temperature level, using a compact conversion structure with a catalytically active medium for this, which will allow to use exhaust gases as a heating coolant, that is, to organize the conversion process directly on board vehicles without additional heat sources.When using a converted gaseous product as a fuel with high calorific value, energy and effective performance and its fuel efficiency are increased. In addition, the presence in the fuel mixture of hydrogen, which is the activating reagent of the combustion process, can contribute to the improvement of the environmental qualities of the power plant [4].

Such power recycling scheme allows you to use of a number of commercial hydrocarbons, including alternative, able to implement efficient reactions of thermal catalytic conversion with the release of an energy-intensive target product. Thermal catalytic conversion is characterized not only by a high degree of utilization of waste heat, but also by improved kinetic and toxic combustion parameters of the produced fuel - synthesis gas. In modern energy-saving transport technologies, such efficient and compact systems have not yet found their practical application.

Since the proposed version of the energy technology scheme for utilization of the secondary energy resources of the power plant implies the need to organize the process of endothermic conversion (thermochemical conversion) of the initial energy carrier, it was named as *thermochemical regeneration of waste heat* (TRWH).

The concept of this type of waste heat utilization of the power plant has been developed on the basis of the fundamental principles of thermochemistry and can be implemented under the operating conditions of any type of power plant of the vehicle. It is based on the law of energy conservation, which in thermochemistry is interpreted as the Hess law and its consequences.

According to this law, the efficiency of thermochemical waste heat recovery directly depends on the chemical nature of the initial energy carrier, which determines the energy index of the endothermic effect of the reaction system of its conversion. It is clear that this indicator is equivalent to the amount of heat taken from an external source (heating coolant) and expended for the conversion process, and going to increase the energy potential of the target product of conversion reactions (synthesis gas).

We illustrate this with the example of the conversion of alternative alcohol fuels, in particular, methyl alcohol, which has been used for a relatively long time as a partial substitute or the main type of fuel for transport [5]. The heat of combustion of methanol $H_{u(m)}$ = 19670 *kJ/kg*.

The heat of combustion of the gaseous products of its thermochemical conversion (synthesis gas) is H_u^K = 23870 *kJ/kg*. Thus, during the combustion of 1 *kg* of these synthesized products obtained from the same mass of liquid methanol, additional thermal energy is released, accumulated during the endothermic conversion of alcohol fuel, equal to $H_u^K - H_{u(m)}$ = 4200 *kJ/kg*. Therefore, in real conditions of the operating cycle, 21.4% of the available energy of the exhaust gases utilized for organizing the endothermic conversion cycle of 1 *kg* of methanol can be returned to the operating cycle and converted into additional work.

3. Efficiency of thermochemical regeneration of waste heat

The use of pre-chemical fuel conversion systems can improve the efficiency of its use, but under certain restrictions. The degree of regeneration depends on the temperature of the conversion process and increases with its decrease. Modern catalysts allow to realize the process of hydrocarbon conversion at certain operating temperatures (usually not below 280- 300°C [6]), thereby determining the minimum possible temperature- limit of the exhaust gases, at which the organization of this process is feasible. If the temperature level of the exhaust gases is lower than the conversion operating temperature, an effective thermochemical regeneration of waste heat process is practically not feasible. Implementation of the TRWH effect at lower temperatures may be feasible in the future, if appropriate low-temperature catalysts are developed in the field of chemical technologies that can initiate conversion reactions of hydrocarbon compounds in such conditions.

The energy technological system for utilization of the secondary energy resources of the power plant, being a kind of "thermochemical adsorber", takes away part of the waste thermal energy and returns it to the power plant for re-participation in the operating cycle (additional work). Therefore, in general, the thermodynamic cycle of this power plant can be considered as *regenerative*.

In the thermodynamic cycle of a traditional power plant (without a regeneration system), the heat Q_1 brought to the working fluid, resulting from the combustion of the fuel, is not fully used. Part of this heat Q_2 is removed from the working fluid and transferred to the "cold body", causing that part of the chemical energy of the fuel that was not implemented for conversion into work. It is known [1] that the thermal efficiency of a cycle for such power plant is defined as:

$$\eta_t = \frac{Q_1 - Q_2}{Q_1} = 1 - \frac{Q_2}{Q_1}$$

In the working cycle with TRWH, part of the heat Q_2 is used to organize the process of endothermic transformation of the initial fuel to increase its energy characteristics: $Q_p = w \cdot Q_2$ (w - degree of regeneration), while the other part (Q_2 - $w \cdot Q_2$) will be transferred to the "cold body", thereby causing the level of irreversible heat loss in the cycle.

Then the indicator of the efficiency cycle with TRWH will be defined as

$$\eta_{t_{TXPOT}} = \frac{Q_1 - (Q_2 - w \cdot Q_2)}{Q_1} = 1 - \frac{Q_2 \cdot (1-w)}{Q_1}$$

It is easy to see that the value of the efficiency index of the cycle with TRWH will be higher than the thermal efficiency of the operating cycle of the power plant without regeneration, and increases with the growth of the degree of regeneration w.

The principal possibility of the effective implementation of TRWH in the operating cycle of the power plant, like any heat engine, will be determined by a number of conditions:

- the initial hydrocarbon product used for the conversion must be able, under the influence of the heat of exhaust gases (if there is an appropriate catalytic medium), to enter into conversion reactions with a pronounced endothermic effect;
- the available energy capabilities of the effluent gas flow from the power plant should correspond to the energy costs for compensation of the endothermic effect of the conversion reaction of the initial hydrocarbon product;
- the composition of the conversion products must reach the state of thermodynamic equilibrium, which is realized when the temperature state of the exhaust gases of the power plant corresponds to the level necessary for the efficient flow of the conversion process.

It should be noted another interesting in the scientific and practical sense aspect of the use of TRWH. With the appropriate organization of the thermochemical conversion of fuel, there is a unique opportunity to regenerate the entire energy potential of the off-gas flow. In addition to the thermal component of this potential discussed above, there is also a chemical component, which is caused by the content in the exhaust gases of incomplete combustion products whose chemical energy due to the imperfection of the combustion process has not been realized in the operating cycle of the power plant. It is possible to regenerate this component of the energy of gases by first passing them through the catalytic medium in order to oxidize the products of incomplete combustion of fuel (*CO, CH,* etc.). The heat released during the exothermic oxidation of these products contributes to an additional increase in the effect of TRWH.

The priority (RF patent No. 2249807) idea of complex regeneration of chemical and thermal energy of exhaust gases of power plant, is reflected in the development of a highly efficient sample of a methanol conversion reactor [7], combined with a catalytic chamber for oxidation of products of incomplete combustion of fuel. The functionality of such a combined apparatus allowed, along with an increase in the degree of regeneration of heat energy from the exhaust gases, to reduce the content of *CO* and *CH*, that is, to improve the environmental quality of the power plant.

According to the results of a series of mathematical studies, the practical implementation of TRWH has been thoroughly studied and carried out in real conditions of the operating cycles of transport power plants. Compact thermocatalytic reactors have been developed, which according to their characteristics are coordinated with the operating conditions of the power plants for ground vehicles. Experimental testing of the effectiveness of TRWH was carried out for power plants (heat engines) operating on alcohol fuel in two thermal power cycles: the cycle with mixed supply of heat (diesel) and the Otto cycle (spark ignition engine) [8,9,10].

For the engine with the first variant of the organization of the working cycle, an increase in the energy efficiency of fuel due to the use of TRWH in the range of 4.5 ... 8.5% was registered [9,10]. The greatest value of the effect corresponded to the modes of high power. Along with the reduction in fuel consumption, due to the fact that the composition of the mixed fuel (synthesis gas) includes the hydrogen-containing component, the improvement of the environmental qualities of the engine was recorded: the content of soot in the exhaust gases decreased by almost 2 times, nitrogen oxides to 16%.

For an engine operating on the Carnot thermal power cycle, the energy-saving effect for individual modes reached up to 15% [8] with a significant improvement in its environmental qualities.

The established manifestation of the energy-saving and ecological effects of the onboard system TRWH makes it relevant for its application for the power plants of ground vehicles.

4. Conclusion

In the course of research on finding solutions to the problem of fuel and energy savings in transport, the following results have been obtained that are of scientific and applied importance.

4.1. Practical significance

Multi-million vehicle complexes of megacities emit billions of joules of thermal energy into the environment, which is irretrievably lost. The use of the proposed energy technology system TRWH will allow utilizing a significant part of this energy to convert it into useful work. This will lead to tangible savings in fuel resources in the field of energy consumption of transport. Due to the technological simplicity of the onboard organization of the conversion process, it can be implemented under the operating conditions of any type of ground vehicles power plants with minimal production and financial costs.

By its influence on the performance of the operating cycle, the TRWH energy technology system has the property of multifunctionality. Along with increased energy savings, it provides the possibility of using environmentally friendly alternative types of fuel. Thus, it allows solving important social problems of energy supply, conservation of traditional fuels and environmental safety in transport.

4.2. Scientific significance

In the course of theoretical research, a number of scientific statements have been formulated, which to date have not yet found their detailed reflection in modern heat power engineering.

1) A statement has been formulated on the possibility of complex thermochemical energy regeneration of the exhaust gases, including the thermal and chemical components of the energy of the gas flow. A similar scheme of the proposed version of the energy technology utilization of secondary energy resources of the power plant in modern research practice has not been studied in detail until now.

2) The principle of functioning of the proposed energy technological system TRWH is based on the principle reflecting only in general form the essence of the classical regenerative cycle of a heat engine. However, its "thermochemical scenario" includes an additional structural stage of the transformation of the chemical energy of the initial fuel, which causes its individual epistemological difference from the classical understanding of the regenerative cycle. In the field of modern research

and technology of power engineering this type of regenerative cycle has not yet found its real reflection.

References

[1] Orlina A S, Kruglova M G 1983 Internal combustion engines. Theory of piston and combined engines (Moscow) *Mashinostroenie* 372 p

[2] Nosach V G, Pereletov I I 1977 Methods to improve fuel efficiency in technological processes *Teplofizika i teplotekhnika* **37** pp 44-47

[3] SHejpak A A, Baldin V P 1985 Utilization steam turbines of autotractor internal combustion engines *Avtomobil'naya promyshlennost'* **12** pp 12–14

[4] Fomin V M, Platunov A S 2011 Hydrogen as a chemical reagent to improve the performance of an automobile engine *Transport na al'ternativnom toplive* **4**(22) pp 30-37

[5] Zvonov V A, CHernyh V I, Zaigraev L S 1995 Feasibility and environmental performance indicators of methanol as a fuel for internal combustion engines *JEkotehnologija i resursosberezhenie* **4** pp 5-8

[6] Fomin V M and Makunin A V 2009 *Thermo chemical recovery of heat contained in exhaust gases of internal combustion engines (a general approach to the problem of recovery of heat contained in exhaust gases)* Theoretical foundations of chemical engineering **43**(5) pp 834-40

[7] Fomin V M, Kamenev V F, Kornilov G S *Engine operation method* (Patent RF №2249807)

[8] Fomin V M, Kamenev V F, Hripach N A 2003 *Improving the efficiency of the internal combustion engine using thermochemical heat recovery of exhaust gases* Problems of energy storage and ecology in mechanical engineering INMASH RAN pp 156-170

[9] Fomin V M, Kamenev V F, Hripach N A 2005 *Theoretical and experimental studies of the work of the internal combustion engine in hydrogen-diesel mixtures* International Scientific Journal for Alternative Energy and Ecology ISJAEE **7** pp 32-42

[10] Fomin V M, Atrash Rami 2012 Improving the performance of a diesel engine on binary biohydrocarbon fuel *Transport na al'ternativnom toplive* **5**(29) pp 36-40

IOP Conf. Series: Materials Science and Engineering **534** (2019) 012033 doi:10.1088/1757-899X/534/1/012033

Planning of driverless vehicle motion trajectory taking into account its dimensional characteristics and under conditions of dynamically changing environment

A Zabolotny

FSUE "NAMI", 2 Avtomotornaya St., 125438, Moscow, Russian Federation

E-mail: alexx_20@mail.ru

Abstract. This paper presents the solution for the task of the planning of the driverless vehicle (DV) driving trajectory for the modern road conditions. The rapidly-exploring random trees (RRT) algorithm consisting in the generation of the treetops under the equiprobability law and search for an optimum trajectory was used as a base. The minimum length acts as a criterion of optimality. The condition for the trajectory generation using the road map model is added to the new algorithm. Thereby, the problem of low processing (response) speed of the optimum trajectory search algorithm was solved.

1. Introduction

Works on the creation of the driverless vehicle began in the 70s of the XX century in Japan. The project "Prometheus" based on the developments of the German scientist Ernst Dickmans acted in the late 80s - early 90s, during which studies were conducted on the management of the driverless vehicles [1]. Work on the creation of the driverless technology of the last century were more experimental than practical. Most of the cars were either remote control prototypes or implemented simple algorithms. Currently, you can monitor the activity aimed at the creation of the driverless vehicles [2]. Such companies as Google, Tesla Motors, Apple, General Motors, Volkswagen, Audi, BMW, Volvo, Nissan and others make the greatest efforts in this field.

The rapid progress in this area is mainly due to the following factors:

- the advent of global navigation systems GPS / GLONASS and the widespread use of small GPS receivers;
- the increase of the productivity of modern computers;
- evolution of vision systems.

Modern technologies allow to create systems with high localization accuracy. Much has already been done in the area of unmanned vehicle architecture design [3].

Thus, the task of a higher level, the synthesis of control algorithms, comes to the fore. Namely - trajectory planning algorithms and algorithms for following it. This article is devoted to the issues related to the construction of the trajectory of movement on a well-known map.

Traditionally, the task of planning a trajectory is solved in two stages: the construction of a graph and the search for a path according to a given criterion in this graph [4]. Existing methods for finding a path on a graph in most cases are modifications of E. Dijkstra's algorithm, which is based on an iterative traversal of nodes of this graph [5]. Usually they can be divided into three categories: exhaustive algorithm, heuristic search, a ray-casting and tracking algorithm. Poor performance is a huge disadvantage of all these algorithms. This parameter is critical for the driverless vehicle in modern conditions of a dynamically changing environment.

This article proposes a modified trajectory planning algorithm based on the method of rapid research using a random tree (RRT). The idea of the RRT algorithm is to form a tree that would evenly explore the search space [6]. The search space is represented as a map in the form of a grid of size M × N, where M is the number of rows, and N is the number of columns [7]. For each node of the grid there is a logical sign of the path possibility.

Distinctive features of the proposed algorithm are:

1. The search space is defined as a vector model of a road map.
2. In the process of trajectory search, limitations imposed by the geometric dimensions of the DV are taken into account.
3. The law of generation of treetops is proposed taking into account the kinematic characteristics of the DV, which allows effectively to explore the search space.

Thus, the trajectory planning time is significantly reduced. As a result, this ensures the non-stop movement of the DV and the avoidance of obstacles.

2. Formulation of the problem

There is a mathematical model of a roadmap, which is described in a two-dimensional Cartesian coordinate system XOY. The boundaries of the roads are described as a sequential set of intervals $\overline{l_\iota} = \{x_{ni}, y_{ni}, d_i, \theta_i\}$ and arcs of circles $\overline{c_\iota} = \{x_{0i}, y_{0i}, R_i, \alpha_i, \varphi_i\}$, the tangents of which are $\overline{l_\iota}$ (Fig. 1),

where i – the object number of the original sequence, organized in such a way that the starting point of the i-th object coincides with the final (i-1)-th.

$x_n \in XOY,\ y_n \in XOY$ – coordinates of the point A_{ln}, which is the "beginning" of the lines segment.

$d \in |\mathbb{R}|$ – line length.

$\theta \in [0; 2\pi]$ – cut angle of line (relative to axis OX).

$x_0 \in XOY,\ y_0 \in XOY$ – coordinates of the circle center A_o.

$R \in |\mathbb{R}|$ – radius of circle.

$\alpha \in [0; 2\pi]$ – the orientation angle of the radius vector drawn from the center of the circle to the "beginning" of the arc A_{cn} (relative to axis OX).

$\varphi \in [-2\pi; 2\pi]$ – angular magnitude of the arc of a circle.

Thus, the map of the road Map can be represented as a set union of n line and m arcs of circles.

$$Map = \left(\sum_{j=0}^{n} \overline{c_\iota}\right) \cup \left(\sum_{j=0}^{m} \overline{l_\iota}\right) \quad (1)$$

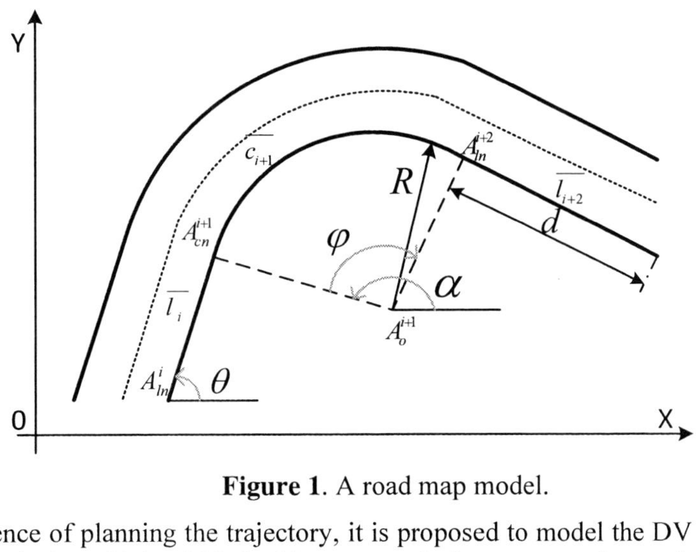

Figure 1. A road map model.

For the convenience of planning the trajectory, it is proposed to model the DV as a circle of radius $r = 0.5 \cdot H$, where H is the vehicle width. In this case, to take into account the restrictions when planning

IOP Conf. Series: Materials Science and Engineering **534** (2019) 012033 doi:10.1088/1757-899X/534/1/012033

a trajectory, you can apply the method of expanding impassable zones by an amount corresponding to r [8]. In this case, the DV can be considered as a material point. In [9], it was shown that the time-optimal path between the initial state vector $\overline{p_s} = \{x_s, y_s, \vartheta_s\}$ and a terminal state vector $\overline{p_f} = \{x_f, y_f, \vartheta_f\}$ consists of an arc of a circle $\overline{R_{ds}} = \{\overline{p_s}, O_{cs}, T_{ex}, r_{ds}, \varphi_{ds}\}$, followed by the line segment $\overline{l_d} = \{T_{ex}, T_{en}\}$, and ends with another arc of the circle $\overline{R_{df}} = \{\overline{p_f}, O_{cf}, T_{en}, r_{df}, \varphi_{df}\}$ (Fig. 2). Where the radii of these arcs r_{ds} and r_{df} are determined by the kinematics of the DV. The trajectory of this type is called the Dubins trajectory.

The previously used notations are described here

$x_s \in XOY$, $y_s \in XOY$ – coordinates of the starting point of the DV A_s.

$\vartheta_s \in [0; 2\pi]$ – initial orientation angle DV (relative to axis OX).

$x_f \in XOY$, $y_f \in XOY$ – coordinates of the desired position A_f.

$\vartheta_f \in [0; 2\pi]$ – terminal orientation angle (relative to axis OX).

$O_{cs}(x_{cs}, y_{cs}) \in XOY$ – center of the first circle.

$O_{cf}(x_{cf}, y_{cf}) \in XOY$ – center of the second circle.

$T_{ex}(x_{ex}, y_{ex}) \in XOY$ – line start point.

$T_{en}(x_{en}, y_{en}) \in XOY$ – straight line end point.

$\varphi_{ds} \in [-2\pi; 2\pi]$, $\varphi_{df} \in [-2\pi; 2\pi]$ – angular values of the first and the second arcs of the trajectory circles.

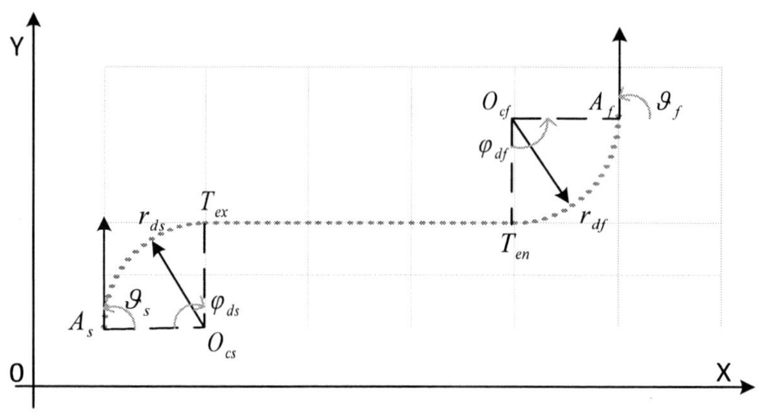

Figure 2. Dubins trajectory.

Dubins generalized trajectory $D \in XOY$ can be represented

$$D = \left(\overline{R_{ds}} \cup \overline{l_d} \cup \overline{R_{df}}\right) \tag{2}$$

The trajectory is considered acceptable if the vehicle, moving on it, does not touch the lets and does not go beyond the limits of the carriageway. For ease of calculation, each obstacle Let_j is conveniently described by a circle with a radius r_{Lj} and center coordinate $O_{Lj}(x_{Lj}, y_{Lj})$, where $j = 1..N$, and N is the total number of obstacles. Taking into account (1), (2) and the writing above, and also on the condition that the DV is located on the carriageway at the beginning of the movement, the criterion for the admissibility of the trajectory D is

$$\left(D \cap \left(\cup_{j=1}^{N} Let_j\right)\right) \cup \left(D \cap (Map)\right) = \emptyset \tag{3}$$

The task of planning a trajectory is to find such an optimal along the length of the trajectory D such that criterion (3) is satisfied. The RRT algorithm is used as the base.

3. Trajectory planning algorithm

The DV technical vision system with a specified time interval searches for obstacles Let_j around the robot. A comprehensive navigation system determines the state vector of its current position in the global space $\overline{p_{tek}} = \{x_{tek}, y_{tek}, \vartheta_{tek}\}$.

The fulfillment of the initial conditions is necessary for the correct operation of the algorithm.

- An initial trajectory $T = \sum_{i=0}^{n} D_i$ is given, which consists of a set of Dubins n-trajectories (2). T is located on the roadway and connects the start and finish points.
- DV is located at the starting point and begins to move along the trajectory at the initial moment.

The criterion (3) is checked in the process of movement when new obstacles appear. If it is not true, then the trajectory planning algorithm is launched. It represents the following sequence of actions:

1) Transformation of the search space for the condition that DV is described as a circle with a radius r.
2) Selection of the terminal state vector $\overline{p_{end}} = \{x_{end}, y_{end}, \vartheta_{end}\}$ for returning the control object from the current position to the initial path T.
3) Generation of treetops A_i ($i = 1..N-1$, where N-1 is a total number of them) according to the law, taking into account the kinematic properties of the DV and the map Map.
4) Formation of a set of trajectory prototypes $Tops$ as the treetops sequences.
5) Construction of the Dubins trajectories (2) between adjacent points A_i и A_{i+1} for each set of treetops $Tops[j]$.
6) Trajectories research and selection of optimal.

3.1. Search space conversion

There are two stages to convert the search space, taking into account the overall characteristics of the DV. Expansion of impassable zones due to the presence of obstacles occurs at the first stage. That is, the magnitude of the obstacle radius r_L is increased by r (half width of DV). Narrowing of the roadway Map occurs in the second stage. The narrowing of the carriageway consists in a parallel transfer of its $Left$ and $Right$ lines l_i by the value of r, and in increasing or decreasing the radius R_i by the value of r (Fig. 3).

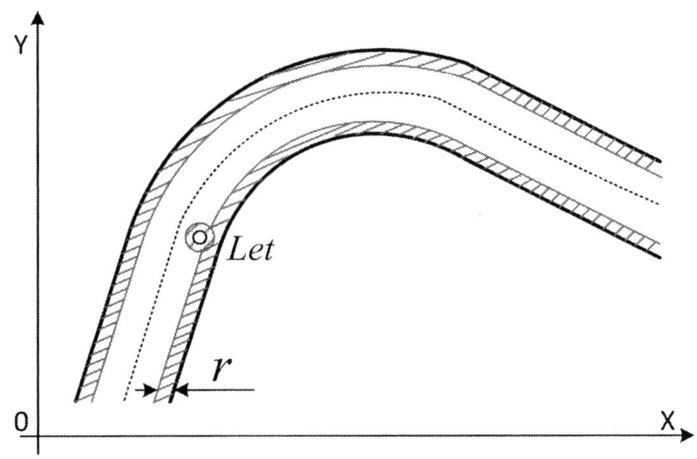

Figure 3. Changing the area of the drivable possibility.

For the $Right$ line, the transformation is $\overline{l_i} = \{x_{ni} - r \cdot \sin\theta_i, y_{ni} + r \cdot \cos\theta_i, d_i, \theta_i\}$.

For $Left$ is $\overline{l_i} = \{x_{ni} + r \cdot \sin\theta_i, y_{ni} - r \cdot \cos\theta_i, d_i, \theta_i\}$.

The calculation of the change in R_i is determined from the condition K (4). This condition is formed on the basis of the geometric meaning of the vector product. It characterizes the direction of rotation of the roadway. If the condition is true, then the roadway changes its direction of travel to the right, otherwise - to the left.

$$K: (\cos(\theta_{i-1})(y_{0i} - y_{ni-1}) - \sin(\theta_{i-1})(x_{0i} - x_{ni-1})) < 0 \tag{4}$$

Based on the above

$$\begin{cases} Right: \begin{cases} if\ K = true, \bar{c_i} = \{x_{0i}, y_{0i}, R_i + r, \alpha_i, \varphi_i\} \\ else \quad \bar{c_i} = \{x_{0i}, y_{0i}, R_i - r, \alpha_i, \varphi_i\} \end{cases}; \\ Left: \begin{cases} if\ K = true, \bar{c_i} = \{x_{0i}, y_{0i}, R_i - r, \alpha_i, \varphi_i\} \\ else \quad \bar{c_i} = \{x_{0i}, y_{0i}, R_i + r, \alpha_i, \varphi_i\} \end{cases}. \end{cases} \tag{5}$$

Thus, DV can be considered as a material point.

3.2. Selection of the terminal state vector

The task of the planning algorithm is to find such a trajectory that would go around the obstacles and return the DV to the original trajectory T. At least two conditions must be met for the desired trajectory to exist. The first is the absence of obstacles in the terminal point area. Secondly, the terminal point must be in sight of the vision system. The orientation angle ϑ_{end} must be equal to the angle of inclination of the trajectory T at the selected point.

3.3. Generate treetops

The following principles are observed when synthesizing the law of treetops generation:

1) uniform study of the search space;
2) full exploration of the search space;
3) quick change of treetops;
4) the ability to take into account the kinematics of the DV motion.

It is proposed to set the treetops in the form of a "grid". The nodes of this "grid" are the treetops. Their schematic layout is shown in Figure 4. The location of the "grid" is determined by the starting point A_1 with coordinates x_{1_gr}, y_{1_gr}. Any other tops $A_i(x_{i\_gr}, y_{i\_gr})$ are easily determined according to (6).

$$\begin{cases} x_{i\_gr} = x_{0\_gr} + k\Delta x_{gr} + b\Delta y_{gr} ctg\theta_{gr} \\ y_{i\_gr} = y_{0\_gr} + b\Delta y_{gr} \end{cases}, \tag{6}$$

where $k = 1..n$, $b = 1..m$, a n и m are number of columns and rows respectively.

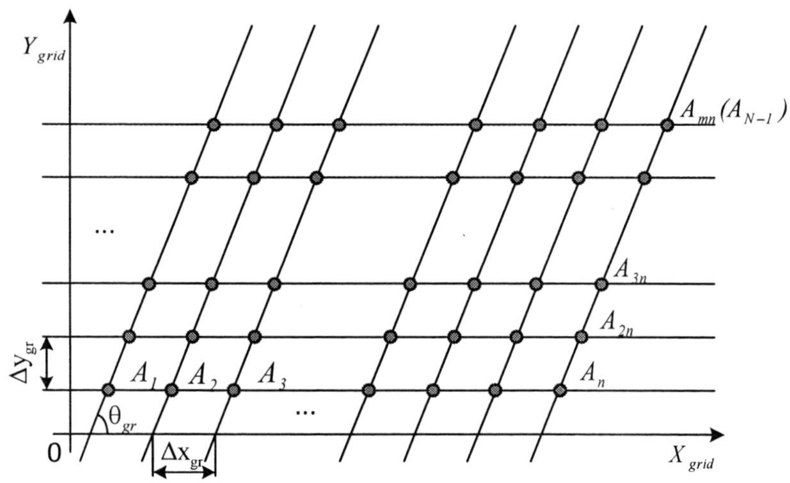

Figure 4. Treetops.

$\Delta x_{gr}, \Delta y_{gr}$ are the parameters that define the "grid" step. They are selected based on the kinematics of the DV motion. They must be such that the DV can move between adjacent points. Changing of the treetops location is done by setting of the angle of the "grid" θ_{gr}. There is a need to search for a trajectory from a treetops set that are on the roadway, for time-optimal a work of the algorithm (Fig. 5). Condition (3) can be used to find these treetops.

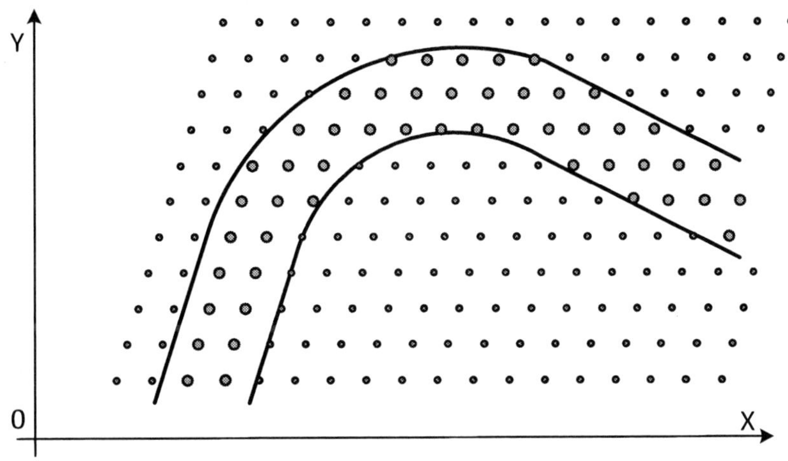

Figure 5. Selection of treetops. Treetops marked with green labels can be used to plan a trajectory.

3.4. Constructing of a trajectory prototypes set

The trajectory prototype is represented as a set of points between which one can construct the Dubins trajectories (2). Each prototype contains a starting vertex A_0 with coordinates (x_{tek}, y_{tek}) and a final vertex A_N with coordinates (x_{end}, y_{end}). These vertices correspond to the vectors of the current and terminal position. It is known that the search space is represented as a "grid" of n columns and m rows. There are rules following which a sequence of treetops is formed for each trajectory prototype (Fig. 6).

1. Each next treetop is selected from a range of grid lines $b + 1..m$, where b is the line number of the previous treetop.
2. For any treetop belonging to the b-line, there must be trajectories for all treetops of the subsequent lines.

$$Tops = \left\{ \begin{array}{l} \{A_0, A_N\}, \\ \{A_0, A_{11}, A_N\}, \{A_0, A_{12}, A_N\} \dots \{A_0, A_{mn}, A_N\}, \\ \{A_0, A_{11}, A_{21}, A_N\}, \{A_0, A_{11}, A_{22}, A_N\} \dots \{A_0, A_{11}, A_{mn}, A_N\}, \\ \dots \\ \{A_0, A_{1k}, A_{21}, A_N\}, \{A_0, A_{1k}, A_{22}, A_N\} \dots \{A_0, A_{1k}, A_{mn}, A_N\}, \\ \dots \\ \{A_0, A_{1n}, A_{21}, A_N\}, \{A_0, A_{1n}, A_{22}, A_N\} \dots \{A_0, A_{1n}, A_{mn}, A_N\}, \\ \{A_0, A_{11}, A_{21}, A_{31}, A_N\}, \{A_0, A_{11}, A_{21}, A_{32}, A_N\} \dots \{A_0, A_{11}, A_{21}, A_{mn}, A_N\}, \\ \dots \\ \{A_0, A_{1k}, A_{2k} \dots A_{bk} \dots A_{mn}, A_N\} \end{array} \right\} \quad (7)$$

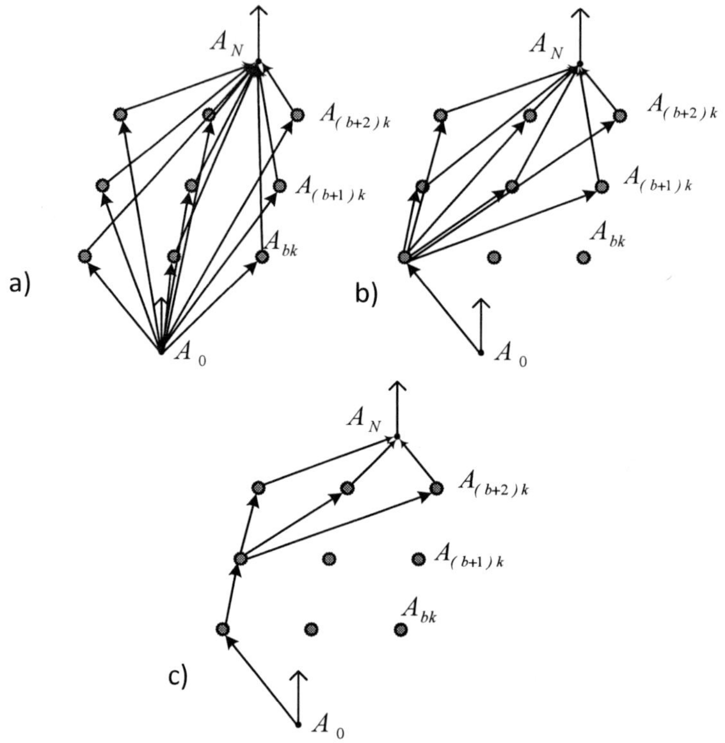

Figure 6. Variants of trajectory prototypes.

3.5. Construction of Dubins trajectories

According to [6], there are four possible Dubins trajectories for a given initial configuration $\overline{p_s}$ and a final configuration $\overline{p_f}$. Option 1 (R-S-R) refers to the case when the DV begins clockwise motion, then follows a movement in a straight line and the second clockwise motion. Option 2 (L-S-L) for the case when the DV begins an anticlockwise motion, then follows a straight line operation and the second anticlockwise motion. The 3rd option (R-S-L) is the case when the DV begins a clockwise motion, then follows a straight line operation and an anticlockwise motion. Option 4 (L-S-R) is the case when the DV begins an anticlockwise motion, then follows a straight line operation and a clockwise motion. As a rule, the trajectory option is selected based on the minimum length. proposed to introduce criteria z_s and z_f that correspond to the first and second circles for determining the minimum-length trajectory. Criteria (8) and (9) characterize the direction of rotation and are formed on basis of the minimum-length trajectory.

$$z_s = \cos\vartheta_s \cdot (y_f - y_s) - \sin\vartheta_s \cdot (x_f - x_s). \quad (8)$$

$$z_f = \cos\vartheta_f \cdot (a_y - y_f) - \sin\vartheta_f \cdot (a_x - x_f), \quad (9)$$

where $[a_x, a_y] = [x', y'] \begin{bmatrix} \cos(fi) & -\sin(fi) \\ \sin(fi) & \cos(fi) \end{bmatrix} + [x_{cs}, y_{cs}]$,

$$fi = pr \cdot arccos\left(\frac{y_f - y_{cs}}{\sqrt{(x_f - x_{cs})^2 + (y_f - y_{cs})^2}}\right),$$

if $(x_f - x_{cs}) \geq 0$ then $pr = 1$, else $pr = -1$,

$$x' = -\frac{z_s}{|z_s|} r_{ds} \sqrt{1 - \frac{r_{ds}}{\sqrt{(x_f - x_{cs})^2 + (y_f - y_{cs})^2}^2}},$$

$$y' = \frac{r_{ds}^2}{\sqrt{(x_f - x_{cs})^2 + (y_f - y_{cs})^2}}.$$

If the criterion (z_s or z_f) is less than zero, then the optimal turn is clockwise (R), otherwise it is counterclockwise (L). It's easy to calculate the coordinates of the circle centers O_{cs} and O_{cf} (10) if the direction of rotation is known.

$$\begin{cases} x_{cs} = x_s + \frac{z_s}{|z_s|} r_{ds} \sin \vartheta_s \\ y_{cs} = y_s - \frac{z_s}{|z_s|} r_{ds} \cos \vartheta_s \\ x_{cf} = x_f + \frac{z_f}{|z_f|} r_{df} \sin \vartheta_f \\ y_{cf} = y_f - \frac{z_f}{|z_f|} r_{df} \cos \vartheta_f \end{cases}. \tag{10}$$

The proposed equation (11), which allows to find the point of change between the motion along a straight line and the circular movement. The condition $r_{ds} = r_{df}$ is accepted to simplify the calculations. It is common in practice.

$$\begin{cases} T_{ex} = T'_{ex} \cdot B + O_{cs} \\ T_{en} = T'_{en} \cdot B + O_{cs} \end{cases}, \tag{11}$$

where $B = \frac{1}{\sqrt{(x_{cf} - x_{cs})^2 + (y_{cf} - y_{cs})^2}} \begin{bmatrix} x_{cf} - x_{cs} & y_{cf} - y_{cs} \\ -y_{cf} + y_{cs} & x_{cf} - x_{cs} \end{bmatrix}$.

If $\frac{z_s}{|z_s|} = \frac{z_f}{|z_f|}$, then

$$T'_{ex} = \begin{cases} 0 \\ -\frac{z_s}{|z_s|} r_{ds} \end{cases},$$

$$T'_{en} = \begin{cases} \sqrt{(x_{cf} - x_{cs})^2 + (y_{cf} - y_{cs})^2} \\ -\frac{z_f}{|z_f|} r_{df} \end{cases}.$$

If $\frac{z_s}{|z_s|} \neq \frac{z_f}{|z_f|}$, then

$$T'_{ex} = \begin{cases} \frac{r_{ds}^2}{0.5\sqrt{(x_{cf} - x_{cs})^2 + (y_{cf} - y_{cs})^2}} \\ -\frac{z_s}{|z_s|} r_{ds} \frac{\sqrt{0.25\left((x_{cf} - x_{cs})^2 + (y_{cf} - y_{cs})^2\right) - r_{ds}^2}}{0.5\sqrt{(x_{cf} - x_{cs})^2 + (y_{cf} - y_{cs})^2}} \end{cases},$$

$$T'_{en} = \begin{cases} \frac{2 - r_{ds} r_{df}}{0.5\sqrt{(x_{cf} - x_{cs})^2 + (y_{cf} - y_{cs})^2}} \\ -\frac{z_f}{|z_f|} r_{df} \frac{\sqrt{0.25\left((x_{cf} - x_{cs})^2 + (y_{cf} - y_{cs})^2\right) - r_{df}^2}}{0.5\sqrt{(x_{cf} - x_{cs})^2 + (y_{cf} - y_{cs})^2}} \end{cases}.$$

4. Conclusion

This paper describes a modified trajectory construction algorithm based on the search space research by method of fast growing random trees. The presented algorithm allows planning the DV trajectory in real time and in a dynamically changing environment. A characteristic feature of the proposed solution is the structuring of the treetops for trajectory planning and the description of the search space as a set of vectors. The kinematic characteristics of DV are taken into account when structuring treetops. This approach can significantly increase the speed of searching for the desired trajectory. The author does not claim a full presentation of the material, but only proposes a methodology for the practical implementation of the trajectory planning problem.

IOP Conf. Series: Materials Science and Engineering **534** (2019) 012033 doi:10.1088/1757-899X/534/1/012033

References

[1] Ivanov A M, Shadrin S S, Karpukhin K E 2013 Intelligent vehicle. Adaptation of subsystem for determination the relative position of moving vehicles (Moscow) *Izvestiya MGTU MAMI* **2** pp 57-62

[2] 12th International Automotive Scientific Forum "Intelligent Transport Systems for Energy Efficiency and Safety" 2016 (Moscow) *AAI* **5** pp 6-7

[3] Batishchev Y, Dudnik V, Gurinov A, Kopkin A 2010 Architecture and possibilities of attitude system of unmanned and ultralight air vehicles based on sensitive elements of new generation *Vestnik of Don State Technical University* **2** pp 229-233

[4] Yakovlev K S, Makarov D A, Baskin E S 2014 UAV path planning technique under flight dynamics constraints (Moscow) *Modelling and Management* **4** pp 3-17

[5] Dijkstra E W 1959 A note on two problems in connexion with graphs *Numerische mathematic* **1** pp 269-271

[6] Randal W. Beard and Timothy W. McLain 2012 Small Unmanned Aircraft: Theory and Practice *Princeton University Press*

[7] Andreychuk A A, Yakovlev K S 2017 The trajectory planning in a two-dimensional space, taking into account the size of the agent (mobile robot, unmanned vehicle) (Kazan) *Fourth All-Russian Scientific and Practical Seminar "Unmanned Vehicles with Elements of Artificial Intelligence"* pp 107-117

[8] Choset H, Lynch K M, Hutchinson S, Kantor G, Burgard W, Kavraki L E, Thrun S 2005 Principles of Robot Motion: Theory, Algorithms, and Implementations (Boston) *MIT Press*

[9] Dubins L E 1957 On curves of minimal length with a constraint on average curvature, and with prescribed initial and terminal positions and tangents *American Journal of Mathematics* **79** pp 497-516

IOP Conf. Series: Materials Science and Engineering **534** (2019) 012034 doi:10.1088/1757-899X/534/1/012034

Creation of electric vehicle ABS operation algorithm with possibility of hybrid braking based on slip-slope approach at wheels slip determining

S Bakhmutov[1], **A Umnitsyn**[1] **and V Ivanov**[2]

[1] FSUE "NAMI", 2 Avtomotornaya St., Moscow, 125438, Russian Federation
[2] TU Ilmenau, Eh renbergstr. 15, 98693 Ilmenau, Germany

E-mail: artem.umnicin@nami.ru

Abstract. This paper represents an Anti-lock Braking System (ABS) operation algorithm implementing hybrid braking; the paper considers braking with a high deceleration of an electric passenger vehicle during the rectilinear driving on the roads with different friction coefficients; a short analysis of the approaches used for evaluation of the target vehicle wheels slip is also performed. The main ABS task consists in maintenance of the target wheel slip coefficient based on the tyre-road friction coefficient. The tyre-road friction coefficient φ_x is determined using the algorithm proposed in the papers by Semmler. The algorithm is about finding the curve peak $\varphi_x - S$ using the least square method. Based on the calculated target slip coefficient, the braking torque at the electric motor shaft operating in the regeneration mode is determined. The electric motor ensures relatively low braking torque at the vehicle wheels preventing the tyres from a significant slip. Reaching the target slip coefficient is ensured by means of friction brake mechanisms. Virtual tests showed that the values of the wheel slip coefficient are close to the required ones. Thus, this system makes it possible to implement hybrid braking with two sources creating the brake torque.

Introduction

The Anti-lock Braking System (ABS) serves to prevent vehicle wheels lock-up in order to maintain steerability as well as to reduce the vehicle braking distance. An increase in the number of papers dedicated to ABS integration with other vehicle systems can recently be observed in foreign sources [1, 2]. One of the special cases of such integration is the use of a traction electric motor, which operates in the energy regeneration mode, within ABS in electric vehicles. It should be noted that, at the moment, there is no unified control strategy accepted for the process of combined braking by means of the friction brake mechanisms and electric motor, which operates in the energy regeneration mode. One of the possible variants of combining two brake torque sources is considered in this paper.

ABS Structure

The vehicle Anti-lock Braking System is a control system with feedback (Figure 1).

IOP Conf. Series: Materials Science and Engineering **534** (2019) 012034 doi:10.1088/1757-899X/534/1/012034

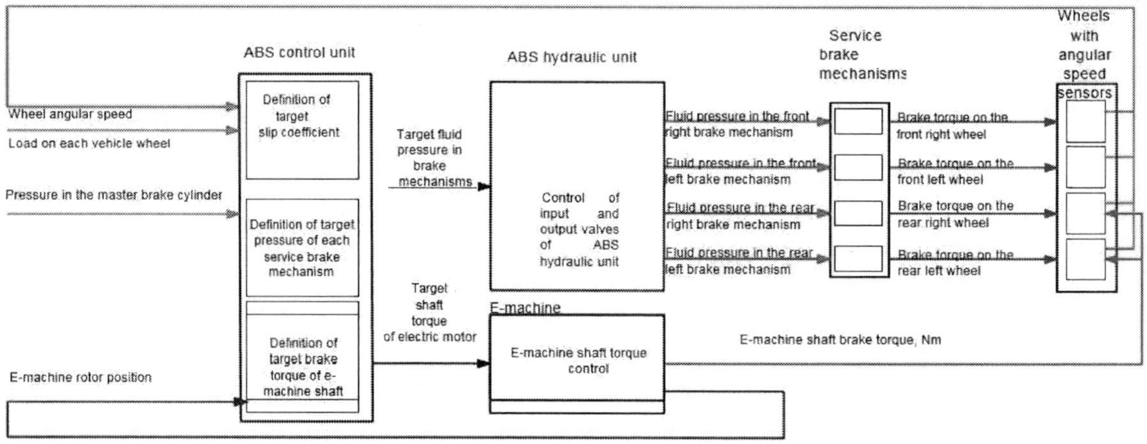

Figure 1. ABS diagram.

The system consists of the following components:

1. Control unit responsible for calculation of the target slip coefficient, as well as for control of the hydraulic unit valves and electric motor of the vehicle.
2. Hydraulic unit, which is an actuator ensuring adjustment of pressure in the service brake mechanisms.
3. Wheels angular speed sensors, which are tracking devices used in order to ensure the feedback.
4. Vertical load sensors, which are tracking devices ensuring acquisition of information about the vertical load on each wheel.

Target wheels slip coefficient determination method

The maximum axial force Fx exerted upon the vehicle is known to be proportional to the tyre-road friction coefficient μ (1).

$$F_x = F_z \cdot \mu, \quad (1)$$

In turn, the tyre-road friction coefficient μ depends on the tyre-road slip coefficient S. The example of this dependence is given in Figure 2.

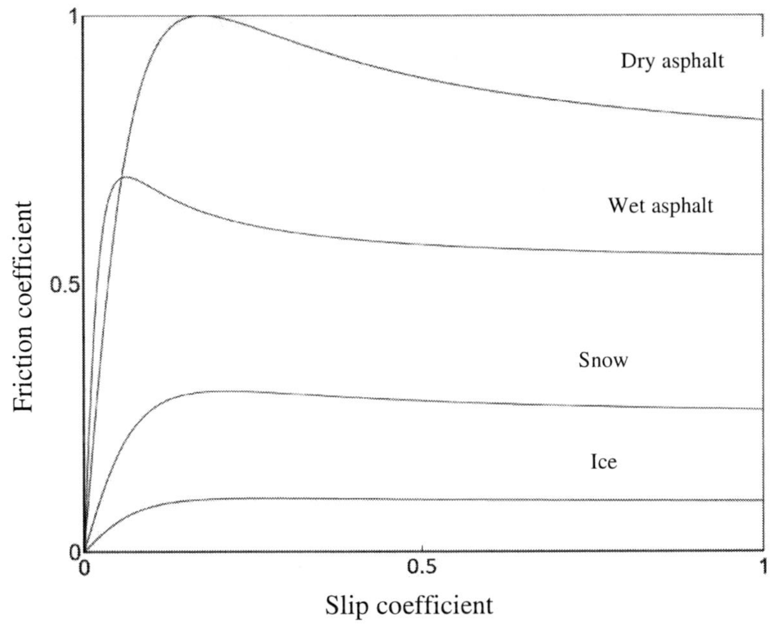

Figure 2. Dependence of friction coefficient on slip coefficient.

For assessment of the tyre-road friction that is the maximum for the given conditions, many papers suggest using the models that describe the interaction of tyres with the bearing surface – mathematical, analytical or empirical ones. The most common tyre models are the models based on the Magic Formula [3], Brush model [4, 5] and others [6]. This approach ensures a high-precision definition of the maximum tyre-road friction coefficient, but it has a number of considerable disadvantages:

1. The models become complicated when it is necessary to consider the slip due to the tyre properties – pressure, tyre tread wear, foreign bodies in the contact area, tyre and bearing surface temperatures.
2. High difficulty of obtaining an accurate tyre model caused by the need to conduct a large number of experiments.
3. Complexity of adjustment of the surface properties estimator during real experiments.
4. Methods of model parametrization based on experimental data demand acquisition and systematization of a large scope of information.

An alternative approach for the definition of the bearing surface friction properties is the slip-slope approach.

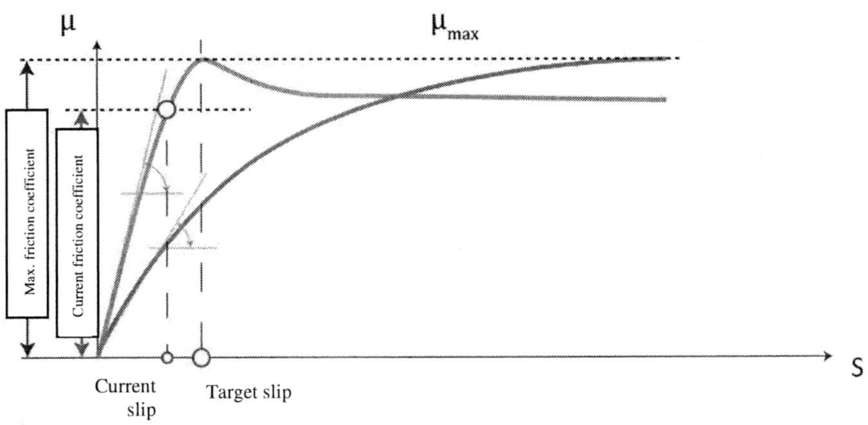

Figure 3. Difference in tangent slope angles at different tyre-road friction coefficients.

As one can see in Figure 3, the tangent slope angle against the curve describing the μ-S function for different friction coefficients of the bearing surface may vary within a quite wide range. This observation became a basis for the slip-slope approach applied for the determination of the maximum tyre-road friction coefficient.

Compared to the methods based on the tyre models only, the methods using the slip-slope approach are characterized by a much smaller number of variables necessary for the correct operation of ABS controllers with the given accuracy as well as by convenience of adjusting the models under the real experiment conditions.

One of the methods for the target slip coefficient determination using the slip-slope approach is the method suggested in Semmler's papers [7]. The process of determination of the target slip coefficient according to this method can be divided into three stages:

1. Determination of a μ-S diagram section containing W points with T time step.
2. Determination of the optimum slip.
3. Setting the value of the target slip.

Determination of the μ-S diagram section is performed by means of approximation of W points characterizing the S slip coefficient and μ tyre-road friction coefficient with the T time step, by a straight line (2).

$$\mu(k) = c_0 + c_1 \cdot S(k), \tag{2}$$

Applying the least square method, it is possible to obtain the value of coefficient c_1 characterizing this line slope. Thus, it is possible to draw a conclusion, whether the average slip coefficient of the considered wheel is higher or lower than the optimum value. In case of c_1 sign reversal, it is supposed that the average slip coefficient of the considered wheel has passed the μ-S curve peak. Thus, the optimum slip coefficient S_{opt} is calculated according to the following formula (3):

$$S_{opt} = \frac{1}{W} \cdot \sum_{i=0}^{W} S(i) \; \forall \; sgn(c_1(k)) \neq sgn(c_1(k-1)), \tag{3}$$

As the target slip shall not considerably deviate from the maximum value, but, at the same time, change of the target slip within the small range about the optimum slip is necessary for obtaining the μ-S curve maximum, the author suggests using the following formula in order to define the target slip S_d (4):

$$S_d(k) = \begin{cases} S_d(k-1) + K_S \cdot sgn\left(S_{opt}(k) - S_d(k-1)\right) \forall sgn(c_1(k)) = sgn(c_1(k-1)) \\ S_{opt}(k) \qquad \forall sgn(c_1(k)) \neq sgn(c_1(k-1)) \end{cases}, \tag{4}$$

Following the tests, this approach showed the exact definition of μ-S curve peak on surfaces with any tyre-road friction coefficient ranging from 0.1 to 1. As a result, this method is well applicable in ABS algorithms.

Vehicle wheel brake torque control algorithm

A rear-wheel-drive electric vehicle of N2 category, the necessary specifications of which are given in Table 1, was selected as an object of investigation for this paper.

Table 1. Specifications of N2 category vehicle.

Fully loaded weight	4200 kg
Weight on front axle	1653 kg
Weight on rear axle	2547 kg
Unladen weight	4060 kg
Weight on front axle	1555 kg
Weight on rear axle	2505 kg
Height of the center of mass	
Unladen condition	998 mm
Fully loaded weight	1001 mm
Tyres	185/75R16C

The characteristics of an e-machine applied in the examined vehicle are shown in Figure 4 [8].

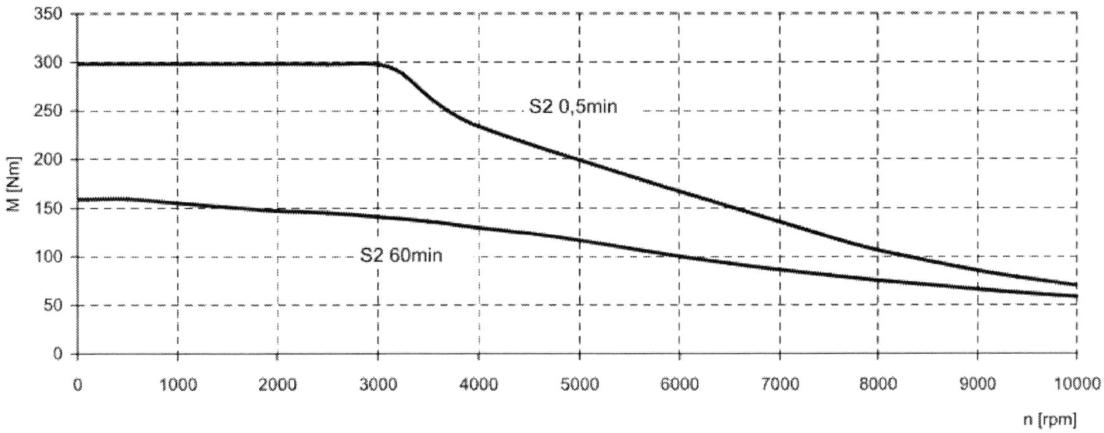

Figure 4. E-machine characteristics.

The brake torque control on the driving wheels is exercised by means of controlling the brake torque from the e-machine operating in the regeneration mode, as well as by controlling the brake mechanisms pressure. The brake torque control on the driven wheels is exercised only by means of brake fluid pressure control in brake mechanisms.

It should be noted that, in this vehicle configuration, the maximum brake torque created by the e-machine does not allow reaching thigh brake efficiency when driving on a surface with friction coefficient higher than 0.3. When driving on a road with the tyre-road friction coefficient less than 0.3, the brake torque from the e-machine is limited and does not reach the maximum value. It is reasonable to exclude PID control of the e-machine shaft torque for the wheel slip control described in [9] in view of the complexity of system adjustment when changing the brake torque source.

The vehicle wheel slip coefficient is adjusted by changing the brake fluid pressure in the service brake mechanisms in the ABS hydraulic unit. The hydraulic unit of the system is a device that controls the pressure in the service brake mechanisms of the vehicle. Each service brake mechanism is connected with the brake master cylinder by means of two valves – input valve and output valve. In the course of service braking characterized by low wheel slip coefficient, the input valve is opened and the output valve is closed. As soon as the wheel slip coefficient surpasses the preset limits, upon a signal from the ABS control unit, the input valve will be closed thus maintaining a constant pressure in the service brake mechanism. If the wheel slip coefficient continues to increase, the output valve will be opened and, as a result, the pressure in the service brake mechanism will decrease. As soon as the wheel slip coefficient reaches the preset value, the output valve will be closed.

The target brake pressure is calculated for each wheel by means of the PID controller. The feedback is performed by monitoring the wheels slip coefficient values. The control error is calculated as a difference between the target slip coefficient (obtained with application of Semmler's method) and actual slip coefficient.

System operation results

For system testing, a number of virtual tests were carried out according to the following scheme:

1. Initial driving speed – 101 km/h, accelerator pedal travel – 0%.
2. After reaching 100 km/h: accelerator pedal travel – 0%, brake pedal travel – 100%.
3. Braking up to 20 km/h.

These tests were carried out on roads with tyre-road friction coefficient μ=0.8 and 0.1.

Figure 5 shows the process of definition of the target front right wheel slip coefficient by the module when driving on the road with μ=0.8.

IOP Conf. Series: Materials Science and Engineering **534** (2019) 012034 doi:10.1088/1757-899X/534/1/012034

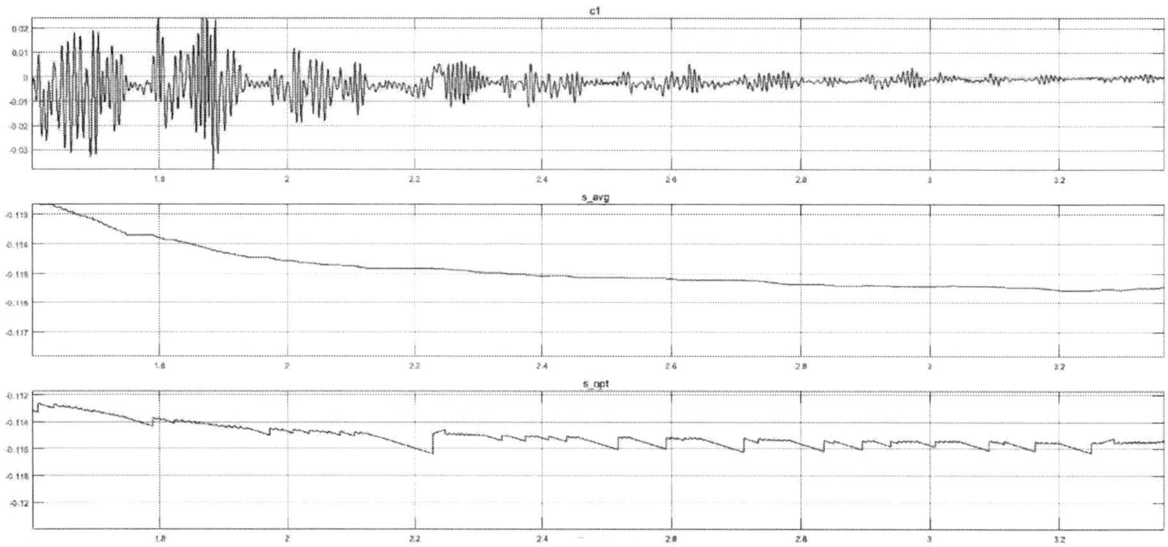

Figure 5. Process of definition of target front right wheel slip coefficient by module at μ=0.8.

Diagrams of wheel slip coefficient vs time on the road with μ=0.8 and 0.1 are shown in Figures 6 and 7, respectively.

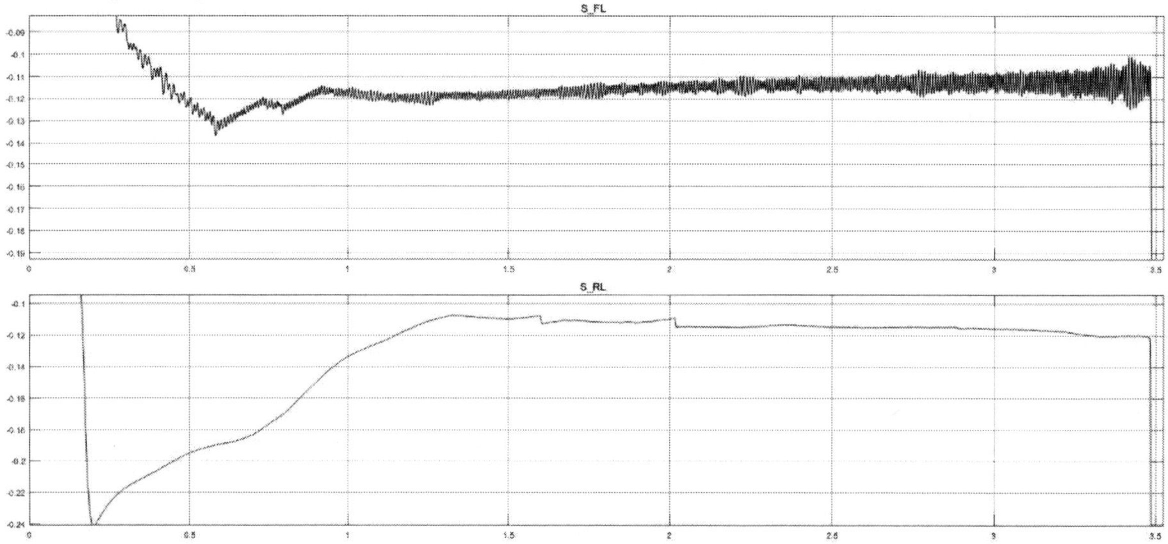

Figure 6. Change of wheel slip coefficient vs time on the road with μ=0.8.

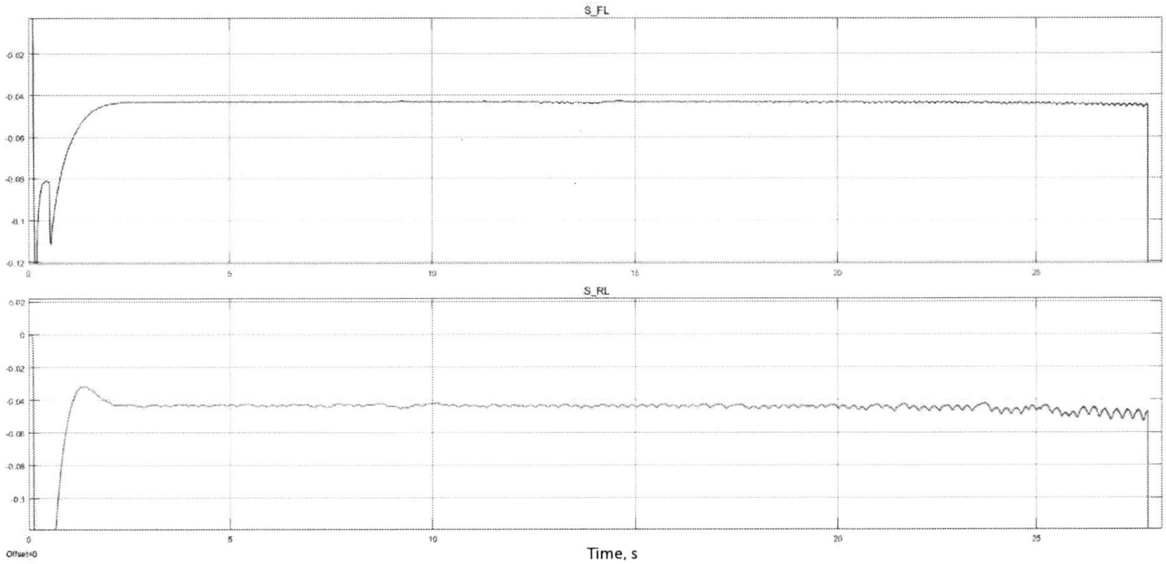

Figure 7. Change of wheel slip coefficient vs time on the road with μ=0.1.

It should be noted that the Anti-lock Braking System presented in this paper allows reduction of the vehicle braking distance by 5-10% in comparison to the "classic" system that does not use PID control. The braking distances for the developed system and for the "classic" system are provided in Table 2.

Table 2. Comparison of braking distance for different ABS algorithms.

Friction coefficient μ	Braking distance, m	
	With the developed system	"Classic" system
0.8	56.6	62.3
0.1	167.5	175.8

Conclusion

Summarizing the abovementioned, it is possible to draw several conclusions:

1. Application of the Slip-slope approach (in particular the Semmler's method) allows reaching the friction coefficient that is close to the maximum on the majority of pavings at any moment, which, in turn, reduces the vehicle braking distance by the value up to 10% in comparison to the "classic" braking algorithm.
2. The easiest adjustment refers to the hybrid braking system, in which the wheel slip is controlled only by one brake torque source.
3. The maximum efficiency of the developed system is reached when braking on roads with a high tyre-road friction coefficient. The braking distance on roads with μ=0.8 decreases by 5.7 meters compared to the "classic system". On roads with μ=0.1, the braking distance decreases by 8.3 meters.

References

[1] Savitski D, Ivanov V, Augsburg K, Pütz T and Barber P 2016 The new paradigm of an anti-lock braking system for a full electric vehicle: Experimental investigation and benchmarking *Proceedings of the Institution of Mechanical Engineers* Part D: Journal of Automobile Engineering

[2] Ivanov V, Augsburg K, Savitski D, Schreiber V, Els S and Dhaens M 2017 Collaborative engineering of integrated chassis control for ground vehicle: Case study of lifelong learning technologies in automotive mechatronics *2017 IEEE International Conference on Mechatronics (ICM)* Churchill VIC pp 449-454

[3] Pacejka H 2012 *Tire and Vehicle Dynamics* (Butterworth-Heinemann: Oxford, UK)
[4] Zhao J, Zhang J and Zhu B 2014 Developmentand Verification of the Tire/Road Friction Estimation Algorithm for Antilock Braking System *Math. Probl. Eng.* (doi:10.1155/2014/786492)
[5] Singh K and Taheri S 2014 Estimation of tire-road friction coefficient and its application in chassis control systems *Syst. Sci. Control Eng.* **3** pp 39–61
[6] Rajamani R 2012 *Vehicle Dynamics and Control* (Springer: New York, NY, USA)
[7] Semmler S 2006 Control of the vehicle brake dynamics with continuously adjustable wheel brakes *VDI Series 12* **632** (Dusseldorf)
[8] Information on http://media3.ev-tv.me/Azure300VDC-400Acurve.pdf
[9] Bakhmutov S V, Ivanov V G, Karpukhin K E and Umnitsyn A A 2018 Creation of operation algorithms for combined operation of anti-lock braking system (ABS) and electric machine included in the combined power plant *IOP Conference Series: Materials Science and Engineering* **315**(1)

Institute of Physics
Dirac House, Temple Back
Bristol BS1 6BE UK

ISSN: 1757-8981
ISBN 978-1-5108-8919-4